T0134412

Lecture Notes on Data Engineering and Communications Technologies

Volume 109

Series Editor

Fatos Xhafa, Technical University of Catalonia, Barcelona, Spain

The aim of the book series is to present cutting edge engineering approaches to data technologies and communications. It will publish latest advances on the engineering task of building and deploying distributed, scalable and reliable data infrastructures and communication systems.

The series will have a prominent applied focus on data technologies and communications with aim to promote the bridging from fundamental research on data science and networking to data engineering and communications that lead to industry products, business knowledge and standardisation.

Indexed by SCOPUS, INSPEC, EI Compendex.

All books published in the series are submitted for consideration in Web of Science.

More information about this series at https://link.springer.com/bookseries/15362

Sanjay Misra · Chamundeswari Arumugam
Editors

Illumination of Artificial Intelligence in Cybersecurity and Forensics

 Springer

Editors
Sanjay Misra 🄳
Østfold University College
Halden, Norway

Chamundeswari Arumugam
Computer Science and Engineering
Sri Sivasubramaniya Nadar College
of Engineering
Chennai, Tamil Nadu, India

ISSN 2367-4512   ISSN 2367-4520   (electronic)
Lecture Notes on Data Engineering and Communications Technologies
ISBN 978-3-030-93452-1   ISBN 978-3-030-93453-8   (eBook)
https://doi.org/10.1007/978-3-030-93453-8

This Springer imprint is published by the registered company Springer Nature Switzerland AG
The registered company address is: Gewerbestrasse 11, 6330 Cham, Switzerland

This Book is Dedicated to Father of Lead Editor Prof. Sanjay Misra

Mr. Om Prakash Misra

27.09.1950–24.12.2019

Preface

A new dimension in terms of illumination of AI is taken up here to showcase the relationship between AI and cybersecurity. This book supplies the related information that are concern in today's scenario in the field of cybersecurity and forensics. It takes you to the current techniques that are practically possible to apply in this domain using machine learning and deep learning concepts. The topic of cybersecurity in terms of IDS, authentication, audit techniques, forensics, IoT, health care, and image recognition are covered under various chapters. The research finding can be used as a base to improve the research thirst in this domain. Many empirical finding is available to explore the applicability of artificial intelligence in this domain. Also, the survey papers indicate many depth information on the applicability of artificial intelligence in this domain.

A pleasure to introduce the book, Illumination of AI in cybersecurity and forensics. Fifteen chapters organized in this book provide an interesting aspect of AI in cyber forensics and forensics. Some of chapters cover the applicability of preprocessing, feature selection, classification, analysis, prediction, dimensionality reduction, optimization, and AI techniques in this domain. The chapters demonstrate and report the dataset applicability in this domain using AI. The various datasets explored to provide the result outcomes are CIC-IDS2017, UNSW-NB15, KDD Cup 99, DARPA, etc. A brief summary of various authors presented in chapters is listed below.

Julián Gómez et al. in their chapter titled, "A Practical Experience Applying Security Audit Techniques in An Industrial Healthcare System", performed a security assessment process. The results of this study include a security audit on an industrial scenario currently in production. An exploitation and vulnerability analysis have been performed, and more that 450 vulnerabilities have been found. This chapter outlines a systematic approach using artificial intelligence to enable the system security team to facilitate the process of conducting a security audit taking into account the sensitivity of their systems.

Joseph Bamidele Awotunde et al., in their work titled "Feature Extraction and Artificial Intelligence-Based Intrusion Detection Model for a Secure Internet of Things Networks", discussed the security issues within IoT-based environments and the application of AI models for security and privacy in IoT-based for a secure network.

The chapter proposes a hybrid AI model framework for intrusion detection in an IoT-based environment and a case study using CIC-IDS2017 and UNSW-NB15 to test the proposed model's performance. The model performed better with an accuracy of 99.45%, with a detection rate of 99.75%. The results from the proposed model show that the classifier performs far better when compared with existing work using the same datasets, thus prove more effective in the classification of intruders and attackers on IoT-based systems.

Chika Yinka-Banjo et al. in their chapter titled, "Intrusion Detection Using Anomaly Detection Algorithm and Snort", focused on Intrusion Detection System (IDS) for a local network to detect network. A statistical approach, as well as a binomial classification, was used for simplicity in classification. The result shows the outlier value for each item considered; a 1 depicts an attack, a 0 depicts normalcy. The results are promising in dictating intrusion and anomalies in an IDS system.

Kousik Barik et al. in their chapter titled, "Research Perspective on Digital Forensic Tools and Investigation Process", provided a comparative analysis of popular tools in each category is tabulated to understand better, making it easy for the user to choose according to their needs. Furthermore, this chapter presents how machine learning, deep learning, and natural language processing, a subset of artificial intelligence, can be effectively used in digital forensic investigation. Finally, the future direction of the challenges and study scope in digital forensics and artificial intelligence is also mentioned for potential researches.

Timibloudi Stephen Enamamu in his chapter titled, "Intelligent Authentication Framework for Internet of Medical Things (IoMT)", explored the use of artificial intelligence to enhance authentication of Internet of Medical Things (IoMT) through a design of a framework. The framework is designed using wearable and or with a mobile device for extracting bioelectrical signals and context awareness data. The framework uses bioelectrical signals for authentication, while artificial intelligent is applied using the contextual data to enhance the patient data integrity. The framework applied different security levels to balance between usability and security on the bases of False Acceptance Rate (FAR) and False Rejection Rate (FRR). Thirty people are used for the evaluation of the different security levels and the security level 1 achieved a result base on usability verse security obtaining FAR of 5.6% and FRR of 9% but when the FAR is at 0%, the FRR stood at 29%. The Intelligent Authentication Framework for Internet of Medical Things (IoMT) will be of advantage in increasing the trust of data extracted for the purpose of user authentication by reducing the FRR percentage.

Stephen Bassi Joseph et al., in their chapter, "Parallel Faces Recognition Attendance System with Anti-Spoofing Using Convolutional Neural Network", proposed a parallel face recognition attendance system based on Convolutional Neural Network a branch of artificial intelligence and OpenCV. Experimental results proved the effectiveness of the proposed technique having shown good performance with recognition accuracy of about 98%, precision of 96%, and a recall of 0.96. This demonstrates that the proposed method is a promising facial recognition technology.

Kenneth Mary Ogbuka et al. in their chapter titled, "A Systematic Literature Review on Face Morphing Attack Detection (MAD)", explored to find MAD methodologies, feature extraction techniques, and performance assessment metrics that can help MAD systems become more robust. To fulfill this study's goal, a Systematic Literature Review was done. A manual search of 9 well-known databases yielded 2089 papers. Based on the study topic, 33 primary studies were eventually considered. A novel taxonomy of the strategies utilized in MAD for feature extraction is one of the research's contributions. The study also discovered that (1) single and differential image-based approaches are the commonly used approaches for MAD; (2) texture and key point feature extraction methods are more widely used than other feature extraction techniques; and (3) Bona-fide Presentation Classification Error Rate and Attack Presentation Classification Error Rate are the commonly used performance metrics for evaluating MAD systems. This chapter addresses open issues and includes additional pertinent information on MAD, making it a valuable resource for researchers developing and evaluating MAD systems.

Kenneth Mary Ogbuka et al., in their chapter titled, "Averaging Dimensionality Reduction and Feature Level Fusion for Post-Processed Morphed Face Image Attack Detection", proposed a MAD technique to perform MAD even after image sharpening operation using averaging dimensionality reduction and feature level fusion of Histogram of Oriented Gradient (HOG) 8 x 8 and 16 x 16 cell size. The 8 x 8 pixels cell size was used to capture small-scale spatial information from the images, while 16 x 16 pixels cell size was used to capture large-scale spatial details from the pictures. The proposed technique achieved a better accuracy of 95.71% compared with the previous work, which reached an accuracy of 85% when used for MAD on sharpened image sources. This result showed that the proposed technique is effective for MAD on sharpened post-processed images.

N. S. Gowri Ganesh et al. in their chapter, "A Systematic Literature Review on Forensics in Cloud, IoT, AI & Blockchain", reviewed the application of forensics using Artificial Intelligence in the field of Cloud computing, IoT, and Blockchain Technology. To fulfill the study's goal, a systematic literature review (SLR) was done. By manually searching six well-known databases, documents were extracted. Based on the study topic, 33 primary studies were eventually considered. The study also discovered that (1) highlights several well-known challenges and open issues in IoT forensics research, as it is dependent on other technologies and is crucial when considering an end-to-end IoT application as an integrated environment with cloud and other technologies. (2) There has been less research dedicated to the use of AI in the field of forensics. (3) Contributions on forensic analysis of attacks in blockchain-based systems is not found.

Mathew Emeka Nwanga et al. in their chapter titled, "Predictive Forensic Based—Characterization of Hidden Elements in Criminal Networks Using Baum-Welch Optimization Technique", contribute to knowledge by providing a terrorist computational model which helps to determine the most probable state and timeframe for the occurrence of terrorist attacks. It also provides the most probable sequence of active internal communications (AICs) that led to such attacks. The Baum–Welch optimization is applied in this research to improve the intelligence and predictive accuracy of the activities of criminal elements. This solution is adaptable to any country around the globe. The result shows that the Foot Soldiers (FS) are most vulnerable with 90% involvement in criminal attacks; the Commander carried out most strategic (high profile) attacks estimated at 2.2%. The private citizens and properties had the highest attack targets (50.6%), whereas the police and military base had 12.3% and 6.7%, respectively. The results show that Boko-Haram carried out the greatest level of attacks at 79.2%, while Fulani extremists are responsible for 20.8% of all acts of terrorism in Nigeria from 2010 to date.

Roseline Oluwaseun Ogundokun et al. presented a chapter tiled, "An Integrated IDS Using ICA-Based Feature Selection and SVM Classification Method". Here, the authors presented a novel method for detecting security violations in IDSs that combines ICA FS and ML classifiers in this work. SVM was employed as a suitability function to identify important characteristics that can aid in properly classifying assaults. The suggested approach is utilized in this context to advance the resulting quality by changing the SVM regulatory parameter values. The goal was to achieve satisfactory results for the IDS datasets, KDD Cup 99, in terms of classifier performance in noticing invasions built on the optimal number of features. In contrast to several state-of-the-art approaches, the suggested model outperforms them in terms of accuracy, sensitivity, detection rate (DR) false alarm, and specificity. IDS may be used to secure wireless payment systems. It is possible to establish secure integrated network management that is error-free, therefore boosting performance.

Yakub Kayode Saheed, in his chapter titled, "A Binary Firefly Algorithm Based Feature Selection Method on High Dimensional Intrusion Detection Data", proposed a binary firefly algorithm (BFFA)-based feature selection for IDS. First performed normalization, subsequently, the BFFA algorithm was used for feature selection stage. Then adopted random forest algorithm for the classification phase. The experiment was performed on high-dimensional University of New South Wales-NB 2015 (UNSW-NB15) dataset with 75% of the data used for training the model and 20% for testing. The findings showed an accuracy of 99.72%, detection rate of 99.84%, precision of 99.27%, recall of 99.84%, and F-score of 99.56%. The results were gauge with the state-of-the-art results and our results were found outstanding.

Suleman Isah Atsu Sani et al. in their chapter titled, "Graphical Based Authentication Method Combined with City Block Distance for Electronic Payment System", proposed a graphical-based authentication method combined with an Artificial Intelligence (AI) domain city block distance algorithm to measure the similarity score between the image passwords at the points of registration and login. To achieve optimal performance of the system and make it robust, an experiment was conducted during the login session using the AI-based city block distance algorithm and other different distance algorithms that include Euclidean, Cosine similarity, and Jaccard. The experimental results show that the proposed city block distance method has the fastest execution time of 0.0318 ms, minimal matching error of 1.55231, and an acceptable login success rate of 64%, compared to when the graphical-based password is combined with other similarity score algorithms. This chapter concludes that the proposed method would be the most reliable authentication for e-payment systems.

S. Hanis et al. in their chapter titled, "Authenticated Encryption to Prevent Cyber-Attacks in Images", provided a secure authenticated encryption algorithm for the storage and transmission of digital images to avoid cyber threats and attacks. The designed algorithm makes use of the deep convolutional generative adversarial network to test if the image is a fake image originated by the intruder. If found fake exclusive OR operations are performed with the random matrices to confuse the intruder. If the image is not fake, then encryption operations are directly performed on the image. The image is split into two 4-bit images and a permutation operation using a logistic map is performed and finally the split images are merged together. Finally, exclusive OR operations are performed on the merged image using the convolution-based round keys generated to generate the concealed image. In addition, authentication is also achieved by calculating the mean of the actual image. The performance analysis shows that the designed technique offers excellent security and also helps in testing the authenticity of the stored images.

Vani Thangapandian in her paper titled, "Machine Learning in Automated Detection of Ransomware: Scope, Benefits and Challenges", proposed a Systematic Literature Review, to discuss the different ransomware detection tools developed so far and highlighted the strengths and weaknesses of Machine Learning-based detection tools. The main focus of this study is on how ransomware attacks are executed and the possible solutions to mitigate such attacks. The main focus of this work is the application of various machine learning and deep learning methods in detecting Ransomware. Many detection models that are developed with high accuracy have been discussed. Out of them, most of the models employ Machine Learning techniques for detection of ransomware as it facilitates automated detection. The proportion of the count (37.5%) of Machine Learning-based models is considerably higher than that of other models (3% each). The vital role of Machine Learning in developing automated detection tool is reviewed from different perspectives and the limitations of Machine Language-based model are also discussed.

This book discusses the AI in cybersecurity on various artifacts related to malware, image authentication, IoT, network, IoMT, facial recognition, criminal network, electronic payment system, etc. Hope this book serve the community who aspire to utilize the AI in cybersecurity and cyber forensics.

Halden, Norway/Ota, Nigeria   Sanjay Misra
Tamil Nadu, India   Chamundeswari Arumugam

Contents

About the Editors

Sanjay Misra is a Professor in Østfold University College, Halden Norway. Before Joining to Østfold University, he was full professor of Computer (software Engineering at Covenant University (400–500 ranked by THE(2019)) Nigeria for more than 9 yrs. He is PhD. in Inf. & Know. Engg (Software Engg) from the Uni of Alcala, Spain & M.Tech. (Software Engg) from MLN National Institute of Tech, India. As of today (21.05.2021)—as per SciVal(SCOPUS- Elsevier) analysis)—he is the most productive researcher (Number 1) https://t.co/fBYnVxbmiL in Nigeria during 2012–17,13–18,14–19 & 15–20 **(in all disciplines)**, in comp science no 1 in the country & no 2 in the whole continent. Total around 500 articles (SCOPUS/WoS) with 500 coauthors worldwide (-110 JCR/SCIE) in the core & appl. area of Soft Engg, Web engg, Health Informatics, Cybersecurity, Intelligent systems, AI, etc. He got several awards for outstanding publications (2014 IET Software Premium Award (UK)), and from TUBITAK (Turkish Higher Education and Atilim University). He has delivered more than 100 keynote/invited talks/public lectures in reputed conferences and institutes (traveled to more than 60 countries). He edited (with colleagues) 58 LNCS, 4 LNEE, 1 LNNS, 2 CCIS, and 10 IEEE proc, 4 books, EIC of IT Personnel, and Project Management, Int J of Human Capital & Inf Technology Professionals—IGI Global & editor in various SCIE journals.

Dr. Chamundeswari Arumugam is a Professor of Computer Science and Engineering Department at SSN College of Engineering, Chennai, India. Her research areas of interests include software estimation, software testing, software project management, cybersecurity, cyber forensics, machine learning, deep learning, and psychology. She has organized International conference IEEE ICCIDS 2019 and reviewed many conference paper and journal. Also she is one of the edited book authors of IGI Global. Guided many undergraduate and postgraduate students to publish research papers. Organized FDP, workshops for faculty members. She is a member of the Computer Society of India (CSI), IEEE and ACM. CSI student branch counsellor at SSN College of Engineering since 2009. She has been organizing many workshops, competitions, project colloquium, and seminars to the students, research scholars, and faculty members under the banner of SSN-CSI students chapter.

A Practical Experience Applying Security Audit Techniques in An Industrial Healthcare System

Julián Gómez, Miguel Á Olivero, J. A. García-García, and María J. Escalona

Abstract Healthcare institutions are an ever-innovative field, where modernization is advancing by leaps and bounds. This modernization, called "digitization", brings with it some concerns that need to be taken into account. The aim of this work is to present a way to combine the advantages of cybersecurity and artificial intelligence to protect what is of most concern: electronic medical records and the privacy of patient data. Health-related data in healthcare systems are subject to strict regulations, such as the EU's General Data Protection Regulation (GDPR), non-compliance with which imposes huge penalties and fines. Healthcare cybersecurity plays an important role in protecting this sensitive data, which is highly valuable to criminals. The methodology used to perform a security assessment process has been orchestrated with frameworks that make the audit process as comprehensive and organized as possible. The results of this study include a security audit on an industrial scenario currently in production. An exploitation and vulnerability analysis has been performed, and more that 450 vulnerabilities has been found. This chapter outlines a systematic approach using artificial intelligence to enable the system security team to facilitate the process of conducting a security audit taking into account the sensitivity of their systems.

Keywords Artificial intelligence · Cybersecurity · Healthcare · Pentesting · Security

J. Gómez (✉) · M. Á. Olivero · J. A. García-García · M. J. Escalona
Web Engineering and Early Testing, (IWT2, Ingeniería Web y Testing Temprano), University of Seville, Seville, Spain

e-mail: julgomrod@alum.us.es

M. Á. Olivero
e-mail: molivero@us.es

J. A. García-García
e-mail: juliangg@us.es

M. J. Escalona
e-mail: mjescalona@us.es

© The Author(s), under exclusive license to Springer Nature Switzerland AG 2022
S. Misra and C. Arumugam (eds.), *Illumination of Artificial Intelligence in Cybersecurity and Forensics*, Lecture Notes on Data Engineering and Communications Technologies 109, https://doi.org/10.1007/978-3-030-93453-8_1

1

1   Introduction

Cybersecurity is a field that is located within computer science but is related to other disciplines such as law, legislation, and security and control forces. According to ISO/IEC 27032:2012 [1], cybersecurity is "the preservation of the confidentiality, integrity, and availability of information in cyberspace. It is good to note that, according to ISO, there is a key difference between "cybersecurity" and "information security": "cybersecurity approaches focus on external threats and the need to use the information for organizational purposes, whereas, in contrast, information security approaches consider all risks, whether from internal or external sources." [2]. Other definitions, such as the NIST [3] definition, explain that cybersecurity is: "The prevention of damage, protection, and restoration of computers, electronic communications systems, electronic communications services, wired communications and electronic communications, including the information they contain, to ensure its availability, integrity, authentication, confidentiality and non-repudiation". In conclusion, the different definitions point out that cybersecurity is responsible for defending information in computer and electronic communication systems. Companies all over the world are pushed to use more and more digital services. In this digitalization, information systems play a key role. One of the biggest risks of this digitization is to do it abruptly, making the systems prone to attacks. These systems often contain sensitive or classified information, such as electronic health records (EHRs), intellectual property, or government or industry secrets. Personal data is always sensitive information, but even more so in the case of healthcare systems. The only way to verify that the system is storing such information reliably and securely is to conduct regular independent security audits.

To successfully manage and conduct these audits, a set of steps or phases must be followed. It is with a structured set of steps that the security audit is complete and comprehensive. These steps or phases are established through security frameworks. These frameworks are work from reliable and renowned sources such as OWASP and OSSTMM. It is the combination of the guides that these security organizations provide what makes the auditing phase as complete and sound as possible.

This chapter shows how a security audit of a real system related to the electronic health record (EHR) has been performed through an industrial case scenario. This audit has been performed by analyzing and exploiting IT vulnerabilities using frameworks that help to orchestrate the security audit. Artificial Intelligence tools have been used to assist in the audit process.

As a side note, this study has conducted its research in a healthcare system. However, the audit process is based on methodologies that are domain agnostic. The audit process can be applied to domain problems of all types: whether the domain is educational, military, or healthcare, as in the system in this paper. Since the audited system is a live system of an industrial healthcare institution, its name will not be disclosed for privacy reasons.

Following the recommendations of the work "A Step by Step Guide for Choosing Project Topics and Writing Research Papers in ICT Related Disciplines" [4], the chapter is organized as follows: Sects. 2 and 3 introduces the background and summarizes some related work. Section 4, presents the execution of the security audit and its results. This section is organized into subsections, each corresponding to one of the phases given by the audit frameworks. Finally, the last section presents a series of conclusions and suggests some lines of research for future work.

2 Background

Every year, ISACA publishes a summary of the state of the art, i.e. the state of the field and the latest innovations in cybersecurity. The report can be found on ISACA's official website [5].

The 2021 report obtains the data after conducting a macro survey in the last quarter of 2021 of industry professionals of different nationalities, ages, and with different years of experience. The pandemic situation experienced during this last year marks a great uncertainty in all aspects of professional life and cybersecurity is no exception. Demand is on the rise and more and more professionals are needed. More than 60% of respondents describe the state of their company's cybersecurity department as "significantly understaffed" and/or "somewhat understaffed." 55% say their company's cybersecurity department is in short supply. 55% say their company's cybersecurity department is understaffed. 55% say there is at least one vacancy in their company. Furthermore, 80% of respondents believe that the demand for professionals, far from decreasing, will increase in the short and medium-term. These vacancies are not being filled, in part, because 60% of professionals think that only 1 or 2 out of every 4 people who want to get a position are well qualified, with applicants failing the most in *soft skills*. Most think this can be solved by training technical personnel who are not directly related to security. Most of the HR department often does not understand the needs of the cybersecurity department.

In a conclusion, the report points out that the business understood from the traditional point of view does not work well. The cybersecurity workforce is scarce and, in the future, will probably remain scarce, because it takes a long time to train a professional from a theoretical point of view. Even more so because the most important thing is experience, which in the vast majority of cases comes after lengthy formal studies. However, applicants are the ones who fail the most in interpersonal and team skills, something that is not taught as much in universities as in the working world. To conclude, ISACA expects 2021 to be the year when companies start hiring as many professionals as there are vacancies to fill. In addition, it suggests not overestimating the effect of "digitized classes" on education, since the most lacking skills are interpersonal: the soft skills.

3   Security Audit Framework

This section will analyze how the security audit itself has been managed. The first step in choosing a methodology is to know what they are and what each one consists of the summary form. Some popular frameworks are OSSTMM, OWASP Web Security Testing Guide, NIST SP 800-115, PTES and ISSAF.

3.1   Standardized Frameworks

As for the security audit framework itself, the OSSTMM and the OWASP Web Security Testing Guide have been chosen. These two frameworks are *de facto* standards in the cybersecurity field. Unlike PTES or ISSAF, the first ones on the list are very popular and are still being updated today. In the case of OSSTMM, it is particularly effective, because, within the existing general and broad methodologies, it is the most comprehensive of all, considering aspects that other guides do not contemplate, such as the human aspect. As for the OWASP Web Security Testing Guide, it is a very popular guide for performing security audits on web applications. Since the system to be audited is a web system, this guide is ideal. Therefore, these are the two frameworks that will be used in the study.

Each of the phases in the following list will receive a subsection explaining what that phase consists of. First, the type of audits to be performed is established with the client.

In addition to the phases, there are different types of audits. Each of the phases will fall into one of the three types explained here:

- **Black box**. This audit is the process that would be followed by someone totally external to the system, who has no prior ideas about how the system is developed from the inside. This type of audit simulates the state that a cybercriminal who is going to attack a system would start from and has to gradually gather information.
- **White box**. This type of audit is the complete opposite of the black box audit, as the auditor now has all the information on how the system is developed inside and can see the source code. However, he has no prior knowledge of what attack vectors he is going to test yet, nor what vulnerabilities there may be.
- **Grey box**. This type of audit relies on the auditor having partial knowledge of the details of the system. In fact, it is the type of audit that most auditors start from: they know what type of system they are going to audit in broad strokes. For example, in the case of the system of this project, the only previous information before the security audit was that the system uses an Odoo 11 web system.

These three types of audits are not mutually exclusive and are performed all at once. Generally, what is done first is a passive scan of the application without knowing anything (black box) and then they go to the specific details (the source code, the configuration files, the platform where the application is deployed...).

Fig. 1 Phases of the security audit

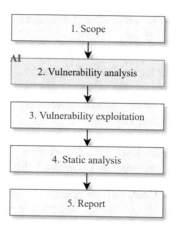

After seeing the types of audit, it will be explained which are the phases to be developed during the security audit. These phases have been established after combining the OSSTMM framework and the OWASP framework.

In addition to this combination, a new phase is proposed. This phase is called "Static Analysis" and aims to analyze the source code of the system to discover new vulnerabilities. More details on this phase can be found in the Sect. 4.

The visual representation of the flow of a security audit can be seen in Fig. 1, with "Static Analysis" being highlighted in blue. The phases are:

1. **Scope**. Definition of the scope and objectives of the audit. See Sect. 4.1.
2. **Social engineering**. Social engineering techniques will be used to try to breach the system starting from the weakest link in the chain: the users. In this paper, this phase will not be discussed due to length restrictions. See Sect. 5.
3. **Vulnerability analysis**. The vulnerabilities will be analyzed if they are corrected. In addition, it will be checked if the system is vulnerable to the vulnerabilities published in the OWASP Top 10. This is the phase that makes use of tools with Artificial Intelligence. See Sect. 4.2.
4. **Vulnerability exploitation**. Some of the vulnerabilities found will be exploited. See Sect. 4.3
5. **Static analysis**. The source code will be passed through code analyzers to find memory bugs and bad *input sanitization*. See Sect. 4.4.
6. **Writing the Report**. An executive report that contains all of the knowledge that has been generated. See Sect. 4.5.

3.2 Artificial Intelligence in This Project

This project presents how a security audit is performed in an industrial healthcare scenario. The tools that adjust most are the tools that help the security auditor in its job, and the existing tools for this objective are limited. Artificial intelligence is used in other fields of cybersecurity such as Malware Detection, Intrusion Detection or Social Engineering.

Nevertheless, some tools fit within Penetration Testing, the scope of the project. Such a set of tools has been obtained from the book titled "Machine Learning for Cybersecurity Cookbook" [6]. The book covers the following tools:

- CAPTCHA breaker
- Neural network-assisted fuzzing
- DeepExploit
- Web server vulnerability scanner using machine learning (GyoiThon)
- Deanonymizing Tor using machine learning
- Internet of Things (IoT) device type identification using machine learning
- Keystroke dynamics
- Malicious URL detector
- Deep-pwning
- Deep learning-based system for the automatic detection of software vulnerabilities (VulDeePecker).

Not all tools had a relation with this project, and thus some were removed. This is the case of VulDeePecker, which analyzes C/C++ code and the system of this project had no C/C++ code whatsoever. This project does not use Tor, nor CAPTCHA or IoT. In the end, only GyoiThon fits with the project, and it was the selected tool.

4 Execution of the Security Audit

Each of the proposed phases of the security audit will be covered in this section, except the social engineering phase, due to limitations in length.

4.1 Scope

This phase is of the kind "Grey box".

The first phase of a security audit is to define the scope. *Scope* refers to exactly which parts of the system are to be audited and under what conditions, not to be confused with the scope of the work itself, which is the performance of a security audit and the proposal of a new phase of the audit process.

To find out what alternatives there are to audit, the best option is to consult checklists that provide as complete a scope as possible. One such checklist is the OWASP ASVS checklist [7]. It provides the steps to follow to investigate a system in depth. Although this checklist is already quite comprehensive, it is good practice to consult other checklists to make the scope even more complete. Briefly summarizing the ASVS:

1. *Process abuse and social engineering*: Attempting to abuse the human element in a way that reveals confidential information or grants access to parts of the system to unauthorized users.
2. *Application identification and resource mapping*: An attempt will be made to execute footprinting techniques to find out the system architecture.
3. *Functionality and attack surface*: Evaluate which parts of the system are critical in its operation and assess how much the system can be compromised by attacking those parts.

- Authentication: Evaluate the identity verification process.
- Authorization: Evaluate the process of granting access and permissions.
- Validation of inputs: Determine which inputs are vulnerable and how they can be used to breach the application.
- Bugs in application logic: Determine if the application has any bugs in its programming.
- Web Server and Framework: Evaluate the web technology used by the server and whether it uses external frameworks that are vulnerable.
- Context-Dependent Encounters: Examine those parts of the system that are unique.

Once the alternatives that exist to audit are known, a meeting with the client to define the scope needs to take place. Some aspects that were **NOT** included in the scope were:

- Wireless networks and LAN.
- If processes of storing sensitive data are followed: such as NIST CSF, ISO 27001 (cybersecurity standard), HIPAA (USA's standard required in healthcare systems) or UK Data Protection act 1998.
- Servers: Firewalls, Groups and file owners, Updates, Backups, Malware scan, Use of antiviruses, IDS.
- Network and protocols: Port scanning, If cryptographic protocols such as SHA256 are followed, Passwords in the database are *hashed & salted*.
- Physical security: Access with card, Cameras and IPS.

Everything that was going to be audited was also discussed. The following list reflects what is **IN** the scope:

- This project's entire system: Login page, Internal page, Custom modules, Known vulnerabilities of Odoo 11
- Basis security protocols analysis such as SSL and HTTPS
- If the following can be obtained or modified: Configuration of any kind, Directory browsing, .htaccess.

4.2   Vulnerability Analysis

This phase is of the kind "Black box". It is in this phase that a tool with Machine Learning has been used.

The analysis of vulnerabilities is one of the most important phases to be carried out in a security audit. These vulnerabilities are present in information systems of all kinds: flaws in application logic, user entries that are not properly escaped or sanitized, redirections ..... The list is long. This section explores how vulnerability scanning can be done, what tools and techniques are available and what it is for.

To make vulnerability scanning as complete as possible, there are numerous guides with lists of vulnerabilities to check, both manually and with an automated tool. One such guide is the OWASP top 10 [8].

This top 10 is a list of the most common vulnerabilities in the web domain. The *superset* of this list is the OWASP ASVS, a much broader list that covers all types of web vulnerabilities. Both lists are important because in the case of the top 10 it is a shortlist that allows you to check if an application suffers from any of these ten vulnerabilities. However, if a more comprehensive list is needed, the OWASP ASVS can be consulted. The OWASP Top 10 has as a summary the following:

1. Injection: Injection flaws, such as SQL, NoSQL, OS, and LDAP injection.
2. Broken Authentication: Application functions related to authentication and session management are often implemented incorrectly, allowing attackers to compromise passwords, keys, or session tokens, or to exploit other implementation flaws to assume other users' identities temporarily or permanently.
3. Sensitive Data Exposure: Many web applications and APIs do not properly protect sensitive data, such as financial, healthcare, and PII.
4. XML External Entities (XXE): Many older or poorly configured XML processors evaluate external entity references within XML documents.
5. Broken Access Control: Restrictions on what authenticated users are allowed to do are often not properly enforced.
6. Security Misconfiguration: Security misconfiguration is the most commonly seen issue. This is commonly a result of insecure default configurations, incomplete or ad hoc configurations, open cloud storage, misconfigured HTTP headers, and verbose error messages containing sensitive information. Not only must all operating systems, frameworks, libraries, and applications be securely configured, but they must be patched/upgraded in a timely fashion.
7. Cross-Site Scripting XSS: XSS flaws occur whenever an application includes untrusted data in a new web page without proper validation or escaping, or updates an existing web page with user-supplied data using a browser API that can create HTML or JavaScript.
8. Insecure Deserialization: Insecure deserialization often leads to remote code execution.
9. Using Components with Known Vulnerabilities: Components, such as libraries, frameworks, and other software modules, run with the same privileges as the application.

10. Insufficient Logging & Monitoring: Insufficient logging and monitoring, coupled with missing or ineffective integration with incident response, allows attackers to further attack systems, maintain persistence, pivot to more systems, and tamper, extract, or destroy data.

Threats to the Coverage

As in other areas of software engineering, the discovery of these vulnerabilities is not a complete process. When an application is in the testing phase and there is no warning in the tool used to test the application, it does not mean that it is known with certainty that the application will not have bugs, but rather that the tool does not find more and that what has been tested does not give bugs, but it does not mean that there cannot be others. Normally, these tools use a *oracle*, which evaluates based on some parameters that are specified whether there are bugs in a part of the code or not. That is why it is said that the *testing* phase is not complete. It is not possible to discover all the bugs in an application and it will never be possible to ensure that software is free of bugs.

The same is true for security auditing. When using an auditing application or even when auditing manually without the help of tools, the certainty that the system will no longer have vulnerabilities is not a given. The program will check for known vulnerabilities and try to be as complete as possible, but there may be vulnerabilities that it cannot discover, for example, because they are of the type **zero day**, vulnerabilities that have not been published and that the person who knows about the vulnerability can exploit. This is why it is so difficult to determine whether an audit is complete or not. The only way to do so is to trust that the security auditor, with the experience and knowledge he has, will report all the vulnerabilities he finds and will always try to do as complete an audit job as possible.

Knowing that one tool cannot discover all vulnerabilities in a system, the following sections are presented.

Selected Tools

Vulnerability scanning tools compete in a hotly contested niche market. There are few free programs, and those that do offer free versions lack many of the features that are needed. These licenses can cost up to 3600 euros per year, as is the case with Nessus, or as affordable as 360 euros per year, as is the case with the Burp Suite. The best strategy is to combine several tools, as they often have different ways of reporting vulnerabilities, or directly report different vulnerabilities.

After comparing Nesus, SolarWinds MSP, OWASP ZAP, Burp Suite and Rapid7 InsightVM, two of the most widespread tools in the cybersecurity field were used to scan for vulnerabilities in a system: OWASP ZAP and Burp Suite. These two programs are dynamic scanners, i.e. they analyze the system in operation and take it as if it were a black box without seeing its source code, which automates the vulnerability discovery process.

Another tool was used, GyoiThon. This tool, according to the description of its GitHub page, is "GyoiThon is Intelligence Gathering tool for Web Server. GyoiThon executes remote access to a target Web server and identifies products operated on the server such as CMS, Web server software, Framework, Programming Language, etc. And, it can execute exploit modules to identified products using Metasploit. GyoiThon fully automatically executes above action."

The tool was used in this phase of the security audit because it helps with the information gathering of the server. Specifically, one of the modules of the tools relies on Machine Learning for doing its research.

4.2.1   OWASP ZAP

The configuration of the tools is not a trivial process, since, for example, the application to be audited has a login system whose access must be described in the tool.

The tool is configured by adding to the "scope" the URL to be audited. Concerning this login, the tool needs an anti-CSRF token, which in a nutshell, is a token that prevents a user from executing malicious content in his web browser from the web page to which he is authenticated.

The tool has two scanners. The passive scanner does not make any changes to the web application, i.e. it only performs GET requests. The passive scanner is also called *spider* and collects information from the application bit by bit. This scanner is also useful because it is fast, although it is not always effective if the web has a lot of AJAX content, for that there is another dedicated scanner. The system of this project, even without much AJAX content, has also been audited by this scanner. In addition, there is another type of scanner: the active scanner. This scanner makes changes to the application, i.e. it will be able to execute more HTTP commands, in addition, to GET, such as POST, PUT, DELETE..... It is a much more aggressive scanner, as it will try to exploit all the security flaws it finds. It should not be run on production systems, as it will attempt to crash the system. In fact, during the execution of this type of scan, the URL provided to me with the copy of the production system crashed due to the high number of requests. If this were to happen on a production system, it could be catastrophic.

Result

The tool, after being run for a couple of hours, finds all the vulnerabilities it can and then generates a report with all the vulnerabilities, their severity and their solution. Perhaps this report is the most important part of running the tool because it is what the development team uses to fix the vulnerabilities.

4.2.2 Burp Suite

The configuration of Burp suite is much simpler than that of ZAP. For login, Burp suite opens a browser window and copies clicks and keyboard entries and then plays them back. There is only one type of scan that removes all the vulnerabilities it finds. However, it does not have the HUD, so the security audit is not as interactive.

Result

As with ZAP, Burp Suite generates a report with all the vulnerabilities it finds. An example of this report can be found in [9].

4.2.3 GyoiThon

The tool has the following options of usage:

```
usage:
    .\gyoithon.py [-s] [-m] [-g] [-e] [-c] [-p] [-l --log_path=<path>]
    [--no-update-vulndb]

    .\gyoithon.py [-d --category=<category> --vendor=<vendor>
    --package=<package>]

    .\gyoithon.py [-i]
    .\gyoithon.py -h | --help
options:
    -s    Optional : Examine cloud service.
    -m    Optional : Analyze HTTP response for identify product/version
    using Machine Learning.
    -g    Optional : Google Custom Search for identify product/version.
    -e    Optional : Explore default path of product.
    -c    Optional : Discover open ports and wrong ssl server
    certification using Censys.
    -p    Optional : Execute exploit module using Metasploit.
    -l    Optional : Analyze log based HTTP response for identify
    product/version.
    -d    Optional : Development of signature and train data.
    -i    Optional : Explore relevant FQDN with the target FQDN.
    -h --help     Show this help message and exit.
```

In this project, the options **s, m, g, e and c** were used in the login page of the system. Regarding the Machine Learning analysis mode, it uses a Naive Bayes to identify the versions of the products, such as the server, the operative system, and the programming language.

Result

The tool generates a report with information of the investigated URL, the vendor name, the version of the product, the type of the product, and perhaps the most useful field, the vulnerabilities associated with the version of the used product. Thanks to this tool, the DevOps of the system were informed that the version of the operating system that was in use had vulnerabilities, and was consequently updated.

4.2.4 CVEs

In addition to having checked everything else, another important aspect to check is the CVEs. These CVEs are registered vulnerabilities that usually have as common information the version of the software they affect and a score that determines the seriousness and severity of the vulnerability.

These CVEs are published in public databases where you can check if there is an exploit associated with that CVE. An exploit is a program that exploits the vulnerability automatically without the person running the code needing to know what the process is to exploit the vulnerability or why it works. Databases, such as *cvdetails.com* [10] or such as *repology.org* [11] have lists of many vulnerabilities found in Odoo.

These CVEs usually carry an explanation of how the vulnerability was found. In this case, at the bottom of the figure, with the arrow pointing in red, we can follow a link to the *GitHub* [12] of the *issue* where aspects such as the description of the vulnerability, the impact, how it was fixed and in which *commits* the fix was made are detailed.

There is no way to audit CVEs associated with systems that do not have *exploits* published. The only way to ensure that the audited system is secure against that CVE is to ask the system administrator to always keep it updated with the latest version. It is for this reason that computer users and system administrators are strongly urged to keep their system up to date, otherwise, the system may be vulnerable to a CVE.

Although most Odoo CVEs do not have *exploits* associated with them, there are some that do, such as the CVE-2017-10803 [13]. Although the *exploit* itself is not associated on the *cvedetails.com* website, we can look for the available *exploit-db.com* [14].

To be more specific, we can search for vulnerabilities affecting the audited version of Odoo 11. One of the most serious vulnerabilities affecting Odoo 11 is, for example, CVE-2018-15632 [15]. This vulnerability, however, does not have an associated exploit. Neither do the other vulnerabilities, thus concluding with CVEs research.

4.3 Vulnerability Exploitation

This phase is of the kind "Black box".

This section aims to further investigate vulnerabilities that have been discovered in the previous sections, as well as to consider as false positives some vulnerabilities that in reality are not. No new tools have been used to investigate the vulnerabilities, except for those already used (OWASP ZAP and Burp suite), as all tests have been performed manually to eliminate false positives.

The analysis tools give the severity and confidence of the vulnerability. Its severity indicates the priority that its remediation should have, and the confidence indicates the certainty that the vulnerability is present or whether it is a false positive. Both tools produce tables that indicate the severity of the vulnerability and, in the case of Burp suite, also indicate how confident Burp suite is about a false positive. The executive report contains a step-by-step reproduction of the exploitation of the vulnerability, the so-called proof of concept.

Scope

The OWASP ZAP report has reported 16 alerts, with a total of 395 instances. The Burp suite has reported 74 vulnerabilities. In total there are more than 450 vulnerabilities to be investigated. Typically, in an enterprise environment, an estimate of how long it would take to investigate all vulnerabilities is usually given based on the number found. Due to limitations in the length of the work presented, two of the most critical vulnerabilities reported by OWASP ZAP will be investigated further and two others by Burp suite.

4.3.1 Vulnerability 1: Redirects

This first vulnerability is present in two forms: as *External redirection* and as *Reflected XSS*. They correspond to the first two vulnerabilities in the OWASP ZAP report and have a high severity and medium confidence. These vulnerabilities have not been flagged by the Burp suite as vulnerabilities. If we look at the URL, the common path of the two vulnerabilities is.

```
https://o****n.com/web/session/logout?redirect=
```

This route itself is very dangerous because it is vulnerable to an XSS and redirection attack. An XSS attack is an attack that relies on an attacker being able to inject Javascript code into a web page. In this case, an attacker could inject code to steal users' cookie information, for example. The redirect attack consists of the website redirecting to another external website.

If the URL is further investigated, the logout page shows a new appearance. Indeed, there are two hidden fields: one with an anti-CSRF token and one with the name *redirect*, which corresponds to the *redirect* of the URL marked as vulnerable. Therefore, it is concluded that the login page also has this vulnerability. If an attacker wanted to exploit it, he would only have to put in the URL a redirect parameter like the following:

```
https://o***n.com/web/login?redirect=https://
github.com/
```

and a URL to which you want to redirect. This way, after users enter their data, they will be redirected to *github.com* (in this example). This is extremely dangerous because, if an attacker wants to exploit it, the end user is unlikely to notice because of the following: the URL has to carry the address of another website, but this can be hidden using a URL shortener, so the URL could become, for example, the following:

```
https://bit.ly/3zxyyTi0
```

This way, the end user will not realize that the *login* page has a hidden field and will be redirected to another website afterward. If the redirected website is a website similar to the one in this project, the user may try to enter their data thinking it is the original website, when in fact it is being stolen. This attack is called water holing.

To fix this vulnerability, it is recommended that the development team try to do the *Login* and *Logout* redirects without relying on hidden form fields, but instead do it with session or internal *cookies*.

4.3.2 Vulnerability 2: DOM XSS

This vulnerability has nine instances. It corresponds to vulnerabilities 2.1–2.9 of the report generated by the Burp suite. All these instances are grouped under the name "Cross-site scripting (DOM-based)", with high severity and tentative confidence. These vulnerabilities have not been flagged by OWASP as vulnerabilities.

The exploitability of these vulnerabilities, if you look at the report notes, depends on the jQuery libraries. These two vulnerabilities are indeed the same and occur due to multiple vulnerabilities that are present in jQuery version 1.11.1. It is recommended to upgrade to the latest version of jQuery and always keep your dependencies up to date.

4.3.3 Vulnerability 3: SQL Injection

This vulnerability has been reported by the Burp suite. It corresponds to the first one in the report and has a high severity and strong confidence. SQL injection is based on the fact that by using characters that cause queries to escape, such as single quotes ('), modifications can be made to the database.

The vulnerable URL is as follows:

```
POST https://o***on.com/web/database/duplicate
```

In the body of the request, the following is added

```
master\_pwd=admin\&name=\%00'\&new\_name=wwu3j
```

Burp suite says it has discovered the vulnerability using the **null byte (0x00)** to get such a bug message:

```
Database duplication error: unterminated quoted
identifier at or near """ LINE 1:
```

```
...b6ax9b6a8sd8deb9hj69h53nma" ENCODING
'unicode' TEMPLATE *^
```

From that error message, Burp suite concludes that the application is vulnerable to SQL injection, in addition to the database being a PostgreSQL. However, after manually checking the strings listed below, it has been determined that the application is not vulnerable to SQL injection and that this is a false positive. If it had indeed been vulnerable, the application would have given information from its database. However, the application continued to give the same error message and no information.

The strings tested for SQL injection are of the kind *%00'SELECT version()*. Since the database is PostgreSQL, we can use *SELECT version()* to guess the version. All strings have also been tested with *URL encoding*.

4.3.4 Vulnerability 4: Path Traversals

Path traversals (also called Directory traversals) are vulnerabilities that allow attackers to break out of the *var/www* folder where the web is hosted on the server and be able to query and modify files on the server. This vulnerability has been reported by OWASP ZAP with eight instances that have high severity and low confidence.

There is a total of eight declared instances, which have different URLs. However, in all URLs the parameter that could be changed to an escape path is the last one. For example, in the first instance, it is:

```
https://o***n.com/web/dataset/call_kw
/mail.message/load_views
```

This should change *load_views* to an escape path. To exploit this vulnerability, escape paths of the kind *../../../etc/passwd* and with encoding such as *..%252f..%252f..%252fetc/passwd* were used.

If the server is vulnerable it should return the file *etc/passwd*. However, after testing the escape paths listed in the snippet above, no way to escape to parent folders containing other files has been found, so it is declared as a false positive.

4.4 Static Analysis

This phase is of the "white box" type. Static code analysis is performed on the code without executing it. The main difference with the programs used in the other sections is that with those programs the analysis was done on the live application, while for this static analysis only the source code is needed.

Static analysis makes it possible to detect vulnerabilities that would otherwise not be easily detected because the functionality will not be executed and will need a stimulus or because a piece of code is not exposed to users.

This static analysis of code is not usually done in audits that follow frameworks, perhaps because they are more associated with its execution in the pipeline of the Continuous Integration tool when the code is uploaded to a repository. However, the power and usefulness provided by these analyses should not be underestimated, as security analysis programs often overlook vulnerabilities that would be impossible to detect without reading the code.

It is assumed that a cybercriminal would not have access to the source code of an application in any case, but that would be relying too much on the fact that such access is impossible to obtain. In both cases, and as we saw in the chapter on social engineering, cybercriminals can gain access to the source code not by a computer but by social means. It is this distrust that leads us to improve the source code of our system, even when it is difficult—not impossible—for a cybercriminal to gain access.

How It Works

Code parsers usually have an abstract syntax tree (AST) underneath them. These ASTs represent a program in a hierarchical way, whose representation is the same as that of compilers. They are said to have an abstract syntax because the representation does not depend on the language being used. Thus, a function would have the same representation in C, Java or Python. In ASTs, the inner nodes represent operators and the last nodes, called leaves, represent variables. To identify vulnerabilities, static analyzers using this technique traverse the tree looking for flaws in the logic of a function.

Tools Used

Multiple lists that compile static tools that are open source. An example of one of the most comprehensive lists is [16]. After obtaining the security tools from the list discussed above, a comparison was made between the following tools: py_find_injection, Sonarqube, Dlint, AttackFlow, pyre and Bandit. The tools selected for the project were Bandit, Dlint and Sonarqube.

4.4.1 Bandit

Bandit is a PyCQA tool [17]. According to its description on GitHub, it is "a tool designed to find common security problems in Python code. To do this Bandit processes each file builds an AST from it and runs the appropriate plugins against the nodes in the AST. Once Bandit has finished scanning all files, it generates a report". The installation of the tool is very simple and its execution is automatic. As an example, the following snippet shows a vulnerability discovered by Bandit.

```
>> Issue: [B110:try_except_pass] Try, Except,
Pass detected.
```

```
Severity: Low    Confidence: High
Location:  ./src/addons/imedea_andrology
/models/andrology_episode.py:317
More Info: https://bandit.readthedocs.io/en
/latest/plugins/b110_try_except_pass.html
```
```
316                 # self.save_pdf_episode(
```

4.4.2 Dlint

Dlint is a static analysis tool of Dlint-py [18]. Like Bandit, its goal is to detect security flaws in Python code. Its installation and configuration are straightforward. As an example, the following snippet shows a vulnerability discovered by Dlint.

```
./src/addons/imedea_project/models/cycle_items
/defrosting.py:3:1: DUO107 insecure use of XML
modules, prefer "defusedxml"
./src/addons/imedea_project/models/cycle_items
/embryonic_study.py:3:1: DUO107 insecure use of
XML modules, prefer "defusedxml"
```

4.4.3 Sonarqube

Sonarqube [19] is a platform that performs static code analysis. Sonarqube has several versions: the community version (open source) and the developer, enterprise and data center versions (paid). One of the most useful features of Sonarqube is the generation of reports. Unfortunately, reports can only be generated natively in the paid versions and in the community version you have to resort to a plugin developed by the community [20]. This plugin is only compatible up to version 8.2 of Sonarqube, from February 2020, when the latest is 8.9, from June 2021. This is why Sonarqube has been run twice in total, once with 8.2 to generate the report that will go later in the Executive Report, and once where it is run without report. The two versions report different vulnerabilities and bugs, so it is useful to run both versions.

4.5 Executive Report

The executive report (also called executive summary) is a document that is delivered to the different departments of a company so that all personnel can understand the work that has been done and the vulnerabilities of the work. For confidentiality reasons, the report that has been generated for this project cannot be publicly disseminated, but the report was delivered to the company. It is advisable to go to public repositories such as [21], which collects several real examples.

There are two main readers of this report: senior management, who are interested in the general high-level details, and developers who have to fix vulnerabilities through a series of technical steps. The report includes these two sections: a general one, in the beginning, as a high-level summary, and then a low-level one, which explains the vulnerabilities that have been found. Vulnerabilities are usually classified by their severity, by the degree of confidence that the vulnerability exists and is not a false positive, and by their category (SQL injection, XSS...). Accompanying this classification should be the Proof of Concept (POC), i.e. the exact steps to exploit all vulnerabilities, as well as some ideas to fix the vulnerability.

In the end, all the reports generated by the automatic tools are included for the record and, in case of doubt, a developer can go to the report of a particular tool and get the information directly, without going through the security auditor's brief.

This report is the part that a security auditor usually likes the least because it is based on creating documentation and not on investigating vulnerabilities. However, it is perhaps the most crucial part of the audit. During the previous sections, there has not been any material produced that is of any use. This report is so important because it is really what the client pays for. Imagine that the auditor, after four months of work, leaves the audit firm just before writing the report. In that case, the client will have nothing, and the work will have been in vain. Moreover, if the auditor had to verbally communicate to the developers all the existing vulnerabilities, there would be no organization and the developers' remediation work would be more arduous. The importance of this report is quite remarkable.

5   Conclusions and Future Work

Companies around the world are going digital, leading them to use more and more healthcare systems. These healthcare systems are not exempt from security holes, which cybercriminals can exploit. Hence the need to perform security audits, especially in the audited system, since being a system in production in a healthcare institution, the data it contains are very sensitive. In this project, an audit has been performed on a system in production implemented in an industrial scenario, more specifically on a healthcare system currently in production. The system uses Odoo 11 ERP to store its patient data.

The study shows how a real security audit is performed on a live industrial healthcare system. More specifically, first of all, in the scope section, it is indicated what can be audited. In the analysis Sect. 4.2 tools are run that identify more than 450 computer vulnerabilities and in the exploitation Sect. 4.3 an attempt is made to drill down to the most critical vulnerabilities. In the static analysis Sect. 4.4 another analysis is performed. This time, the tools are run only on the custom code of the modules. Finally, all the knowledge generated is collected in the Executive Report Sect. 4.5.

The study uses a tool that uses Machine Learning to make the analysis. The tool looks for the version of the products that a system uses (such as the operative system, information of the server engine, the programming language...) and checks if there are CVEs associated with that version of a product. It is with this tool that uses Artificial Intelligence that some vulnerabilities that could not have been discovered, were discovered and resolved.

In light of the results, it can be concluded that critical and highly confidential systems, should be subject to periodic security audits, to to protect such confidentiality. Future work might be oriented in various directions according to the examined perspectives and identified gaps, such as the social engineering one. A social engineering audit, which explores the human factor of a system, maybe as important as a technical audit of the system since there is no point in having a technically fortified system if employees later reveal sensitive information. Finally, it is remarkable to say that the lack of tools that use artificial intelligence to perform penetration analysis is a limiting factor to be taken into account. Little by little, more and more tools are appearing, and artificial intelligence is undoubtedly joining the field of cybersecurity[1]. The intersection of artificial intelligence and cybersecurity is an effective symbiosis, but one that still needs more research.

Acknowledgements This work has been partially supported by the NICO project (PID2019-105455GB-C31) of the Spanish the Ministry of Science, Economy and University and NDT4.0 (US-1251532) of the Andalusian Regional Ministry of Economy and Knowledge. This research has been developed with the help of G7innovation.

References

1. ISO (2012) Information technology—Security techniques—Guidelines for cybersecurity. https://www.iso.org/standard/44375.html
2. ISO (2018) Information technology—Security techniques—Guidelines for cybersecurity. https://www.iso.org/obp/ui/#iso:std:iso-iec:tr:27103:ed-1:v1:en
3. NIST (2021) Cybersecurity. https://csrc.nist.gov/glossary/term/cybersecurity
4. Misra S (2021) A step by step guide for choosing project topics and writing research papers in ICT related disciplines. In: Misra S, Muhammad-Bello D (eds) Information and communication technology and applications. Springer International Publishing, Cham, pp 727-744. ISBN: 978-3-030-69143-1
5. ISACA (2021) State of cybersecurity. https://www.isaca.org/go/state-of-cybersecurity-2021
6. Tsukerman E (2019) Machine learning for cybersecurity cookbook. Packt
7. OWASP (2019) OWASP application securityverification standard. https://owasp.org/www-project-application-security-verification-standard/
8. OWASP (2017) OWASP top ten. https://owasp.org/www-project-top-ten/
9. Burp suite (2019) Burp suite scanner sample report. https://portswigger.net/burp/samplereport/burpscannersamplereport
10. CVEdetails (2021) CVEdetails.com. https://www.cvedetails.com/vulnerability-list.php
11. repology (2021) repology.org. https://repology.org/project/odoo/cve
12. Valov P (2020) SoCyber. Security advisory CVE-2018-15632. GitHub. https://github.com/odoo/odoo/issues/63700
13. Ayrx (2017) Vulnerability details: CVE-2017-10803. CVEdetails. https://www.cvedetails.com/cve/CVE-2017-10803/
14. exploit-db (2021) exploit-db.com. https://www.exploit-db.com/exploits/44064

---

[1] https://g7innovation.com/en.

15. Valov P (2020) SoCyber. Vulnerability details: CVE-2018-15632. CVEdetails. https://www.cvedetails.com/cve/CVE-2018-15632/
16. analysis-tools-dev. github.com/analysis-tools-dev/static-analysis. GitHub (2021). https://github.com/analysis-tools-dev/static-analysis
17. PyCQA (2021) bandit. GitHub. https://github.com/PyCQA/bandit
18. dlint-py (2021) dlint. GitHub. https://github.com/dlint-py/dlint
19. sonarqube (2021) Sonarqube. https://www.sonarqube.org/
20. cnescatlab (2021) cnescatlab/sonar-cnes-report. GitHub. https://github.com/cnescatlab/sonarcnes-report
21. Github:juliocesarfort (2021) juliocesarfort/public-pentesting-reports. GitHub. https://github.com/juliocesarfort/public-pentesting-reports

Feature Extraction and Artificial Intelligence-Based Intrusion Detection Model for a Secure Internet of Things Networks

Joseph Bamidele Awotunde◉ and Sanjay Misra◉

Abstract Security has been a concern in recent years, especially in the Internet of Things (IoT) system environment, where security and privacy are of great importance. Our lives have significantly transformed positively with the emergence of cutting-edge technologies like big data, edge and cloud computing, artificial intelligence (AI) with the help of the Internet, coupled with the generations of symmetric and asymmetric data distribution using highly valued real-time applications. Yet, these cut-edge technologies come with daily disastrous ever-increasing cyberattacks on sensitive data in the IoT-based environment. Hence, there is a continued need for groundbreaking strengths of AI-based models to develop and implement intrusion detection systems (IDSs) to arras and mitigate these ugly cyber-threats with IoT-based systems. Therefore, this chapter discusses the security issues within IoT-based environments and the application of AI models for security and privacy in IoT-based for a secure network. The chapter proposes a hybrid AI-model framework for intrusion detection in an IoT-based environment and a case study using CIC-IDS2017and UNSW-NB15 to test the proposmodel's performance. The model performed better with an accuracy of 99.45%, with a detection rate of 99.75%. The results from the proposed model show that the classifier performs far better when compared with existing work using the same datasets, thus prove more effective in the classification of intruders and attackers on IoT-based systems.

Keywords Security and privacy · Internet of Things · Intrusion detection systems · Artificial intelligence · Deep learning · Cloud computing · Edge computing · Symmetric · Asymmetric data

J. B. Awotunde (✉)
Department of Computer Science, University of Ilorin, Ilorin, Nigeria
e-mail: awotunde.jb@unilorin.edu.ng

S. Misra
Department of Computer Science and Communication, Østfold University College, Halden, Norway

© The Author(s), under exclusive license to Springer Nature Switzerland AG 2022
S. Misra and C. Arumugam (eds.), *Illumination of Artificial Intelligence in Cybersecurity and Forensics*, Lecture Notes on Data Engineering and Communications Technologies 109, https://doi.org/10.1007/978-3-030-93453-8_2

1   Introduction

The emergence of innovative technologies like the Internet of Things with storage resources of cloud computing has resulted in the generation of big data called big data. This has led to the witness of massive data generation by humans through IoT-based devices and sensors [1], thus changing the world of businesses in various aspects and society in general [2]. The authors in [3] argue that these cutting-edge technologies recently drive the global market with the connecting and productive big data managed by big data analytics. Hence, these infrastructures have created attractions from the business industries and the government and resulted in the illegal accessibility of these valuable and sensitive data globally [4]. But these big data in real-world applications have been categorized into asymmetric and symmetric data distribution.

The symmetric data comes from the relationship of social networks users, and the asymmetric data comes as a result of regular network traffic with the probability of dissemination of various malicious within network protocols. There are still great hidden patterns and knowledge within real-world applications, irrespective of the missing information. Hence, it resulted in an effective and efficient way of purifying various valuable patterns from these huge data generated from the IoT-based systems, and this becomes significant in such an environment [5].

These ubiquitous technologies have really impacted people's lives in respective of background, race, and every aspect of society, and these have resulted into various kinds of attacks on these pervasive technologies, and the growing dependency on the use of Internet facilities have led to a continuous risk against the nodes and protocols of the network [6]. Thus, the ubiquitous technologies need incorporated and tangible security solutions for proper security and privacy platform. The most important features of cyberspace security are confidentiality, integrity, and availability (CIA). Anything cut short of these features by negotiating the CIA or bypassing these technologies' security components is called cybercrime or network intrusion [7, 8].

With the rapid growth of ubiquitous technologies and the inception of the internet, several kinds of cybercrime or attacks have grammatically evolved globally. Not minding the tireless efforts of various experts in cybersecurity developing various defense techniques, intruders have not relented and have always found a way of targeted, valuable resources by launching automated, cultured, and adaptable cyber-attacks. These attackers have causes remarkable mayhem to individuals, governments, and even various businesses worldwide [9]. A report form authors of [10] have shown that by 2021 from cybersecurity over six trillion US dollar may be lost due to various cybercrimes, and these several cutting-edge attacks could have resulted to loss of billion dollar worldwide. These result from over five million cybercrimes recorded daily through computers that have been compromised, thus a whopping 1.5 trillion US dollars. Consequently, due to the intrinsic ability of Intrusion Detection Systems (IDSs) to detect an intrusion in real-time, the methodology in recent times has witnessed increasing popularity [11].

The IDS is a difficult field that deals with detecting cyber-threats such as hostile activities or policy violations on data networks by examining the information included in the data packets that have been transmitted [12]. The data packets' contents are converted into a vector of continuous and categorical variables such as size, addresses, and flags, among other things that denotes the existence of a network link. This vector can be compared to pre-registered vectors associated with normal traffic or attacks like signature-based intrusion detection (ID), looking for comparable patterns [12]. To detect attacks, the vector might be utilized as an input to statistical or machine learning classification methods.

For instance, the authors of [13] provide a good summary of the importance of security features in cloud computing platform surveillance. They also presented a three-level cloud-based IDS that employed rules to express event Calculus's definition and monitoring aspects. Additionally, the suggested technique made advantage of the hypervisor framework to focus on application supervision during runtime and facilitate automatic reconfiguring of these programs. Finally, the article claimed to have considerably enhanced the security of cloud computing. The ID is the process of measuring and reviewing events occurring in a computer system or network for evidence of intrusion [14].

Furthermore, they describe an intrusion as an attempt to circumvent a network's or computer system's security safeguards, thereby jeopardizing the system's CIA. Finally, based on network packets, network flow, system logs, and rootkit analysis [15, 16], authors define an IDS as a piece of hardware or software that monitors various malicious actions within computer systems and networks. The misused detection (knowledge or signature-based) and anomaly-based approaches are the two basic approaches of detecting intrusions within computer systems or networks. However, the hybrid-based strategy has exploded in popularity in the last decade, combining the benefits of the two ways outlined above to create a more robust and effective system [17].

There are a variety of traditional methods for ID, such as access control systems, firewalls, and encryption. These attack detection systems have some drawbacks, especially when systems are subjected to a large number of attacks, such as denial of service (DOS) attacks. Furthermore, the systems can achieve higher false positive and negative detection rates. Researchers have applied AI models for ID in recent years with the goal of boosting attack detection rates over traditional attack detection methodologies. Since simple machine learning algorithms have significant drawbacks, and security threats are on the rise. The newest versions of AI learning models are needed, especially for the selection of features and intrusion analysis. Therefore, this chapter presents the security issues within IoT-based environments and discusses the state-of-the-art AI models for ID in an IoT-based environment for a secure network. The chapter also proposes a particle swarm optimization (PSO) model to extract relevant features from the datasets, and Convolution Neural Network (CNN) was used to classify the intruder within the IoT-based environment. The proposed system can automatically perform the selection of significant features that can be used for the classification of the datasets. Hence, the major contributions of this chapter are:

(i)   The chapter proposed a novel feature extraction based on PSO algorithms, and CNN was used to identify and detect an attacker within an IoT-based network. The combined algorithms were utilized to make use of their capabilities while avoiding computational overhead expenses.

(ii)   The proposed system was evaluated using two widespread and recent datasets by analyzed the capture packet file within a network and using various performance metrics like accuracy, precision, recall, F1-score, and ROC, respectively.

(iii)   The model was compared with the state-of-the-art methods basic of conventional AI-based models. The findings show that the model outperforms recent work that uses the same datasets.

(iv)   The IDS model also achieves huge scalability with meaningful reduction of the training time and giving low probability of false alarms rate with an overall high degree of accuracy when compared with existing methods.

The remaining part of this chapter is as follows: Sect. 2 presents the security issues in the Internet of Things environments, Sect. 3 discusses the applications of Artificial Intelligence for security and privacy in Internet of Things systems. Section 4 presents the methodologies used, and Sect. 5 discusses the results with a comparative analysis of the chapter. Finally, Sect. 6 concluded the chapter.

2   The Security Issues Within IoT-Based Environments

Because of the growing number of services and users in IoT networks, the security of IoT systems has become a critical concern [7]. Smart things become more effective when IoT systems and smart surroundings are integrated. The consequences of IoT security flaws, on the other hand, are extremely harmful in vital smart contexts such as health and industry [18]. Applications and services will be at risk in IoT-based intelligent devices without adequate security mechanisms. Information security in IoT systems demands more research to meet these challenges [19, 20]. The CIA are three fundamental security principles of applications and services in IoT-based embedded systems. IoT-based smart houses, for example, suffer security and privacy issues that cut across all layers of the IoT framework [21].

The security of IoT systems and the complexities and interoperability of IoT settings are significant impediments to the establishment of intelligent devices in the physical world [22]. Attacks on IoT networks, such as DoS or DDoS attacks, have an impact on IoT services and consequently on the services provided by embedded systems. Researchers look at the IoT's security concerns from a variety of perspectives, including the security susceptibility of IoT routing protocols [23, 24]. This chapter will concentrate on IDSs for IoT-based systems, regardless of protocol.

The security vulnerabilities that arise in the various IoT layers are the source of IoT security concerns. The physical layer faces obstacles such as physical damage, hardware failure, and power limits. The network layer faces issues such as DoS assaults, sniffers, backdoor attacks, and illegal users. The application layer faces issues such as malicious code attacks, application vulnerabilities, and software flaws [25]. According to [26], any IoT system's security issues can be divided into four categories: Threats to authentication and physical security, as well as dangers to confidentiality, data integrity, and privacy.

For IoT-based users around the world, cyber security is a top priority. However, some concerns go beyond conventional cyber threats and can result in severe security breaches. Figure 1 displays the security and privacy in IoT-based systems.

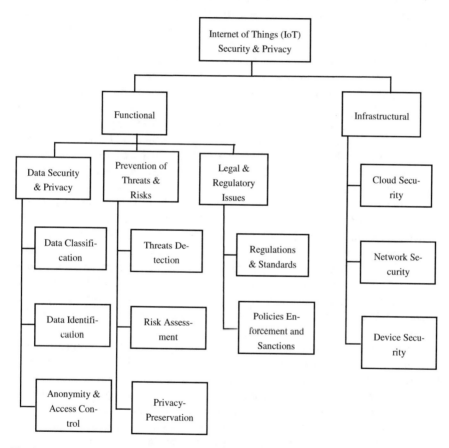

Fig. 1 The security and privacy in internet of things systems

2.1 The Following Are Some Examples of Malicious Threats

Security Problem in RFID

The RFID system isn't without flaws. It has a wide range of applications. RFID is vulnerable to a variety of security threats and problems, all of which must be handled and addressed in WHD (wearable health care devices). Confidentiality and key management are essential. One of the most widely used techniques for automatically identifying things or individuals is radio frequency identification (RFID). The RFID application, which is based on a combination of tags and readers, is widely employed in a variety of industries, including distribution networks, engineering, and transportation control systems. Despite its many advantages, however, the technology raises a lot of hurdles and concerns, particularly in terms of security and privacy, which are deterring more researchers. RFID systems, like other devices and networks, are susceptible to both physical and electronic attacks. As technology advances and becomes more accessible, hackers who wish to steal private information, get access to restricted areas, or bring a system down for personal benefit are becoming more common. Spying occurs when an unauthorized RFID reader listens in on conversations between a label and a reader and obtains classified information. The hacker must also understand the basic protocols, tags, and reader information in order for this technique to work.

All that is required for the assault on force research is a hacker's brain and a cell phone. According to top specialists, power analysis assaults on RFID devices can be mounted by monitoring the energy usage levels of RFID tags. When examining the power pollution levels of smart cards, researchers discovered an intrusion method, specifically the difference in supply voltages between valid and incorrect passwords. RFID tags and readers, like other goods, can be reverse-engineered; however, to get optimal performance, a thorough grasp of the protocols and features is required. Attackers will deconstruct the chip to figure out how it works in order to accept files from it.

During signal transmission, a man-in-the-middle attack occurs. Similar to eavesdropping, the attacker waits for communication between a label and a user before intercepting and modifying the data. While posing as a standard RFID component, the attacker intercepts the unique indication and then sends incorrect data. A Denial of Service attack is any RFID device malfunction that is linked to an attack. Physical attacks are widespread, including utilizing noise interference to jam the device, obstructing radio signals, and even erasing or deactivating RFID labels.

Cloning and hacking are two distinct procedures that are frequently carried out at the same time. Cloning is the process of transferring data from an original tag and applying it to a modified tag to get access to a restricted area or object. Because the intruder must know the label's details to reproduce it, this type of assault has been utilized in access control and inventory management operations. Viruses may not have adequate storage capacity in RFID tags right now, but they could represent a substantial threat to an RFID system in the future. When a virus programmed on an RFID tag by an unknown source is read at a plant, it has the potential to bring the

RFID device to a halt. The virus moves from the sticker to the reader, then to corporate servers and apps, resulting in the failure of associated devices, RFID modules, and networks.

Distributed Denial of Service Attacks (DDoS)

Since the introduction of non-legacy IoT devices, DDoS attacks have become increasingly dangerous. Attackers may now use the weak security implementation of IoT devices to gain control of them and use them to launch an attack on the targeted system or network. The number of attacks has been shown to increase as the cost of adding additional IoT devices rises. A DDoS attack's principal goal is to deny legitimate users access to channel and latency facilities, resulting in service interruption [27, 28]. The invader begins with non-legacy IoT systems like CCTV cameras, camcorders, baby tracking devices, and wearable gadgets, which have insufficient built-in security and other flaws, including low computational power and energy density.

IoT systems are not just difficult to attack, but they are also cheap. Attackers can obtain control of compromised IoT devices for free or at a fraction of the cost of hosting a server rather than investing in and maintaining expensive networks to launch powerful DDoS attacks. Companies do not maintain track of a device's security credentials until after it has been released to the public. Hackers take advantage of several authentication flaws in the code. The makers do not release security fixes for these devices that correct the flawed software. If an attacker gains control of a compromised IoT device, he or she is free to change the device's security credentials. Suppose the infected computer is ever tracked for the duration of the attack. In that case, the device's vendor or manufacturer will be unable to retune the safety permits and reclaim control from the invader. The invader plans to exploit the system to cause as much harm as possible to the victim for as long as feasible.

Mobile Devices for Internet of Things Services

Mobile devices with secure credential storage, increased storage capacity, wireless networking interfaces, and computer power can now be utilized in healthcare to collect crucial health parameters, as in Body Area Networks, and manage healthcare. The importance of privacy and protection in IoT-based systems cannot be overstated [29]. The IoT-based system's users must be aware of the potential of security vulnerabilities and information manipulation and practices becoming accessible on mobile devices as more devices and sensors become available [30]. The use of tablets and handheld devices by various users of IoT-based platforms elevates the potential of security breaches on both sides of the IoT-based settings. An intruder will install sophisticated malware in cellphones, which will remain dormant until the person in possession of the malware-infected computer enters a specific place, at which point the virus will be activated. As a result, an adversary can employ malware to tarnish a hospital's reputation by activating it whenever users of malware-infected PCs visit the facility.

IoT-based systems are becoming increasingly dangerous, and any disruption or abuse could result in significant financial loss or even life-threatening difficulties.

Weak authentication mechanisms could allow a malicious attacker to get access to sensitive data and shut down all hospital systems. As a result, it's critical to ensure the safety of patients, linked devices, and hospital networks and make the operating ecosystem immune to such attacks [31, 32].

An attacker can steal a client's medical record using a Man-in-the-Middle (MITM) attack on the communication network. This allows the intruder to quickly collect plain-text data from internet traffic and change a message. EMRs may also be obtained by the opponent using malicious software portable apps used by patients. The public uploading of the EMR on the network will jeopardize patients' privacy, especially for those who do not want their health problems publicized. Reverse technology is the process of creating things out of thin air. An attacker can use malicious software on mobile devices to interact with medical equipment and supply incorrect data through the application layer of the medical devices. Control system errors can lead to a physician making the wrong decision, which can have major ramifications for the patient's health [33, 34].

Unintentional Misconduct

IoT-based security is not always compromised by unscrupulous individuals seeking to harm others. Twelve percent of security issues in IoT systems were caused by unintended human behavior that resulted in a breach of patient data protection. These errors might range from misplacing a patient's file to malfunctioning security equipment. They can also happen when old computers with patient data are discarded [1, 2]. Hackers, network invaders, former workers, and others have the ability to steal or access information, disrupt operations, and harm systems. An intruder gains access to an IoT-based system via an external network and steals patient records in this pure technology hazard. As a result, it's an unsolved issue on the horizon [35]. During an emergency, hospitals encourage doctors to shatter the glass (BTG) approach, which allows them to bypass entry authorisation. The IoT system's normal work cycle is disturbed in BTG scenarios. This BTG method permits doctors or other staff employees to abuse or divulge sensitive information about patients without their knowledge or consent. Health-care providers preserve records to protect against deliberate or inadvertent information misuse. In BTG scenarios, the new method is both preventative and unsuccessful [36, 37].

Insider Abuse

In 2013, insider misuse was responsible for 15% of all security breaches in the healthcare industry [38]. This word refers to circumstances in which firm employees steal goods or information or participate in other criminal conduct. Surprisingly, the amount of persons who work in the healthcare profession only to infiltrate the system and obtain access to patient health information stands out as an example of insider misappropriation. This information is typically stolen in order to get access to funds or commit tax fraud. Insider threats are becoming more and more of a worry for businesses. If these attacks are carried out, insiders' in-depth knowledge of security procedures and monitoring protocols puts firms in jeopardy. As a result, finding insiders is a significant task that has captivated the interest of scholars for over a decade. The authentication of the approved sensor nodes might be compromised, or the culprit could steal token or other information from the networks and start an attack on the entire system.

Detecting anomalies suggestive of unusual and malicious insider behavior [39], recognizing elements in attacks [40], and recognizing behavioral causes [40] have all been thoroughly discussed [41]. In an effort by the CMUCERT Insider Threat project, a pioneering assess insider threats age, sabotage, and intellectual property (IP) theft [42]. The study used a paradigm called System Dynamics to identify and characterize important paths, which the majority of insiders follow in a series of isolating questionable behavior and MERIT (Management and Information on the Risks of Security Breach) copies. In addition, the authors distinguish between insiders who unknowingly aid an attack or expose the IoT to unnecessary harm and stake-holders who act deliberately by breaking standards (to allow their daily activities) or becoming irresponsible (phishing targets) [43].

Insider risks occur when personnel within an IoT use their privileged access to compromise the system's security, credibility, or availability [44]. The severity of the insider threat is well acknowledged, as evidenced by numerous real-life incidents and detailed studies [39, 44, 45]. We believe that in an era of IoT, where everything is a device capable of connecting, preserving, and exchanging important corporate data, the danger will become far more difficult to control for IoTs; this is a view-point shared by many others [46]. In some circumstances, it makes no sense to let these gadgets be "insiders" recognize anything as having the potential for permitted entrance because standard perimeters are getting increasingly vague. As a result, it's critical to understand how to deal with the threat of insiders in IoT contexts. Regrettably, no comprehensive investigation of this threat has been conducted so far.

Data Integrity Attack

In a Data Integrity attack, an attacker can tamper with a patient's data, further deceive the recipients by introducing inaccurate patient information, and then submit the erroneous information. Erroneous treatment, patient status, and emergency calls to specific people may all be the outcome of these threatening attacks. Data manipulation has the potential to result in a patient's death. Denial of Service (DoS) attacks are widespread at all layers of the network and can be carried out in a number of ways.

Data integrity is one of the most important security concerns in the Internet of Things since it affects both data storage and transfer. In the Internet of Things, data is constantly sent, with some of it being deposited and exchanged by third-party vendors who provide utilities to users. Throughout the life of the data, it must be kept confidential. Multiple service access interfaces may result in security issues. Data deposited in the schemes can be amended or deleted by attackers. Malicious apps, for example, could be installed and cause data loss. Smart city systems must mitigate this danger in order to assure data privacy. Data that does not meet the applicable requirements should be discarded using acceptable ways during the data lifecycle in IoT, which includes various phases. Data dependability is a serious concern because IoT-based data is robust in design and large in size [47].

Denial of Service Attack (DoS)

In a DoS attack, an intruder floods the system's data exchange with unidentified traffic, rendering services unavailable to others and preventing other nodes from transmitting data until the busy channel is recognized [48]. In a DoS assault, the attacker usually takes advantage of the activity by altering a certain number of flags in control ledges. Due to the labels in control packets, it is difficult to trace such an attack because nodes in the IEEE 802.11 standard do not counter-check everything. Patient data could be accessed in a DoS attack if there is no certification or authority to examine data [49]. The DoS assault frequently keeps the device's data channel busy, preventing any other data from reaching the network's other sensors. Data connection across networks is disrupted or unavailable as a result of DoS attacks. This type of attack puts system or healthcare facility accessibility and network operation, and sensor responsibilities in jeopardy.

The most common and easiest-to-enforce DoS attacks are on IoT networks. They are described as an incursion that can compromise the network's or systems' capacity to achieve their intended goals in a variety of ways. The Internet of Things has been heavily condemned from its beginning for the lack of attention devoted to safety issues in the design and deployment of its hardware, apps, and infrastructure parts [1–3, 7]. This sloppy approach has resulted in a slew of vulnerabilities that hackers and cybercriminals have successfully exploited to infiltrate IoT elements and utilize them for a variety of purposes, including staging Denial of Service and DoS attacks [50]. Users cannot access network facilities or data due to DoS assaults and DoS (DDoS) sharing. A DDoS assault is defined as a DoS attack that has been compromised by several nodes. Given their often sophisticated and economically appealing exterior form, many IoT devices are built from low-cost generic hardware parts. Security vulnerabilities are almost often built-in to these processors and software, making it impossible for owners and administrators to keep track of them. Furthermore, the wireless problem's facilities and teamwork for firmware and software updates are still immature. As a result, updating or repairing these unprotected IoT PCs is difficult.

Router Attack

Data routing is crucial for healthcare-based systems since it enables the supply of intelligence over the internet and simplifies connection mobility in huge facilities. Routing, on the other hand, is complicated by the fact that wireless networks are transparent. In this invasion, the attacker focuses on data transferred between sensors in various wireless sensor nodes. This is because the safe transmission of medical records to the intended recipient, who could be a physician or a specialist, is the most important prerequisite of a wireless health care system. Few implementations employ multi-trust guiding in this attack, steering basic and key facts displaying patients' daily care rankings. Multi-trust guidance is critical for growing the system's incorporation district and, as a result, providing stability at the expense of complexity.

By facilitating data flow, routers play a critical role in network communications. Protocol flaws, router software oddities, and weak authentication can all be exploited by router assaults. Two types of attacks that can arise are distributed denial of service and brute force assaults. Attacks have an immediate impact on network services and business processes. The TCP protocol employs synchronization packets known as TCP/SYN packets for link requests between computers and servers. The originator's computer When an SYN flood attack occurs, a large number of TCP/SYN packets with a forged URL are sent out. The channel's destination node is unable to connect to the root because the path is unreachable. If a router is unable to verify a TCP message, it will quickly run out of resources [51]. This is a sort of denial of service since the breadth of the assault will deplete the router's resources.

A brute force attack occurs when a hacker tries to guess a password in order to gain access to a router. The invader will utilize software with a dictionary of terms to crack the password. Depending on the strength of the password and the combinations used to discover a match, the attack could take a short time if it is relatively weak. This type of attack isn't limited to business routers; if a hacker is within range of the router, it can also happen at home. Unauthorized access to routers can be gained by a dissatisfied employee who has access to the network topology, router login and password information, and knowledge of the network topology. To avoid this problem, passwords should be changed regularly, and rigorous access controls should be implemented. Routers must have robust and up-to-date software with solid configurations to decrease their vulnerability to assaults.

Select Forwarding Attack (SFA)

In order to carry out the attack, the attacker must get access to one or more sensors. As a result, community-oriented particular forwarding is the name given to this type of forwarding. In this technique, an attacker gains access to a sensor and drops data packets, sending them to nearby sensors to arouse suspicion. This attack significantly impacts the device, especially if the sensor is located close to the base. As a result of the packet loss generated by the SF attack, pinpointing the source of packet loss can be challenging. As a result of the partial data received by the receiver, the attack is extremely dangerous to any patient or smart medical health system.

Attacking wireless communication with selective forwarding has a major impact on network efficiency and wastes substantial energy. Previous countermeasures assumed that all peers within the communication range could notice the attacker's wrongdoing. Previous techniques have struggled to accurately detect misbehaviors because smart networks require a minimum signal-to-noise ratio to adequately gather frames and because nodes incursion is unavoidable in densely spread wireless sensor networks. In a selective forwarding assault (SFA), an intruder impersonates a normal node throughout the transmission period and selectively dismisses traffic from neighboring nodes [52]. Non-critical data can be delivered properly, but vital data can be destroyed, such as information obtained from an adversary in a military application. Because it is immoral to lose confidential information in monitoring [53], it will do major harm to WSN. Detecting and isolating SFA is a major research topic in the field of WSN defense.

Sensor Attack

Due to accidental sensor malfunction in suspicious behavior on a cellular network perpetrated by external attackers. Sensor control necessitates the usage of cellular network limits since the sensor might be exhausted and switched off. In this situation, an attacker might simply replace a malicious sensor in the network and carry out harmful activities with ease. As a result, if the patient data is not dispersed evenly among numerous sensors, the hacker has complete control over the data. As a result of the lack of a legal permission format, false data can be injected or served.

Replay Attack

When an intruder gains unauthorized access to a computer, a reverse attack might occur. When the sender stops sending data, the attacker runs a test on the system and sends a signal to the receiver. The attacker then takes over as the primary source. The attacker's main goal in these attacks is to generate network assurance. The attacker sends a notification to the receiver that is primarily used in the validation process. A replay attack is defined as a security breach in which data is processed without authorization and then rerouted to the recipient with the intent of luring the latter into doing something illegal, such as misidentifying or authenticating themselves or a duplicate operation. Any threat has some sort of effect on the system. The most significant effects on a health monitoring system include unauthorized access, data alteration, rejection of continuous surveillance, data goal route adjustment, and data reduction.

In this type of attack, an unauthorized person gains access to the Smart Health system, captures network traffic, and transmits the message to the receiver as the original sender [54]. The attacker wants to earn the trust of the system. A replay attack is a security breach in which any data is kept without permission and then sent again to the intended recipient. By gaining unauthorized access and then stealing critical medical information, this attack might severely impact an IoT-based system [55].

3   Applications of Artificial Intelligence for Security and Privacy in Internet of Things Systems

The development of smart devices with sensing and acting capabilities has increased the IoT platform's functionality. Because so many devices are connected to the network, a tremendous amount of data is generated [56]. In an IoT world, processing and computing is a difficult problem; thus, AI and other new technologies come to the rescue to handle the IoT security challenge. As illustrated in Fig. 2, IoT and AI can be used together to enhance overall analysis, productivity improvement, and overall accuracy.

Recent breakthroughs in AI may enhance the accuracy of security solutions, reducing the threats posed by the current cyberattacks [57, 58]. While using AI

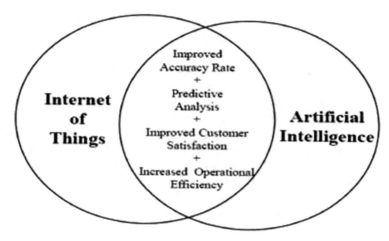

Fig. 2 The common proficiencies of IoT and AI

techniques like classification and clustering is not new, their importance has lately been highlighted as AI models (e.g., deep learning) grow. Historically, the majority of AI-based security study based on predicting attack patterns and their distinctive properties. It can, nevertheless, be intrinsically vulnerable to new sorts of advanced attacks with unique properties. A recent tendency has been to apply the concept of anomaly detection to construct more generic ML algorithms to overcome this restriction of present ML systems and effective security countermeasures against unexpected assaults not identified by conventional attack patterns [59].

Most applications, such as antivirus scanners, NIS, spam detectors, and fraud detection systems could benefit from AI models. In general, such systems leverage AI models to analyze massive volumes of data generated by network traffic, host processes, and human users to identify suspicious activity [60]. There is a general belief that using AI for security applications will become commonplace in the near future. However, security solutions based on AI may be subject to a new sort of complex attack known as adversarial AI [61, 62]. The adversary can effectively alter the contents of the input to AI models to circumvent classifiers designed to detect them in numerous security domains (e.g., e-mail spam detection). Furthermore, moving normal samples to the abnormal sample class and/or vice versa could compromise the training data set used to build classifiers.

The authors in [63] demonstrated how AI might aid IoT in processing large amounts of unstructured and contradictory data in real-time, making the system more realistic. In this study [64], the authors suggest the large margin cosine estimation (LMCE) technique for detecting the adversary in IoT-enabled systems. In paper [65], the work on malware detection in IoT systems using AI is discussed. Similarly, in the article [66], the authors suggested a model for making the system tamper-proof by combining Blockchain and AI in IoT design.

The authors of [67] proposed a critical infrastructure intrusion detection system that uses an ANN classifier with backpropagation and Levenberg-Marquard features to detect abnormal network behavior. In a related effort, the authors used an ANN model for DoS/DDoS detection in IoTs in [68], and provided a decentralized IDS for IoT devices based on artificial immunity in [69]. In [70], another group of researchers introduced the Possibility Risk Identification centered Intrusion Detection System (PRI-IDS) approach to detect replay attacks using Modbus TCP/IP protocol network traffic to detect replay assaults. On the other hand, these systems had a high rate of false alarms and had difficulty detecting certain novel threats.

The authors developed IDs in wireless networks in [71], and the Aegean AWID dataset was utilized to validate the system's accuracy. A PC, two laptops, one tablet, two cellphones, and a smart TV were used to collect the AWID dataset using a SOHO 802.11 wireless network protocol. On the other hand, the collection only includes traces from the MAC layer frame and excludes IoT device telemetry data. The authors of [72] created a BoT-IoT dataset based on a realistic IoT network architecture. DDoS, DoS, service scan, keylogging, and data exfiltration are examples of attacks that include both legal and hostile traffic. The network traffic reported by the simulated IoT-based model utilizing the BoT-IoT dataset was above 72 million.

The author has provided a scaled-down version of the dataset with roughly 3.6 million records for evaluation purposes. In a similar study [73], an IoT-based dataset was employed for ADS detection in a network of IoT devices based on DoS threats. SNMP/TCMP flooding, Ping of Death, and TCP SYN flooding were used to capture data in a smart home scenario utilizing traditional and DoS assaults. However, because the dataset was not taken using an IoT-based device, it was free of XSS-Cross-site-site Scripting and malware threats. Reference [74] proposed Deep RNN for IOT IDS, which included a traffic analysis engine and categorization. The pieces of traffic information are preprocessed in a format that can be processed. Finally, a backpropagation algorithm is used to train the deep NN classifier. The classifier is divided into two categories based on system traffic: normal and attack, and an alarm is triggered if an attack is identified.

4   Methods and Materials

4.1   Particle Swarm Optimization (PSO) Model for Feature Extraction

The algorithm is an evolutionary model inspired by the predatory of birds' behaviors and was proposed [75]. The process of findings optimal fitness solutions for particles can be mimic using the methods of birds finding foods. The local optimal fitness value and the current best global fitness value of particles without knowing the optimal fitness value can provide the speed of motion for each particle. This provides the overall particle swarm to move in the direction of the best possible solution.

The two parameters of each particle can be mathematically represented by.

The position is denoted by

$$x_i^k = \left[x_{i1}^k, x_{i2}^k, x_{i3}^k, \ldots, x_{id}^k \right] \tag{1}$$

And the velocity by

$$x_v^k = \left[v_{i1}^k, v_{i2}^k, v_{i3}^k, \ldots, v_{id}^k \right] \tag{2}$$

The position and velocity transform during the iteration update formula of each particle to:

$$v_{id}^{k+1} = \omega v_{id}^k + c_1 r_1 \left(pbest_{id} - x_{id}^k \right) + c_2 r_2 \left(gbest_{id} - x_{id}^k \right) \tag{3}$$

$$x_{id}^{k+1} = x_{id}^k + v_{id}^{k+1}, \tag{4}$$

where the local optimal position is represented by $pbest_{id}$ of the i-th, the global optimal of all particles in the population is represented by $gbest_{id}$, ω is the inertia weight, k the number of the current iteration, v_{id}^{k+1} is the d-th part of the velocity of the i-th particle of k iteration, c_1 and c_2 represented the cognitive and social parameters called acceleration coefficients, r_1 and r_2 uniformly distributed over the interval [0, 1] are two random numbers, the d-th component of the position of the i-th particle is represented by x_{id}^{k+1} particle in the $k+1$ iteration, and the velocity of the i-th particle in the k iteration of the d-th the component is represented by the x_{id}^k.

Particles can readily escape the current local ideal value when the inertia weight is too great, but they are not directly coupled in the final iteration. When the inertia weight is too low, the particles, on the other hand, are easily sucked into the local ideal value. Therefore, it is necessary to adjust the inertia weight adaptively. Hence, a dynamic inertia weight called the APSO algorithm is introduced. This model was used to adjusts the inertia weight adaptively with the fitness values. The algorithm can be represented mathematically as follows:

$$\omega = \left\{ \omega_{min} - (\omega_{max} - \omega_{min}) * \frac{(f_{cur} - f_{min})}{(f_{avg} - f_{min})}, \quad f_{cur} \leq f_{avg}, \quad \omega_{max}, \quad f_{cur} > f_{avg}, \right. \tag{5}$$

where the current particle fitness value is represented by f_{cur}, the current population average fitness value is represented by f_{avg}, and the smallest particles fitness values represented by f_{min} in the current population.

4.2 The Convolutional Neural Network Algorithm

The newest version of neural network with a multi-layer structure is called CNN composed of the various two-dimensional plane in each layer of the network [76, 77].

To activate the weighted sum of the elements in the previous layer, the output of each neuron is obtained.

The $C1$ layer can be represented by $C1_{ij}^{out}$ given output of the j-th neuron on the i-th feature plane can be mathematically given as:

$$C1_{ij}^{out} = F\left(\sum_{i=1}^{flx5} w_t^{in} \times f_{-raw_t^{in}}\right), \tag{6}$$

where the feature of the position of the characteristic plane is represented by $f_{-raw_t^{in}}$ corresponding to the convolution kernel weight in the input layer, w_t^{in} represents the weight of the t-th position of the convolution kernel, and the length of the filter is represented by fl. Three types of nonlinear activation functions were used the sigmoid, tanh, and relu represents by $Fi(\cdot)(i = 1, 2, 3)$ and can be mathematically denoted as follows:

$$F_1(x) = \text{sigmoid}(x) = \frac{1}{(1 + \exp^{-x})}, \tag{7}$$

$$F_2(x) = \tanh(x) = \frac{(\exp^x - \exp^{-x})}{(\exp^x + \exp^{-x})}, \tag{8}$$

$$F_2(x) = \text{relu}(x) = \max(0, x), \tag{9}$$

In the F2 layer, the output of the m-th neuron $full1_m^{out}$ is as follows:

$$full1_m^{out} = F\left(\sum_{n=1}^{n_{keep}} w_{min}^{keep} \times keep_n + b_m^{f2}\right), \tag{10}$$

where b_m^{f2} is the offset of the t-th neuron of the F2 layer, the connection weight between n-th neurons is w_{min}^{keep} and remaining working neuron after the processing of the previous layer, and $keep_n$ is the n-th neuron of the remaining working neuron.

Initially, the number of F3 neurons and activation mode is specified. In the same way that the F2 layer connects to the preceding layer, each neuron in this layer connects to the previous layer. The third and fourth layers of the fully connected layer are intended to improve learning of nonlinear combinations of compressed elements and learn the innovative functions generated by the convolution layer using the weight network connection. In the F4 layer, F3's output value is transmitted to an output layer in the last layer. The number of multi-classification task categories governs the output layer's number of neurons.

$$\text{softmax}(y)_i = \frac{\exp^{y_i}}{\sum_{i=1}^{kind} \exp^{y_i}}, \tag{11}$$

Table 1 The detailed summary of the CIC-IDS2017 dataset

Classes	Instances	Training set	Testing set
DDoS	60,477	42,335	18,142
DoS	111,082	77,757	33,325
Botnet	1,504	1,053	451
Probe	67,929	47,550	20,379
SSH_patator	4,715	3,301	1,414
FTP_patator	6,348	4,444	1,904
Web attack	4,743	3,320	1,423
Normal	990,814	693,570	297,244
Total	1,247,612	873,328	374,284

where the output value of the i-th neuron is y_i in the output layer, and the number of the network attack types is represented by *kind*.

4.3 The CIC-IDS2017 Dataset Characteristics

The Canadian Institute for Cybersecurity has released the CIC-IDS2017 dataset [78], which is unique, complex, and exhaustive, meeting the eleven most important criteria. For example, attack diversity, which includes 80 network velocity components, is a broad feature set and the necessary processes for compiling an accurate and consistent benchmark dataset. Furthermore, the authors cleverly structured the dataset to collect network traffic for five days, from Monday to Friday, which includes innocuous activity on Monday, but not on Tuesday, Wednesday, or Thursday. The first day is considered normal, and the next days are filled with cutting-edge attack traffic like DDoS, Brute Force, Heart-bleed, and Infiltration. Finally, taking into account the whole computational complexity of the CIC-IDS2017 dataset, a subset of this dataset was created by selecting 565,053 instances at random for testing purposes. Table 1 provides the statistical summary of the CIC-IDS2017 dataset.

4.4 Performance Analysis

The proposed hybrid model was evaluated using various metric performances and compared to other current models using the same dataset with the following performance metrics like accuracy, precision, recall, F1-score. To solve the confusion matrix, the statistical indices true positive (TP), true negative (TN), false positive (FP), and false-negative (FN) were generated, as indicated in Eqs. (12)–(25).

$$\text{Accuracy: } \frac{\text{TP} + \text{TN}}{\text{TP} + \text{FP} + \text{FN} + \text{TN}} \tag{12}$$

$$\text{Precision: } \frac{TP}{TP + FP} \tag{13}$$

$$\text{Sensitivity or Recall: } \frac{TP}{TP + FN} \tag{14}$$

$$\text{Specificity: } \frac{TN}{TN + FP} \tag{15}$$

$$\text{F1-score: } \frac{2 * \text{Precision} * \text{Recall}}{\text{Precision} + \text{Recall}} \tag{16}$$

$$\text{TPR: } \frac{TP}{TP + FN} \tag{17}$$

$$\text{FPR: } \frac{FP}{FP + FN} \tag{18}$$

5 Results and Discussion

The CIC-IDS 2017 dataset was used to test the effectiveness of the proposed system, PSO classifier was used for features extraction to reduce the features of the dataset to only the most relevant attributes, and CNN was used to classified the attack on the dataset. The PSO was used for dimensionality reduction and reduced the features to 11 relevant features. The total number of instances in the dataset is 1,247,612, and this was divided to 70%(873,328) training and 30%(374,284) testing due to the huge amount of data involved, thus help to work with reduced numbers of instances on the dataset. The performance of the proposed method is shown in Table 2 using various metrics.

Table 2 shows the performance evaluation of the proposed model using two classes of attacks and normal, and the results show a better performance with the PSO-CNN model with an accuracy of 99.45%, the sensitivity of 98.99%, specificity of 97.62%, the precision of 98.49%, and F1-score of 99.07% with time (sec) of 69 stamps respectively. The proposed model achieved optimal results on the CIC-IDS 2017 dataset used to test the proposed model's performance for detecting intrusion. Figure 3 displays the overall performance of the proposed model PSO-CNN.

Table 2 The performance evaluation of the proposed model

Models	Accuracy	Sensitivity	Specificity	Precision	F1-score	Time (S)
CNN	95.32	96.07	95.05	96.12	95.54	69
PSO-CNN	99.98	98.99	99.62	99.49	99.73	69

PERFORMANCE EVALUATION OF THE PSO-CNN MODEL

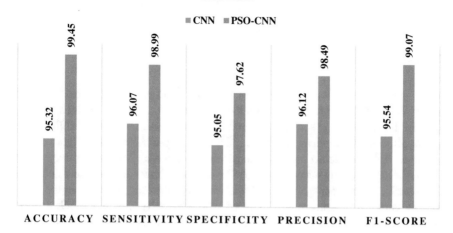

Fig. 3 The performance evaluation of PSO-CNN model

The comparison of the proposed model with the existing model.

The PSO-CNN model was compared with some selected methods that used the same dataset for performance effectiveness. The two models also used different feature selection classifiers with an ensemble classifier, and the results are presented in Table 3. The proposed model recorded a far-fetched performance in term of the metrics used for evaluation.

Table 3 shows the comparison results of the proposed model with existing classifiers, and the finding reveals that the model performs better with the results recorded. For instance, the accuracy of the proposed model is 99.98, which is almost the same as the KODE proposed by authors in [4] with 99.9% accuracy, but the model performed better in precision and F1-score with 99.49% and 99.73%, respectively. The model

Table 3 The comparison of the proposed model with two existing methods using CIC-IDS 2017 datasets

Model	Accuracy	FAR	Precision	F-score	DR
K-means [4]	99.72	0.011	0.992	0.992	0.997
One-class SVM [4]	98.92	0.011	0.982	0.990	0.989
DBSCAN [4]	97.76	0.012	0.986	0.985	0.977
EM [4]	95.32	0.013	0.960	0.949	0.952
KODE [4]	99.99	0.011	0.992	0.993	0.997
CNN	98.38	0.009	0.981	0.995	0.989
Proposed model	99.98	0.009	0.995	0.997	0.999

records a low model building time of 69 against the 217.2 s in the KODE model and 0.009 false alarm rate against o.012 in the KODE model.

6 Conclusion

The loss of non-creditworthy customers has created a huge amount of loss for banks and other sectors; thus fraud detection has become useful in the financial segments. But the detection and prediction of fraud in financial sectors are very difficult due to the diversity of applicant behaviors. This study provided an intelligent model based on ANN for detecting credit and loan fraud in a highly competitive market for credit leaden limits management. ANN simplifies how banks would detect loan fraud within credit management and will make an efficient judgment in the event of a reduction in loaning supply if faced with a negative liquidity shock. Hence, concentrate on the primary goal of increasing bank profits. The results show that ANN greatly detects fraud among loan lenders and loan administrators. Therefore, the bank profit is increased by implementing the advised loan choice based on real facts. The results reveal that our proposed method outperforms other state-of-the-art methods using real transaction data from a financial institution. Future work could apply a genetic algorithm for better feature selection, which would improve the system's performance and a hybrid technique for a better result.

References

1. Awotunde JB, Ogundokun RO, Misra S (2021) Cloud and IoMT-based big data analytics system during COVID-19 pandemic. Internet Things 2021:181–201
2. Awotunde JB, Adeniyi AE, Ogundokun RO, Ajamu GJ, Adebayo PO (2021) MIoT-based big data analytics architecture, opportunities and challenges for enhanced telemedicine systems. In: Enhanced telemedicine and e-health: advanced IoT enabled soft computing framework, pp 199–220
3. Abiodun MK, Awotunde JB, Ogundokun RO, Adeniyi EA, Arowolo MO (2021) Security and information assurance for IoT-based big data. In: artificial intelligence for cyber security: methods, issues and possible horizons or opportunities. Springer, Cham, pp 189–211
4. Jaw E, Wang X (2021) Feature selection and ensemble-based intrusion detection system: an efficient and comprehensive approach. Symmetry 13(10):1764
5. Khan MA, Karim M, Kim Y (2019) A scalable and hybrid intrusion detection system based on the convolutional-LSTM network. Symmetry 11(4):583
6. Meryem A, Ouahidi BE (2020) Hybrid intrusion detection system using machine learning. Netw Secur 2020(5):8–19
7. Awotunde JB, Chakraborty C, Adeniyi AE (2021) Intrusion detection in industrial internet of things network-based on deep learning model with rule-based feature selection. Wirel Commun Mob Comput 2021:7154587
8. Xu C, Shen J, Du X, Zhang F (2018) An intrusion detection system using a deep neural network with gated recurrent units. IEEE Access 6:48697–48707
9. Sarker IH, Kayes ASM, Badsha S, Alqahtani H, Watters P, Ng A (2020) Cybersecurity data science: an overview from machine learning perspective. J Big data 7(1):1–29

10. Damaševičius R, Venčkauskas A, Toldinas J, Grigaliūnas Š (2021) Ensemble-based classification using neural networks and machine learning models for windows PE malware detection. Electronics 10(4):485
11. Dang QV (2019) Studying machine learning techniques for intrusion detection systems. In: International conference on future data and security engineering. Springer, Cham, pp 411–426
12. Lopez-Martin M, Sanchez-Esguevillas A, Arribas JI, Carro B (2021) Supervised contrastive learning over prototype-label embeddings for network intrusion detection. Inf Fusion
13. Muñoz A, Maña A, González J (2013) Dynamic security properties monitoring architecture for cloud computing. In: Security engineering for cloud computing: approaches and tools. IGI Global, pp 1–18
14. Kagara BN, Siraj MM (2020) A review on network intrusion detection system using machine learning. Int J Innov Comput 10(1)
15. Bhosale KS, Nenova M, Iliev G (2020) Intrusion detection in communication networks using different classifiers. In: Techno-societal 2018. Springer, Cham, pp 19–28
16. Liu H, Lang B (2019) Machine learning and deep learning methods for intrusion detection systems: a survey. Appl Sci 9(20):4396
17. Saleh AI, Talaat FM, Labib LM (2019) A hybrid intrusion detection system (HIDS) based on prioritized k-nearest neighbors and optimized SVM classifiers. Artif Intell Rev 51(3):403–443
18. Awotunde JB, Jimoh RG, Folorunso SO, Adeniyi EA, Abiodun KM, Banjo OO (2021) Privacy and security concerns in IoT-based healthcare systems. Internet Things 2021:105–134
19. Weber M, Boban M (2016) Security challenges of the internet of things. In: 2016 39th international convention on information and communication technology, electronics and microelectronics (MIPRO). IEEE, pp 638–643
20. Sfar AR, Natalizio E, Challal Y, Chtourou Z (2018) A roadmap for security challenges in the internet of things. Digit Commun Netw 4(2):118–137
21. Ali B, Awad AI (2018) Cyber and physical security vulnerability assessment for IoT-based smart homes. Sensors 18(3):817
22. Bajeh AO, Mojeed HA, Ameen AO, Abikoye OC, Salihu SA, Abdulraheem M et al (2021) Internet of robotic things: its domain, methodologies, and applications. Adv Sci Technol Innov 203–217
23. Granjal J, Monteiro E, Silva JS (2015) Security for the internet of things: a survey of existing protocols and open research issues. IEEE Commun Surv Tutor 17(3):1294–1312
24. Görmüş S, Aydın H, Ulutaş G (2018) Security for the internet of things: a survey of existing mechanisms, protocols and open research issues. J Fac Eng Archit Gazi Univ 33(4):1247–1272
25. Kumar SA, Vealey T, Srivastava H (2016) Security in internet of things: challenges, solutions and future directions. In: 2016 49th Hawaii international conference on system sciences (HICSS). IEEE, pp 5772–5781
26. Liu X, Zhao M, Li S, Zhang F, Trappe W (2017) A security framework for the internet of things in the future internet architecture. Fut Internet 9(3):27
27. Bhardwaj A, Mangat V, Vig R, Halder S, Conti M (2021) Distributed denial of service attacks in cloud: state-of-the-art of scientific and commercial solutions. Comput Sci Rev 39:100332
28. Bhati A, Bouras A, Qidwai UA, Belhi A (2020) Deep learning based identification of DDoS attacks in industrial application. In: 2020 fourth world conference on smart trends in systems, security and sustainability (WorldS4). IEEE, pp 190–196
29. Taha AEM, Rashwan AM, Hassanein HS (2020) Secure communications for resource-constrained IoT devices. Sensors 20(13):3637
30. Alaba FA, Othman M, Hashem IAT, Alotaibi F (2017) Internet of things security: a survey. J Netw Comput Appl 88:10–28
31. Thomasian NM, Adashi EY (2021) Cybersecurity in the internet of medical things. Health Policy Technol 100549
32. Alsubaei F, Abuhussein A, Shandilya V, Shiva S (2019) IoMT-SAF: internet of medical things security assessment framework. Internet Things 8:100123

33. Hyman WA (2018) Errors in the use of medical equipment. In: Human error in medicine. CRC Press, pp 327–347
34. Royce CS, Hayes MM, Schwartzstein RM (2019) Teaching critical thinking: a case for instruction in cognitive biases to reduce diagnostic errors and improve patient safety. Acad Med 94(2):187–194
35. National Research Council, C. O. R. P. O. R. A. T. E. (1997) Standards, conformity assessment, and trade into the 21st century. Standard View 5(3):99–102
36. Satyanaga A, Kim Y, Hamdany AH, Nistor MM, Sham AWL, Rahardjo H (2021) Preventive measures for rainfall-induced slope failures in Singapore. In: Climate and land use impacts on natural and artificial systems. Elsevier, pp 205–223
37. Hao F, Xiao Q, Chon K (2020) COVID-19 and China's hotel industry: impacts, a disaster management framework, and post-pandemic agenda. Int J Hosp Manag 90:102636
38. Chernyshev M, Zeadally S, Baig Z (2019) Healthcare data breaches: implications for digital forensic readiness. J Med Syst 43(1):1–12
39. Cappelli DM, Moore AP, Trzeciak RF (2012) The CERT guide to insider threats: how to prevent, detect, and respond to information technology crimes (Theft, Sabotage, Fraud). Addison-Wesley
40. Maasberg M, Zhang X, Ko M, Miller SR, Beebe NL (2020) An analysis of motive and observable behavioral indicators associated with insider cyber-sabotage and other attacks. IEEE Eng Manag Rev 48(2):151–165
41. Cotenescu V, Eftimie S (2017) Insider threat detection and mitigation techniques. Sci Bull "Mircea Cel Batran" Naval Acad 20(1):552
42. Glancy F, Biros DP, Liang N, Luse A (2020) Classification of malicious insiders and the association of the forms of attacks. J Crim Psychol
43. Yuan S, Wu X (2021) Deep learning for insider threat detection: review, challenges and opportunities. Comput Secur 102221
44. Nurse JR, Buckley O, Legg PA, Goldsmith M, Creese S, Wright GR, Whitty M (2014) Understanding insider threat: a framework for characterising attacks. In: 2014 IEEE security and privacy workshops. IEEE, pp 214–228
45. Sarkar KR (2010) Assessing insider threats to information security using technical, behavioural and organisational measures. Inf Secur Tech Rep 15(3):112–133
46. Nurse JR, Erola A, Agrafiotis I, Goldsmith M, Creese S (2015) Smart insiders: exploring the threat from insiders using the internet-of-things. In: 2015 international workshop on secure internet of things (SIoT). IEEE, pp 5–14
47. Altulyan M, Yao L, Kanhere SS, Wang X, Huang C (2020) A unified framework for data integrity protection in people-centric smart cities. Multimed Tools Appl 79(7):4989–5002
48. Abdelrahman AM, Rodrigues JJ, Mahmoud MM, Saleem K, Das AK, Korotaev V, Kozlov SA (2021) Software-defined networking security for private data center networks and clouds: vulnerabilities, attacks, countermeasures, and solutions. Int J Commun Syst 34(4):e4706
49. Butt SA, Jamal T, Azad MA, Ali A, Safa NS (2019) A multivariant secure framework for smart mobile health application. Trans Emerg Telecommun Technol e3684
50. Ayo FE, Folorunso SO, Abayomi-Alli AA, Adekunle AO, Awotunde JB (2020) Network intrusion detection is based on deep learning model optimized with rule-based hybrid feature selection. Inf Secur J: Glob Perspect 1–17
51. Sivaraman V, Venkatakrishnan SB, Ruan K, Negi P, Yang L, Mittal R et al (2020) High throughput cryptocurrency routing in payment channel networks. In: 17th {USENIX} symposium on networked systems design and implementation ({NSDI} 20), pp 777–796
52. Zhang Q, Zhang W (2019) Accurate detection of selective forwarding attack in wireless sensor networks. Int J Distrib Sens Netw 15(1):1550147718824008
53. Liu A, Dong M, Ota K, Long J (2015) PHACK: an efficient scheme for selective forwarding attack detection in WSNs. Sensors 15(12):30942–30963
54. Rughoobur P, Nagowah, L. (2017, December). A lightweight replay attack detection framework for battery depended IoT devices designed for healthcare. In: 2017 International conference on Infocom technologies and unmanned systems (trends and future directions) (ICTUS). IEEE, pp 811–817

55. Liu X, Qian C, Hatcher WG, Xu H, Liao W, Yu W (2019) Secure internet of things (IoT)-based smart-world critical infrastructures: survey, case study and research opportunities. IEEE Access 7:79523–79544
56. Mohanta BK, Jena D, Satapathy U, Patnaik S (2020) Survey on IoT security: challenges and solution using machine learning, artificial intelligence and blockchain technology. Internet Things 11:100227
57. Mosteanu NR (2020) Artificial intelligence and cyber security—Face to face with cyber attack—A maltese case of risk management approach. Ecoforum J 9(2)
58. Singh S, Sharma PK, Yoon B, Shojafar M, Cho GH, Ra IH (2020) Convergence of blockchain and artificial intelligence in IoT network for the sustainable smart city. Sustain Cities Soc 63:102364
59. Ogundokun RO, Awotunde JB, Misra S, Abikoye OC, Folarin O (2021) Application of machine learning for ransomware detection in IoT devices. In: Studies in computational intelligence, vol 972, pp 393–420
60. Lee JH, Kim H (2017) Security and privacy challenges in the internet of things [security and privacy matters]. IEEE Consum Electron Mag 6(3):134–136
61. AbdulRaheem M, Balogun GB, Abiodun MK, Taofeek-Ibrahim FA, Tomori AR, Oladipo ID, Awotunde JB (2021, October) An enhanced lightweight speck system for cloud-based smart healthcare. Commun Comput Inf Sci 1455:363–376
62. Vorobeychik Y, Kantarcioglu M (2018) Adversarial machine learning. Synth Lect Artif Intell Mach Learn 12(3):1–169
63. Ghosh A, Chakraborty D, Law A (2018) Artificial intelligence in Internet of things. CAAI Trans Intell Technol 3(4):208–218
64. Wang S, Qiao Z (2019) Robust pervasive detection for adversarial samples of artificial intelligence in IoT environments. IEEE Access 7:88693–88704
65. Zolotukhin M, Hämäläinen T (2018) On artificial intelligent malware tolerant networking for IoT. In: 2018 IEEE conference on network function virtualization and software defined networks (NFV-SDN). IEEE, pp 1–6
66. Singh SK, Rathore S, Park JH (2020) Blockiotintelligence: a blockchain-enabled intelligent IoT architecture with artificial intelligence. Futur Gener Comput Syst 110:721–743
67. Linda O, Vollmer T, Manic M (2009) Neural network-based intrusion detection system for critical infrastructures. In: 2009 international joint conference on neural networks. IEEE, pp 1827–1834
68. Ogundokun RO, Awotunde JB, Sadiku P, Adeniyi EA, Abiodun M, Dauda OI (2021) An enhanced intrusion detection system using particle swarm optimization feature extraction technique. Procedia Comput Sci 193:504–512
69. Chen R, Liu CM, Chen C (2012) An artificial immune-based distributed intrusion detection model for the internet of things. In: Advanced materials research, vol 366. Trans Tech Publications Ltd., pp 165–168
70. Marsden T, Moustafa N, Sitnikova E, Creech G (2017) Probability risk identification based intrusion detection system for SCADA systems. In: International conference on mobile networks and management. Springer, Cham, pp 353–363
71. Kolias C, Kambourakis G, Stavrou A, Gritzalis S (2015) Intrusion detection in 802.11 networks: empirical evaluation of threats and a public dataset. IEEE Commun Surv Tutor 18(1):184–208
72. Koroniotis N, Moustafa N, Sitnikova E, Turnbull B (2019) Towards the development of realistic botnet dataset in the internet of things for network forensic analytics: bot-IoT dataset. Futur Gener Comput Syst 100:779–796
73. Hamza A, Gharakheili HH, Benson TA, Sivaraman V (2019) Detecting volumetric attacks on lot devices via SDN-based monitoring of mud activity. In: Proceedings of the 2019 ACM symposium on SDN research, pp 36–48
74. Almiani M, AbuGhazleh A, Al-Rahayfeh A, Atiewi S, Razaque A (2020) Deep recurrent neural network for IoT intrusion detection system. Simul Modell Pract Theory 101:102031
75. Eberhart R, Kennedy J (1995) A new optimizer using particle swarm theory. In: MHS'95. proceedings of the sixth international symposium on micro machine and human science. IEEE, pp 39–43

76. Hu F, Zhou M, Yan P, Li D, Lai W, Bian K, Dai R (2019) Identification of mine water inrush using laser-induced fluorescence spectroscopy combined with one-dimensional convolutional neural network. RSC Adv 9(14):7673–7679
77. Awotunde JB, Ogundokun RO, Jimoh RG, Misra S, Aro TO (2021) Machine learning algorithm for cryptocurrencies price prediction. Stud Comput Intell 2021(972):421–447
78. Sharafaldin I, Lashkari AH, Ghorbani AA (2018) Toward generating a new intrusion detection dataset and intrusion traffic characterization. ICISSp 1:108–116

Intrusion Detection Using Anomaly Detection Algorithm and Snort

Chika Yinka-Banjo, Pwamoreno Alli, Sanjay Misra, Jonathan Oluranti, and Ravin Ahuja

Abstract Many organizations and businesses are all delving into crafting out an online presence for themselves. This could either be in the form of websites or mobile apps. Many advantages come from an online presence; however, there are some drastic disadvantages that, if left unchecked, could disrupt any business or organization. Chief amongst these disadvantages is the aspect of security. However, many of the techniques that some organizations utilize to guard against unwanted access have been inadequate, and as a result, many unauthorized system break-ins have been reported. This is not made any better by the fact that certain applications used in hacking or system breach are now commonplace. Therefore, the focus of this work is to take an Intrusion Detection System (IDS) for a local network to detect network intrusion. A statistical approach, as well as a binomial classification, was used for simplicity in classification. The result shows the outlier value for each item considered; a 1 depicts an attack, a 0 depicts normalcy. The results are promising in dictating intrusion and anomalies in an IDS system.

Keywords Network security · Intrusion detection system · Snort · Computer network · Cyber security · Anomaly detection · Outlier detection

C. Yinka-Banjo (✉) · P. Alli
Department of Computer Science, University of Lagos, Yaba, Lagos, Akoka, Nigeria
e-mail: cyinkabanjo@unilag.edu.ng

S. Misra
Department of Computer and Communication, Østfold University College, Halden, Norway

J. Oluranti
Center of ICT/ICE, CUCRID, Covenant University, Ota, Nigeria
e-mail: jonathan.oluranti@covenantuniversity.edu.ng

R. Ahuja
Shri Vishwakarma Skill University, Gurgaon, Hariyana, India

© The Author(s), under exclusive license to Springer Nature Switzerland AG 2022
S. Misra and C. Arumugam (eds.), *Illumination of Artificial Intelligence in Cybersecurity and Forensics*, Lecture Notes on Data Engineering and Communications Technologies 109, https://doi.org/10.1007/978-3-030-93453-8_3

1 Introduction

With the development of the Internet and its latent capacity, an ever-increasing number of individuals are getting associated with the Web each day to exploit its advantages. On one side, the Internet provides enormous potential to the business regarding arriving at the clients. Simultaneously it additionally poses many safety hazards to the business over the network. With the advent of cyber-attacks, data safety has become a significant issue everywhere in the world [14]. This shows that no business or organization with an online presence is exempt from the possibility of reporting a security incident. This further buttress the importance of security policies to detect the various forms of cyber-attacks and ways by which such attacks can be prevented from occurring in the future [1]. The all-round security of the network is paramount but some areas of these require special attention, such as different network access levels for the members of the organization. The network access level for an accountant would drastically differ from that of a cleaner or security personnel. This will further improve network security as well as bandwidth utilization [3]. This helps and also shows the need for a proper system to detect various unauthorized access to the organizations' resources on specific aspects of the network before irreparable damages are done.

Today, many entities found on the web are striving to maintain confidentiality, integrity, and availability of their resources (services they provide) and various techniques and processes that can be used to safeguard their systems and networks from intrusion. Though these techniques provide some level of security, like with anything, they have areas in which they are lacking [2, 14].

1. Many organizations employ the use of firewalls. Firewalls are security policies that generally block out or control the traffic to and from a network.

   A firewall security strategy directs which traffic is approved to pass in each bearing. Likewise, it forces limitations on incoming and outgoing network bundles to and from private networks. The firewall sifts access between networks to prevent intrusions and does not alert the admin from within the network [14].
2. Information can be hidden from users who are not authorized to view or access such types of information. This is done via encryption; user access level is taken into consideration when employing such a policy.
3. Another technique is authentication which is used to verify the clients or users of a particular network system. There might be many individuals on the network who either use easy passwords or are generally careless with such information that intruders can gain access to the network.
4. Antiviruses also aid in keeping the integrity of a network intact, but this doesn't provide any meaningful data as to if any intrusion has occurred or not.

As more computers foster an online presence, many of their resources become susceptible to unauthorized access, attacks amongst other things. This then requires a means of combating and counteracting these attacks. Organizations, go online for the undeniable advantages that come from an online presence. However, many of these organizations do not know how best to defend against attacks. Some also employ outdated means of protection or inefficient techniques. The safety and integrity of these systems is of the utmost importance, which is why aggressive monitoring and vigilance of the system are required. An unprotected system will lead to major disadvantages for the organizations, and this directly defeats the purpose of an online presence. This is where an intrusion detection system plays a vital role [5, 9]. An intrusion prevention system may not be enough to aid the organizations, as stated earlier, but with an efficient intrusion detection system, both the attacks and further deviating attacks can be calibrated. Most organizations would always opt for an optimal network as well as a good environment free from threats and various malware. A decisive and efficient intrusion detection system will aid the simplicity of the network as well as network management. The most prominent significance of the study is to reduce the results of unauthorization in the network system. This, in turn, allows network administrators to focus on more industrious events.

The aim of this paper is to do suitable research into intrusion detection systems for most online organizations. This work primarily serves as a benchmark for which many other online organizations can form a basic intrusion detection system that is efficient and reliable.

The main objectives of the work are to:

- To build a reliable framework that detects various threats to a system.
- To make the intrusion detection system accurate to prevent false positives and also not easily cheated.
- Proper validation of the intrusion detection system.
- To identify anomalous data.

The paper is structured in 5 sections. The background and related work are provided in section two, methodology in Sect. 3, results and discussion in Sect. 4, conclusion drawn, and future work in Sect. 5.

2 Background and Literature Review

Intrusion detection systems are discussed by several authors [11, 16, 23].

As a general rule of thumb, Intrusion Detection Systems will identify intrusions or an attempt at intruding into a system by monitoring the following:

- Traffic on the network: The Intrusion Detection System combs through every form of traffic coming into the network and basically flags possible threats.
- Activity and location: Some locations can be flagged if a numerous number of attacks have been noted to originate from those areas.

In situations of breaches, the device could be configured in such a way that notifications could be sent as well as alarms raised in the advent of attacks. Certain measures could also be taken to curtail these attacks by ending sessions and replacing them with other connections as a stopgap to further intrusion. These intrusion activities could be stored for further review.

Intrusion Detection Systems have modes of searching and monitoring networks. As stated above, these IDS services can examine traffic for threats and send this information collated to a database or system console [15].

The following section discusses the various types of computer attacks and the need and types of intrusion detection systems.

2.1 Computer Attacks

Computer attacks are summed up as malicious activities aimed at a computer system for a certain end goal. There are also various unauthorized activities aimed at gaining some sort of advantage or leverage over the attacked system. These unauthorized activities could have goals of stealing, altering, exposing, disabling, or even destroying an existing system. The need to have countermeasures cannot be overemphasized. Total security is a unique idea—it doesn't exist anywhere. All networks are powerless against insider or outsider assaults and eavesdropping. No one wants to risk having the information displayed to the casual onlooker or open malicious intrusions [24].

Hackers have unique ways and techniques by which they bypass the security of a system. Understanding these will help in preventing such vulnerabilities from leaking into other systems. These attacks are numerous in number, and every day hackers find innovative ways of breaching the securities of a system. Some of these attacks include:

1. Denial of Service (DOS): attacks that overcome or flood the resources of the network and make it so it cannot provide resources for service requests. Another variation of this is the distributed denial of service (DDOS) attack. Like the Denial-of-Service attack, it also targets a network resource, but it is executed from many other host devices which are filled with malicious programs. There different types of DOS and DDOS attacks; some are:

    - Teardrop attack
    - Botnets

2. Phishing and spear-phishing attacks: This is a process of sending emails to unsuspecting individuals that appear to be from legitimate sources with the end goal of gaining vital information or coercing others to do a particular thing. This makes use of social engineering and technical trickery. It could attach a file that loads certain malware onto the recipient's device. It could also link to an illegal site that forcefully downloads malware. Spear phishing is a pinpoint type of phishing activity that is detailed by the attacker for the intended recipient. It is very hard to notice and even more difficult to defend against.

3. Social engineering: This form of attack deceives authenticated users into giving out pertinent information that may end up giving attackers access to the network and its varying resources. Some of the information given out could be passwords and other important resources.

4. Worms: These are generally self-preserving programs that duplicate and spread through the network. They have different modes of propagation and derail the performance of the system.

5. Masquerading: This is essentially when an attacker makes use of a fake identity to gain access to a network. This could be anything from a fake address to a fake user.

6. Password attacks: Passwords are generally a widely accepted means of authentication, and as such, attackers focus on gaining authenticated user's passwords. They are various ways they go about doing this. They could use brute force, which is randomly guessing and hoping that one works.

7. SQL injection: This is an issue with database-dependent or driven sites. This is done when the attacker executes a query to the database through the input data from a client to a server.

2.2 Need for Intrusion Detection Systems

There are several important processes towards securing a network; however, an intrusion system gives an immersive process by not only detecting but documenting hostile activities [16, 22]. It scans the network when other security measures are easily passed through.

IDS's are also invaluable when compared to firewalls that just close off communication and allow certain trusted parts of the network traffic through without detecting inside attacks. It is important to note that an IDS is not a firewall. Firewalls are mostly used at the borders of a network, and they are sort of gate-keepers as they usually only control traffic coming in or exiting the network. This has its shortcomings; however, as many times, attackers can be from within the network. It can therefore stand to reason that there is a valid need for intrusion detection, and some of the needs include:

- To detect and log attacks. The IDS detects as well as provides adequate information concerning the attack that a regular firewall would not be able to. The IDS, however, takes further by alerting the admin to the threat.
- The intrusion system also takes down failed attempts.
- They catch insider attacks as well as those from outside the network.

Now at this point, it is also important to note that though the IDS is invaluable in what it brings to the table, it is not without its shortcomings, like with all things. Therefore, the following are notable [17, 19]:

- Intrusion Detection Systems do not prevent intrusions.

- They may not be able to detect advanced attacks as these things are continually improved on by the attackers.
- They will have errors and imperfect intrusion detection.

Setting aside these drawbacks, it is safe to say that an IDS is a device that greatly enhances the overall security and safety of a network.

2.3   Types of Intrusion Detection Systems.

The failure or success of an attack largely depends on the architecture of the IDS in question. Various IDS types exist and could meet the requirement of some organizations.

An intrusion detection system utilizes numerous methods to distinguish incidents. Most IDS advances utilize different detection methods, either independently or incorporated, to give more expansive and precise identification [12]. They went on to classify IDS into three distinct groups:

1. Signature-based detection.
2. Anomaly-based detection.
3. Stateful protocol analysis.

2.3.1   Signature-Based Detection

This is usually some sort of noticeable pattern that has relations to a known threat. In this form of detection, the aim is actively searching for signatures in contrast to noticed events to tell potential threat occurrences apart.

Pattern Matching identifies the assaults utilizing "signatures" or by a few activities they perform. So it is additionally called a signature-based IDS [23].

Signature-based IDS alludes to the detection of assaults by searching for explicit patterns, for example, byte groupings in network traffic or realized malignant guidance groupings utilized by malware [16].

In misuse detection, assaults follow very well-defined designs that take advantage of the system shortcomings and application software. Since these assaults follow well-defined patterns and signatures, they are typically encoded ahead of time and used to compare against user conduct. It suggests that misuse detection requires explicit information on given intrusive conduct. In a signature-based detection, a foreordained attack pattern in the form of signatures is also used to decide the network attacks [28].

Signature-based detection generally has its advantages; some are:

- It is easy to use
- Clearly defined attacks generally yield low false positives.

It also has some disadvantages, chief amongst them are:

- Impractical for most inside attacks
- Alerts can be raised despite the outcome.

2.3.2 Anomaly-Based Detection

This generally analyzes user and system behavior. When an oddity in the behavioral pattern arises (the anomaly), the IDS takes it as an intrusion. However, this highlights a drawback with anomaly detection as it can result in a large number of false positives because not all changes in behavioral patterns are intrusion-based.

Jabez and Muthukumar [10] aimed to create an IDS based entirely on anomaly detection with outlier detection that would be definite and not easily bypassed by little variations in the observable patterns and would also be low in false positives generated and adaptive.

Regular data objects have a thick neighborhood, whereas "outliers" are far separated from their neighbors. The outliers are the items of the outer layers. The significant thought behind this methodology is to allocate a data example to an outlier degree called Neighborhood Outlier Factor (NOF) and track down the rare information whose behavior is very distinct when contrasted with enormous amounts of normal information [10].

As stated earlier, false positives are a major drawback of this model, and it can be said that while the system is still computing what constitutes a normal behavior from an anomaly, several other attacks could have been made. It also highlights that behaviors change over time, and there is a need to update the system according so as not to throw unwanted alarms for new normal behaviors [15, 16].

2.3.3 Stateful Protocol Analysis

Stateful protocol analysis is an interaction of contrasting foreordained profiles of commonly acknowledged meanings of favorable convention exercises for every convention state against noticed occasions to recognize deviations. Stateful convention examination depends on seller-created widespread profiles that indicate how specific conventions ought to and ought not to be utilized [12].

Their IDS is capable of comprehending and tracking the state of the network, transport, and other protocols that can be said to have a state. It generally depends on the developed profiles that tell how any particular protocol should and shouldn't be used.

Its drawback is in the fact that it is very limited in what it can do as regards to monitoring requests. Stateful protocol analysis is limited to probing a single request or reply. Since different attacks cannot be noticed by just looking at a single request, an attack could involve many requests [6–8, 18, 20, 25–27].

3   Methodology

The conceptual framework for an intrusion detection system is presented in this section. The flow of packets over a secured network is achieved. It looks at unauthorized resource access from both outsides and within the network perimeter. Lastly, various possible intrusion prevention techniques that will aid the proficiency of the IDS are considered.

3.1   Data Collection and Analysis

This work made use of the open-source IDS (Snort). It was configured to log various traffic flowing into a private network. The data that was collected is then used to determine the importance of an intrusion detection system over a secured network. Snort is used because:

- It is open source and can be used since it is cost-efficient.
- The application is lightweight.
- Snort also can be used in prevention.
- Snort is scalable.

  The functional requirements include:

- The system setup comes with some prevention system to add efficiency to the IDS.
- A packet capture driver.

  The non-functional requirements include:

- Provision of security to network systems.
- Stipulating guidelines on the use of certain resources.
- Educating users on the essential know-how necessary for network security.

3.2   Snort

Snort is an open-source intrusion detection system that aids in the defense of a system. Snort has the capability to log all incoming traffic into the system. It also aids prevention with the use of rules which can be used to alert the administrator to the presence of some sort of intrusion. These rules are usually particular to the needs of the system and what it wants to monitor [21]. Typically, events like port scanning, access to various services and looking into the browsing of a web server, and the likes can be detected using snort. From there, decisions can be made on the actions to be taken as regards the various events.

```
C:\Snort\bin>snort -V

     ,,_      -*> Snort! <*-
   o"  )~     Version 2.9.17-WIN64 GRE (Build 199)
   ''''       By Martin Roesch & The Snort Team: http://www.snort.org/contact#team
              Copyright (C) 2014-2020 Cisco and/or its affiliates. All rights reserved.
              Copyright (C) 1998-2013 Sourcefire, Inc., et al.
              Using PCRE version: 8.10 2010-06-25
              Using ZLIB version: 1.2.11
```

Fig. 1 Snort validation showing the version

3.2.1 Installation and Validation

Snort for windows is configured using a typical packet capture driver, usually WinPcap; here, we use npcap. Snort can be downloaded from the site; http://www. snort.org.

After which, various steps are taken in the installation process and validation of the system.

3.2.2 Validation of Snort

Typically, after the installation of snort and the packet capture driver, we can then validate the snort installation by running the snort command; snort −V as seen in Fig. 1; this validates if snort has been installed successfully by returning the version number of snort that we are running. This command, however, is not to be confused with a snort −v, which is a way of running a packet dump mode. The packet dump mode is where snort outputs traffic in the system.

3.2.3 Configuration of Snort

After the snort setup has been validated, we begin the configuration process of snort by going to the directory and accessing the 'snort.conf' file.

Here, we specify the home network we wish to monitor and treat any other address as external. The home network is the network we wish to monitor and protect. We also have to correctly specify our paths to these directories properly for a smooth running of snort. The paths are specified in a Linux machine way, which would not do anything for us until we re-specify it in a way windows can understand. This, however, is optional as you can choose to run a virtual machine that uses Linux on it and as so much of these come pre-specified already.

54 C. Yinka-Banjo et al.

3.2.4 Snort Rules

Snort comes with a feature we can use to classify rules; these rules are customized to reflect the needs of the network generally. A snort has a certain order or syntax that it follows, and it is easily recognizable.

Table1 shows a rough template of the thinking to write snort rules. The action column shows the possible type of actions that should be taken when a rule is triggered. The protocol column shows on which protocol type was the rule triggered. The source address is generally set to any, but specific addresses can also be specified; the same goes for the source port as well. The direction can be single-directional or bidirectional. The destination address is basically the IP address that such packets were destined for. The destination port, like the address, is the port where the packets are destined for. The rule option offers a way to tell where any intrusion was detected.

Figure 2 shows some of the rules we wrote to test out the functionality of snort and if it would process the incoming packets. These rules follow the syntax shown in Table 1; it is a basic rule; however, more advanced rules can be written with familiarity when used for testing snort. The results are seen in Fig. 3.

Figure 3, shows the working power of snort to monitor a network. This was done in a few seconds after activating snort on our home network. The idea is to be able to test for various scenarios of attacks. This can also aid us in gaining some sort of idea into the kind of attacks our system is weak against and thereby proffer better policies to better address those issues. So as we see from Fig. 2, we had tested the TCP, ICMP, and UDP protocols.

Table 1 Table showing how snort rules are written

Action	Protocol	Source address	Source port	Direction	Destination address	Destination port	Rule option
Alert	Tcp	Any	Any	<>	Ip address	8001	Msg
Log	Udp						Logto
Pass	Icmp						Ipoption
Drop							Seq
Reject							Itype
Sdrop							Icode
							Id

```
alert icmp any any -> any any (msg:"Testing ICMP alert"; sid:1000001;)
      .........  .......      .......  ..........................
alert udp any any -> any any (msg:"Testing UDP alert"; sid:1000002;)
      .......  .......     .......  ..........................
alert tcp any any -> any any (msg:"Testing TCP alert"; sid:1000003;)
```

Fig. 2 Rules written to test snort

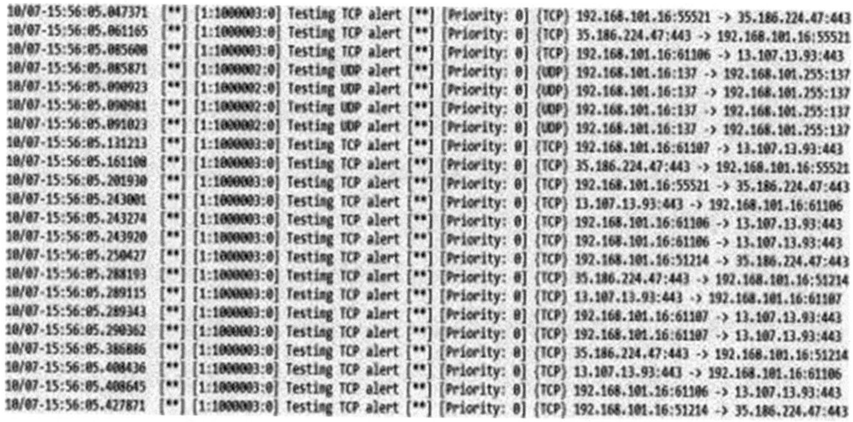

```
10/07-15:56:05.047371  [**] [1:1000003:0] Testing TCP alert [**] [Priority: 0] {TCP} 192.168.101.16:55521 -> 35.186.224.47:443
10/07-15:56:05.061165  [**] [1:1000003:0] Testing TCP alert [**] [Priority: 0] {TCP} 35.186.224.47:443 -> 192.168.101.16:55521
10/07-15:56:05.085608  [**] [1:1000003:0] Testing TCP alert [**] [Priority: 0] {TCP} 192.168.101.16:61106 -> 13.107.13.93:443
10/07-15:56:05.085871  [**] [1:1000002:0] Testing UDP alert [**] [Priority: 0] {UDP} 192.168.101.16:137 -> 192.168.101.255:137
10/07-15:56:05.090923  [**] [1:1000002:0] Testing UDP alert [**] [Priority: 0] {UDP} 192.168.101.16:137 -> 192.168.101.255:137
10/07-15:56:05.090981  [**] [1:1000002:0] Testing UDP alert [**] [Priority: 0] {UDP} 192.168.101.16:137 -> 192.168.101.255:137
10/07-15:56:05.091023  [**] [1:1000002:0] Testing UDP alert [**] [Priority: 0] {UDP} 192.168.101.16:137 -> 192.168.101.255:137
10/07-15:56:05.131213  [**] [1:1000003:0] Testing TCP alert [**] [Priority: 0] {TCP} 192.168.101.16:61107 -> 13.107.13.93:443
10/07-15:56:05.161108  [**] [1:1000003:0] Testing TCP alert [**] [Priority: 0] {TCP} 35.186.224.47:443 -> 192.168.101.16:55521
10/07-15:56:05.201930  [**] [1:1000003:0] Testing TCP alert [**] [Priority: 0] {TCP} 192.168.101.16:55521 -> 35.186.224.47:443
10/07-15:56:05.243001  [**] [1:1000003:0] Testing TCP alert [**] [Priority: 0] {TCP} 13.107.13.93:443 -> 192.168.101.16:61106
10/07-15:56:05.243274  [**] [1:1000003:0] Testing TCP alert [**] [Priority: 0] {TCP} 192.168.101.16:61106 -> 13.107.13.93:443
10/07-15:56:05.243920  [**] [1:1000003:0] Testing TCP alert [**] [Priority: 0] {TCP} 192.168.101.16:61106 -> 13.107.13.93:443
10/07-15:56:05.250427  [**] [1:1000003:0] Testing TCP alert [**] [Priority: 0] {TCP} 192.168.101.16:51214 -> 35.186.224.47:443
10/07-15:56:05.288193  [**] [1:1000003:0] Testing TCP alert [**] [Priority: 0] {TCP} 35.186.224.47:443 -> 192.168.101.16:51214
10/07-15:56:05.289115  [**] [1:1000003:0] Testing TCP alert [**] [Priority: 0] {TCP} 13.107.13.93:443 -> 192.168.101.16:61107
10/07-15:56:05.289343  [**] [1:1000003:0] Testing TCP alert [**] [Priority: 0] {TCP} 192.168.101.16:61107 -> 13.107.13.93:443
10/07-15:56:05.290362  [**] [1:1000003:0] Testing TCP alert [**] [Priority: 0] {TCP} 192.168.101.16:61107 -> 13.107.13.93:443
10/07-15:56:05.386886  [**] [1:1000003:0] Testing TCP alert [**] [Priority: 0] {TCP} 35.186.224.47:443 -> 192.168.101.16:51214
10/07-15:56:05.408436  [**] [1:1000003:0] Testing TCP alert [**] [Priority: 0] {TCP} 13.107.13.93:443 -> 192.168.101.16:61106
10/07-15:56:05.408645  [**] [1:1000003:0] Testing TCP alert [**] [Priority: 0] {TCP} 192.168.101.16:61106 -> 13.107.13.93:443
10/07-15:56:05.427871  [**] [1:1000003:0] Testing TCP alert [**] [Priority: 0] {TCP} 192.168.101.16:51214 -> 35.186.224.47:443
```

Fig. 3 Snort packet processing in testing mode

Figure 4 shows the network channels snort can aid us in monitoring, so we can specify the channel we wish to monitor for any anomalous behaviors as well as any typical intrusion attacks. The network channel is based on the type of connection your home network is making use of. Ours typically was just connecting to a local router, which was the index 4 in Fig. 4. Now with index four selected, we can write basic rules that can be used to test and evaluate the channel. From Fig. 4, it is important to note that the disabled channels cannot be accessed by snort. This makes it even more important to know the channels we wish to monitor before setting up a snort session.

From Fig. 5, we see timestamps for a typical IDS logging the traffic on a network. This specifies the type of traffic coming into the network and the IP addresses for the devices in communication. The time stamps show us the time at which the IDS logs the traffic. The IP addresses show the communication going on under the hood, and typically a description follows if the need arises.

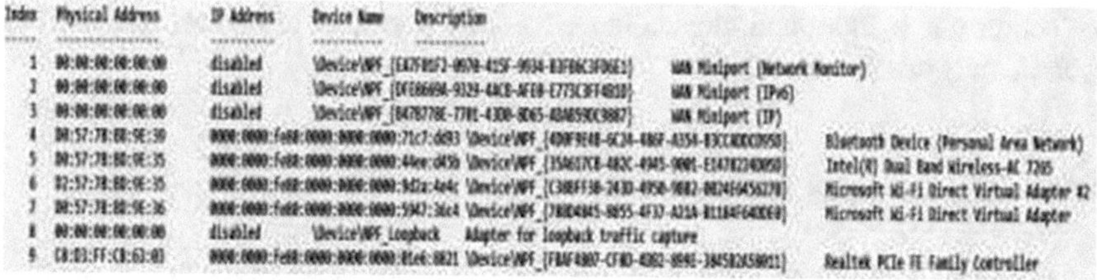

Index	Physical Address	IP Address	Device Name	Description
1	00:00:00:00:00:00	disabled	\Device\NPF_{EA7F85F2-0670-415F-9934-E3F86C3F06E1}	WAN Miniport (Network Monitor)
2	00:00:00:00:00:00	disabled	\Device\NPF_{DFE8669A-9329-4AC8-AFE8-E773C3FF4810}	WAN Miniport (IPv6)
3	00:00:00:00:00:00	disabled	\Device\NPF_{8478778C-77A1-4300-8D65-4DA059DC9887}	WAN Miniport (IP)
4	04:57:78:80:9E:39	0000:0000:fe80:0000:0000:0000:71c7:dd93	\Device\NPF_{40DF9E48-6C34-486F-A354-E3CCADCC9D58}	Bluetooth Device (Personal Area Network)
5	04:57:78:80:9E:35	0000:fe80:0000:0000:0000:44ee:d45b	\Device\NPF_{15A617C8-482C-4945-9001-E147B2240050}	Intel(R) Dual Band Wireless-AC 7265
6	02:57:78:80:9E:35	0000:0000:fe80:0000:0000:0000:9d2a:4a4c	\Device\NPF_{C88FF38-243D-4950-9882-BD245645627B}	Microsoft Wi-Fi Direct Virtual Adapter #2
7	04:57:78:80:9E:36	0000:0000:fe80:0000:0000:0000:5947:3dc4	\Device\NPF_{78DD4845-8055-4F37-A25A-811BAF640DE9}	Microsoft Wi-Fi Direct Virtual Adapter
8	00:00:00:00:00:00	disabled	\Device\NPF_Loopback	Adapter for loopback traffic capture
9	C0:03:FF:C0:63:00	0000:0000:fe80:0000:0000:0000:81e6:0821	\Device\NPF_{FBAF4807-CF0D-4D82-894E-3845B2A58921}	Realtek PCIe FE Family Controller

Fig. 4 Snort configuration initialized in IDS mode

```
1 16:56:12 00:01:26 telnet 1754 23 192.168.1.30 192.168.0.20 0 -

2 16:56:15 00:00:13 ftp 1755 21 192.168.1.30 192.168.0.20 0 -

3 16:56:17 00:00:01 smtp 43493 25 192.168.0.40 192.168.1.30 0 -

4 16:56:17 00:00:00 auth 1756 113 192.168.1.30 192.168.0.40 0 -
                   .........
5 16:56:19 00:00:01 smtp 43494 25 192.168.0.40 192.168.1.30 0 -

6 16:56:19 00:00:00 auth 1761 113 192.168.1.30 192.168.0.40 0 -
                   .........
7 16:56:19 00:00:01 ftp-data 20 1762 192.168.0.20 192.168.1.30 0 -

8 16:56:22 00:00:00 ftp-data 20 1767 192.168.0.20 192.168.1.30 0 -

9 16:56:24 00:00:02 ftp-data 20 1768 192.168.0.20 192.168.1.30 0 -

10 16:56:25 00:01:01 telnet 1769 23 192.168.1.30 192.168.0.20 0 -
```

Fig. 5 An excerpt from a dataset log of the TCP dump

3.3 Anomaly Detection in IDS

Anomaly detection is mainly taking a statistical approach to some data collated for the purpose of outlier analysis. This can be made easier by the automation of its processes which helps in reducing the overall time. There are many instances where anomaly detection can be applied, including disease detection, credit card fraud detection, and intrusion detection. This is done mainly by looking at the data and bringing out the various anomalies in each of these instances. Intrusion Detection Systems (IDS) depend on numerical models, algorithms, and structural arrangements proposed for effectively recognizing inappropriate, wrong, or odd movement inside the network frameworks. Intrusion Detection Systems can be named having a place with two fundamental groupings depending on the detection strategy utilized: anomaly and signature-based detection. Anomaly detection methods, which we center our workaround, depending on a solid portrayal of normal and not in a specific network situation [13].

3.3.1 The Formulas and Algorithm

This algorithm will focus mainly on simplicity as we will use the mean and variance to calculate the probability for each dataset. If the probability is found on the high side, it is normal. If, on the other hand, it is found below, then that data sample can be considered an anomaly. This, however, will vary from different datasets. The steps we will use follows this process.

3.3.2 Algorithm1: The Anomaly Detection Algorithm

STEP 1: Compute the mean

$$\mu = \frac{1}{m} \sum_{i=1}^{m} x^i \tag{i}$$

where m is the length of the dataset and x^i is a single data called a random sample.

Sample Example:

```
dataLength = len(data)

summ = np.sum(data, axis=0)
mean = summ/dataLength
mean
```

```
0        84.107555
1     45566.743000
```

This shows an actual implementation of the sample mean for the two parameters we considered, and the result of both is displayed.

STEP 2: Compute the variance

$$\sigma^2 = \frac{1}{m} \sum_{i=1}^{m} \left(x^i - \mu\right)^2 \tag{ii}$$

Sample Example:

```
vr = np.sum((data - mean)**2, axis=0)
variance = vr/dataLength
variance
```

```
0     1.311212e+04
1     3.446051e+13
```

This shows the computed variance where the random samples and sample mean are used to provide the value "vr" which is then used to calculate the variance. The results for the two parameters are displayed as well.

STEP 3: Compute the probability for each data example

$$p(x;\ \mu, \Sigma) = \frac{1}{(2\pi)^{n/2}\ |\Sigma|^{\frac{1}{2}}} \exp(-\frac{1}{2}(x - \mu)) \tag{iii}$$

Sample Example:

```
k = len(mean)
X = data - mean
prob = 1/((2*np.pi)**(k/2)*(np.linalg.det(var_dia)**0.5))* np.exp(-0.5* np.sum(X @ np.linalg.pinv(var_dia) * X,axis=1))
prob
```

```
0           1.830893e-10
1           1.952422e-10
2           2.234907e-10
3           1.864972e-10
4           2.117513e-10

            ...
125968      1.618295e-10
125969      1.830892e-10
125970      1.819398e-10
125971      2.064926e-10
125972      1.819393e-10
```

This shows the computed probability that takes the random samples, sample mean as well as the variance–covariance matrix (Σ). The variance–covariance matrix is also a diagonal function where the result is found by finding its determinant.

```
var_dia = np.diag(variance)
var_dia
```

```
array([[1.31121171e+04, 0.00000000e+00],
       [0.00000000e+00, 3.44605146e+13]])
```

This shows the actual computed variance-covariance matrix. This is basically gotten by taking the values of the variance and making a matrix or in the case a NumPy array of the variance.

The precision and recall can be calculated as such:

$$\text{precision} = \frac{\text{True Positives}}{\text{True Positives} + \text{False Positives}} \tag{iv}$$

$$\text{recall} = \frac{\text{True Positives}}{\text{True Positives} + \text{False Negatives}} \tag{v}$$

From this, we can compute the F1 score:

$$F1 = \frac{2PR}{P + R} \tag{vi}$$

where P is precision and R is recall.

The anomaly detection algorithm used above is based on a statistical approach; this approach takes a random sample (x^i) and calculates the sample mean. The sample mean is the summed average of the random samples. We then compute the sample variance (σ^2) for the model based on the random samples as well as the sample mean. The probability distribution (p) denotes a multivariate normal distribution taking the computed random samples, the sample means, variance–covariance matrix Σ is then used to give a probability distribution for every random sample.

We then proceed with the precision and recall for the distribution. In pattern recognition, retrieval, and classification, precision and recall are metrics used to determine how accurate and useful the data collected is. The precision finds the positives that are actually positive, and recall finds the number of positives retrieved out of all positive data examples. From this, we can then compute the $F1$ score, which is basically a weighted average of precision and recall.

3.4 Dataset

The KDDCUP99 and DARPA 1998–1999 models are the two most commonly used datasets for network intrusion detection research, and despite their flaws, they are still in use. The problem with using live data created is that it doesn't cover all forms of data, which means it can't be used to assess our algorithm's efficiency.

3.4.1 DARPA

The DARPA network Intrusion dataset has been criticized for its inability to detect zero-base attacks and the lack of false positives. The DARPA dataset is considered easy to replicate because it follows no pattern and can be quickly simulated. The data is used to teach students studying network intrusion how to comprehend the process of intrusions and how to identify them [4].

Figure 6 shows a serviced plot made to show attacks in a "tcpdump_inside" sensor as plotted by the Massachusetts Institute of Technology. It shows attack sessions which the arrows point to, and it can see some such attacks like Ipsweep, Probe using admin, and the likes. The x-axis shows the time, and the y-axis shows the TCP or UDP service. The hash marks show network sessions at those times and services.

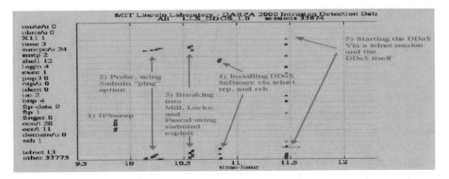

Fig. 6 tcpdump_inside attack from DARPA intrusion detection evaluation (MIT)

3.4.2 KDD

When combining the domain of network intrusions with machine learning methods, the KDD99 dataset is the most often utilized dataset. The dataset frequently contains a large number of irrelevant or redundant elements that cannot be employed in any sort of computation and can skew and impair the detection process. However, for network-based anomaly detection systems, the KDD is still commonly employed. This is what we took into account in our research [4].

3.4.3 Data Features

- **Basic Features**

The basic feature shows the feature of each network connection vector in the dataset and what they actually mean, and the values they hold. They are nine in number and give basic information about the session, like how long it took for a connection and the likes (Table 2).

- **Content Features**

The content features typically show actions, as opposed to a basic feature that tells us about the session. It can show how many successful logins were attempted on a given connection or how many were failed or how operational access were attempted (Table 3).

- **Traffic Features**

The traffic features are divided into time-related features as well as host-based related features. The traffic features generally show how many times a connection was initiated. In the time aspect, it shows how often such attempts were initiated (Table 4).

Table 2 Basic features of individual TCP connections

Feature name	Description
Duration	Length of time of connection
Protocol_type	Type of protocol e.g. TCP, UDP etc.
Service	Network service on network e.g. http, telnet etc.
Src_bytes	Number of data bytes from source to destination
Dst_bytes	Number of bytes from destination to source
Flag	Normal or error status of connection
Land	If the source and destination Ip are equal the value becomes 1 otherwise it is 0
Wrong_fragment	Total number of wrong fragments in the connection
Urgent	Number of urgent packets in the connection

Table 3 Content features within a connection by domain knowledge

Feature name	Description
Hot	Number of hot indicators
Num_failed_logins	Number of failed login attempts
Logged_in	1 if successfully logged in, 0 if not
Num_compromised	Number of compromised conditions
Root_shell	1 if root shell is obtained, 0 if not
Su_attempted	1 if Su root command attempted, 0 if not
Num_root	Number of root access
Num_file_creations	Number of file creation operations
Num_shells	Number of shell prompts
Num_access_files	Number of operations on access control files
Num outbound_cmds	Number of outband commands in an ftp session
Is_hot_login	1 if the login belongs to the hot list, 0 if not
Is_guest_login	1 if the login is a guest login, 0 if not

3.5   Monitoring and Data Collation

The process starts by allowing the IDS to monitor the network we wish to observe for any intrusion. The IDS (snort) goes through every piece of traffic from TCP, ICMP, and UDP traffic as we specified and monitors for deviation from expected. These deviations are flagged by the IDS and logged in a log folder in the snort directory. The sessions for how long the IDS will work are completely based on the discretion of the admin. The admin can have it running for as long as they wish.

Table 4 Traffic features and their parameters

Feature name	Description
Count	Number of connections to the same host as the current connection in the past two seconds
Serror_rate	Percentage of connections that have SYN errors
Rerror_rate	Percentage of connections that have REJ errors
Same_srv_rate	Percentage of connections to the same service
Diff_srv_rate	Percentage of connections to different services
Srv_count	Number of connections to the same service as the current connection in the past two seconds
Srv_serror_rate	Percentage of connections that have SYN errors
Srv_rerror_rate	Percentage of connections that have REJ errors
Srv_diff_host_rate	Percentage connections to different hosts

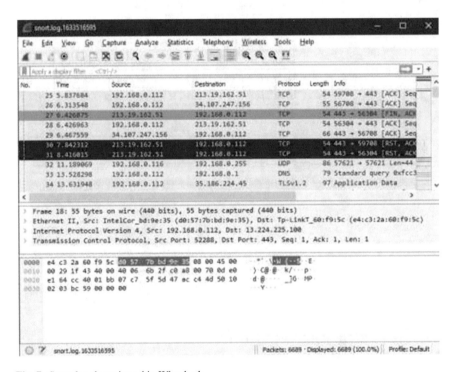

Fig. 7 Snort log data viewed in Wireshark

The data is generated by accessing the snort log file and opening it in Wireshark. The log file can now be read as well as exported out of Wireshark as an excel file that can now be cleaned and utilized.

det_host_srv_count	det_host_same_srv_rate	det_host_diff_srv_rate	det_host_same_src_port_rate	det_host_srv_diff_host_rate	det_host_serror_rate	det_host_srv_serror_rate	det_host_rerror_rate	det_host_srv_rerror_rate	xAttack
25	0.17	0.03	0.17	0.00	0.00	0.00	0.05	0.00	normal
1	0.00	0.60	0.88	0.00	0.00	0.00	0.00	0.00	normal
26	0.10	0.05	0.00	0.00	0.00	1.00	1.00	0.00	dos
255	1.00	0.00	0.00	0.04	0.28	0.01	0.00	0.01	normal
255	1.00	0.00	0.00	0.00	0.00	0.00	0.00	0.00	normal

Fig. 8 Training dataset showing the data items

Figure 7 illustrates some of the data items acquired by snort; it is also color-coded to make it easier to see the protocols. Green indicates TCP traffic, dark blue indicates DNS traffic, light blue indicates UDP traffic, and black indicates TCP traffic with issues. These color codes can be altered according to the user's preferences. However, because it was created in a reasonably safe context, this file does not generally cater to all types of attacks. This is the main reason why KDD99 was chosen over easy-to-replicate sample data since it is more commonly used and can produce a more trustworthy result.

Wireshark is a network protocol analyzer that captures packets from a network connection from the Internet, a single PC, or a large network of computers. A packet sniffer is another name for it. It facilitates the organization and interpretation of packets acquired, in our instance, using sniff. It primarily performs three functions: packet capture, filtering, and visualization.

3.5.1 Data Handling

The dataset contained a variation of attack categories, from the typical Denial of Service (DOS) attacks to probing attacks as well as U2R and R2L unauthorized access attacks. The idea, however, was aimed at easily manipulating the data, and as such, a binomial approach was chosen as opposed to considering every form of the attack. This, of course, further highlights a drawback with the binomial approach, as unauthorized access attacks are not as anomalous as the likes of denial of service as well as probing.

The dataset was divided into two forms (Train and Test), and modifications were made to aid the binomial approach I employed in the data consideration. The images below show the training dataset (Fig. 8).

4 Presentation and Discussion of Results

We used a binomial approach to consider the data, but before then, we looked at some relationships between the data obtained.

In Fig. 9, the boxplot shows the contrast between the three protocol types considered in this work, and their count, which has some correlation with their attacks, shows the outliers in each protocol box. In the diagram, we further see that both the ICMP and UDP protocols have significant values ranging from a count of about

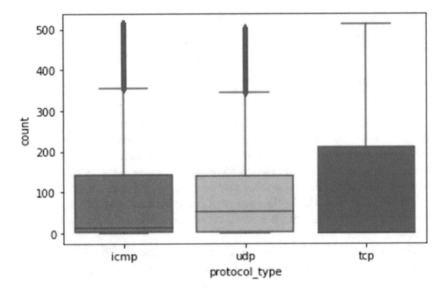

Fig. 9 Boxplot showing the protocols and their outliers

300–500. The TCP, however, shows no signs of there being an outlier present. This, however, does not mean outliers do not exist in the TCP protocol type. Further investigations into these protocol types will show if some of the systems recognize them as either true positive, false positives, or false negatives. The boxplot gives us, however, a rough idea of where to start our investigations from. Some of the data are directly correlated, and some are not; some exist some redundant entries of null as well, which actually provide no insight or information whatsoever.

From Fig. 10 and Table 5, we can see the distribution of the attacks and how their count ranges wildly. Even within the attacks, we see outliers. This shows to a certain degree the unreliability of simple plotting processes. So it further buttresses the idea of using a computable model to ascertain with some level of accuracy what an outlier is and what it is not. The count and the source byte data rows are then isolated and used for the computation which is used in training the model.

4.1 Implementation

We first start by computing the mean for the data, and this is done by dividing the sum of the count as well as the source byte by the number of data elements in each of them. The mean is then used to calculate the variance, which is then made a 2×2 matrix, and the probability of each training example can be computed. There is generally no predefined probabilistic value that we are working towards. Our dataset will determine the probability value we should consider.

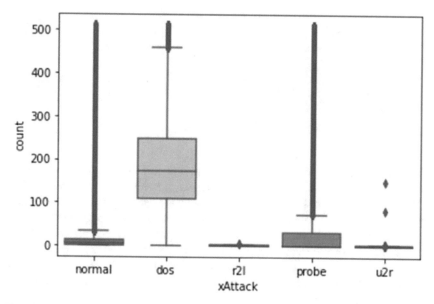

Fig. 10 Boxplot showing the attacks

Table 5 Attacks with the effects as well as some examples

Attack class	Effects	Attack type
DOS	Generally, depletes resources and is unable to handle legitimate requests	Teardrop, Worm, Neptune
Probe	Mainly used to monitor and gain information about a victim	Satan, Ipsweep
R2L	Attacker gains access to a remote machine	Xlock, Spy
U2R	A normal account login used to try and gain access or administrator privileges	Buffer overflow, Perl

It is also important to compute the precision and recall; these enable us to calculate the F1 score for each data in the training model. However, in computing the scores, we must talk about the terms used to calculate the precision and recall. These terms are true positives, false negatives, false positives. True positives are cases where the algorithm detects a data item as an anomaly, and it turns out to be so. False positives are cases that point out a data item as an anomaly, but it is not. False negatives are actual anomalous data items that are detected as not being anomalous.

So these are then used to determine the score from which a threshold probability value is chosen as the standard by which we determine if other data items are anomalous or not (Fig. 11).

After the probability function is computed, we then moved on to testing the data by cross-validating it with the test dataset. During this process, we use a predefined set of data where the values of the various attacks are categorized as "Normal"

```
def probability(data):
    summ = np.sum(data, axis=0)
    dataLength = len(data)
    mean = summ/dataLength
    vr = np.sum((data - mean)**2, axis=0)
    variance = vr/dataLength
    var_dia = np.diag(variance)
    k = len(mean)
    X = data - mean
    prob = 1/((2*np.pi)**(k/2)*(np.linalg.det(var_dia)**0.5))* np.exp(-0.5* np.sum(X @ np.linalg.pinv(var_dia) * X,axis=1))
    return prob
```

Fig. 11 Function to calculate the probability of each data item

which represents "0" or attack, which is: represented with "1". Then, we write a classification function to determine how well the algorithm can process each form of the data item. This basically entails classifying the data items in the three classes spoken of before: the true positives, false positives, and false negatives. A list of probabilities that are lower or equal to the mean probability is initiated. They are used as benchmark values to compute the F1 score. After the F1 score is computed, the values are put into a list where the scores should typically range from 0 to 1. Here a score of 1 generally depicts a perfect score. Figure 12 shows the computed classification, F1 score, and f score for the testing dataset.

The classification function calculates the true positives, false positives, and false negatives based on the probability of each data item being either a normal data item or an anomalous data item (Fig. 13).

The F1 score is then calculated based on the range of values in the probability list we defined. Finally, the threshold value is determined and compared to the rest of the other values. Anything below a threshold can be considered normal, and anything above the threshold can be considered anomalous.

In Table 6, we see some of the values for the F1 score gotten from the dataset. The dataset is relatively large, and as such inserting, all the values of the F1 would be very tedious. Some from the range of values listed in the F1 score, we specify our threshold value.

```
def classification(ep, prob):
    tp, fp, fn = 0, 0, 0
    for i in range(len(y)):
        if prob[i] <= ep and y[i][0] == 1:
            tp += 1
        elif prob[i] <= ep and y[i][0] == 0:
            fp += 1
        elif prob[i] > ep and y[i][0] == 1:
            fn += 1
    return tp, fp, fn
```

Fig. 12 Classification function of the data items in the testing dataset

```
def f1(ep, prob):
    tp, fp, fn = classification(ep, prob)
    precision = tp/(tp + fp)
    recall = tp/(tp + fn)
    f1 = 2*precision*recall/(precision + recall)
    return f1
```

Fig. 13 F1 function of the data items in the testing dataset

Table 6 List of F1 score values of the testing dataset

S/N	F1 score
1	0.27707943300347687
2	0.26449864498644987
3	0.08994966715375873
4	0.07473251028806585
5	0.22250603093514967
6	0.11128775834658187
7	0.03487775688151821
8	0.1390092879256966
9	0.3187889581478183
10	0.23656215005599104
11	0.07473251028806585
12	0.2668649452480735
13	0.11128775834658187
14	0.19478058025951303

In Table 7, the results for the dataset show that more work is still needed in classification. The binomial approach has its drawbacks, especially regarding unauthorized access attacks (U2R and R2L). This is because the count of the attacks is relatively low and can be depicted as normal data items as well.

In summary, an IDS provides vital information of what goes on under the hood of the system. Its benefits cannot be downplayed though at times, it leaves more to be desired from not just detection but also prevention. The drawbacks of IDS leave a lot of room for improvement in the world of security. False positives and negatives are also huge drawbacks that can affect the overall effectiveness of the solution you aim to get. They present erroneous values that deviate from the true reality of things when computed.

Table 7 Count and src-bytes showing the respective outlier values

S/N	Count	Src_bytes	Outlier value
1	110	0	1
2	1	312	0
3	5	245	0
4	4	298	0
5	1	740	0
6	3	304	0
7	4	0	0
8	4	42,340	0
9	1	32	0
10	9	249	0
11	206	0	1
12	16	208	0
13	24	225	0
14	1	0	0
15	216	0	1
16	472	0	1
17	15	227	0
18	10	36	0
19	2	283,618	0
20	238	0	1

5   Conclusion and Future Work

In conclusion, a good IDS must also come in tandem with a good prevention policy, or else it really serves no purpose. There are also areas where having a rigid detection policy will aid the overall security of the system. There are also many other forms of attacks that are to be looked into. These attacks could be novel malicious attacks that can be considered and have a well-defined policy to handle them. Another thing we had to contend with was data overload. The amount of data we could efficiently analyze is very important. At what point do we say this is redundant to the challenges of modern-day security and how beneficial the data would be to our goals.

First, we consider ways to optimize the binomial classification better to take into consideration those unauthorized attack types. Typically an IDS has many germane uses as stated in previous chapters. However, it is important to still note that an IDS system doesn't prevent such intrusions from taking place. This further buttresses the need to have a system that can not only detect but also prevent such attacks from ever happening. So ideally, an IDS should typically come with an Intrusion Prevention System (IPS). This is where we focus on seeing how we can incorporate an IDS and IPS to optimize the system further and prevent attacks.

References

1. Abdulbasit A, Alexei L, Clare D (2011) A misuse-based network intrusion detection system using temporal logic and stream processing. In: International conference on network and system security, pp 1–8
2. Aleksandar M, Marco V, Samuel K, Alberto A, Bryan DP (2017) Evaluating computer intrusions detection systems: a survey of common practices. Res Group Stand Perform Eval Corpor 48(1), Article 12. https://doi.org/10.1145/2808691
3. Alex D (2012) Intrusion detection using VProbes. Mass Inst Technol 1(2):28–31
4. Alia Y, Eric A (2018) Network intrusion dataset used in network security education. Int J Integr Technol Educ 7(3):43–50
5. Alireza H, Hossein S, Ahmad K (2006) A new framework: anomaly detection with snort intrusion detection system. In: Workshop on information technology and its disciplines
6. Bellovin SM (2001) Computer security—An end state? Commun ACM 44:131–132
7. Gopallkrishna NP, Kushank J, Nandan L, Narendra K, Yashasvi Z, Rohan S, Jyoti C (2014) Network intrusion detection system. Int J Eng Res Appl 4(4):69–72
8. Hamdan OA, Rafidah N, Zaidan BB, Zaidan AA (2010) Intrusion detection system. J Comput 2(2):130–133
9. Ibrahim K, Kemal H (2013) Open source intrusion detection system using snort. In: The 4th international symposium on sustainable development, pp 1–6
10. Jabez J, Muthukumar B (2015) Intrusion detection system (IDS): anomaly detection using outlier detection approach. Int Conf Intell Comput Commun Converg 48:338–346
11. Jaiganesh V, Sumathi P, Vinitha A (2013) Classification algorithm in intrusion detection system: a survey. Int J Comput Technol Appl 4(5):746–750
12. Lata KI (2013) Novel algorithm for intrusion detection system. Int J Adv Res Comput Commun Eng 2(5):2104–2110
13. Lukasz S, Marcin G, Tomasz A (2013) Anomaly detection preprocessor for snort ids system. In: Image processing & communications challenges. Springer, Heidelberg, pp 225–232
14. Manu B (2016) A survey on secure network: intrusion detection and prevention approaches. Am J Inf Syst 4(3):69–88. https://doi.org/10.12691/ajis-4-3-2
15. Mohammad JM, Mina S, Marjan KR (2010) Intrusion detection in database systems. Springer, Heidelberg, pp 93–101
16. Mohit T, Raj K, Akash B, Jai K (2017) Intrusion detection system. Int J Tech Res Appl 5(2):38–44
17. Muthu KR, Bala STV (2013) Intrusion detection system in web services. Int J Sci Res 2(2):224–228
18. Naga SLM, Radhika Y (2018) Detection and analysis of network intrusions using data mining approaches. Int J Appl Eng Res 13(6):4059–4066
19. Paresh G, Vishal G, Atish J, Sneha B (2018) Intrusion detection system using data mining. Int Res J Eng Technol 5(3):58–61
20. Rahul Y, Kapil V (2017) Snort-J48 algorithm based intrusion detection and response system (IDRS) for cloud computing. Int J Res Sci Eng 3(2):465–470
21. Rishabh G, Soumya S, Shubham V, Swasti S (2017) Intrusion detection system using snort. Int Res J Eng Technol 4(4):2100–2104
22. Sahar S, Mohamed H, Taymoor NM (2011) Hybrid multi-level intrusion detection system. Int J Comput Sci Inf Secur 9(5):23–29
23. Shivani A, Priyanka W, Shivam P, Sangram N, Sunil D (2020) Intrusion detection system. Int J Sci Res Sci Eng Technol 7(3):13–16. https://doi.org/10.32628/IJSRSET207293
24. Snehal B, Priyanka J (2010) Wireless intrusion detection system. Int J Comput Appl 5(8):975–8887
25. Tanmay P, Piyush I, Omar K, Ashish N, Sheetal B (2017) Smart intrusion detection system. Int Res J Eng Technol (IRJET) 4(4):3404–3406
26. Tariq A, Abdullah A (2014) Hybrid approach using intrusion detection system. Int J Comput Netw Commun Secur 2(2):87–92

27. Vijayarani S, Maria SS (2015) Intrusion detection system—A study. Int J Secur Priv Trust Manag (IJSPTM) 4(1). https://doi.org/10.5121/ijsptm.2015.4104
28. Vinod K, Om PS (2012) Signature based intrusion detection system using snort. Int J Comput Appl Inf Technol I(III):35–40

Research Perspective on Digital Forensic Tools and Investigation Process

Kousik Barik, A. Abirami, Karabi Konar, and Saptarshi Das

Abstract The digital forensic tool plays a crucial role in protecting from share trading fraud, financial fraud, identity theft, and money laundering. Investigators in Digital Forensic analysis have the right to use various forensic tools for investigation. Regretfully, several contemporary digital forensics technologies are lacking in a number of areas. The research aims to study different phases of digital forensic methods and various issues encountered during the investigation process. In addition, the study also focuses on the mindful analysis of different kinds of digital forensic tools. The methodology includes building, providing defense to any scene, proper review, effective communication, and identification. It also involves the identification of the digital investigation opportunities and recognizing all the policies for controlling proof. The tools and procedures that can be utilized to investigate digital crime are discussed in this article. Desktop forensic tool, Live Forensic tool, Operating System Forensic tool, and Email Forensic tool are some of the covered categories. AI-based network logs are evaluated using a variety of machine learning techniques and compared with various metrics such as accuracy, precision, recall, and F1-Score. A comparative analysis of popular tools in each category is tabulated to understand better, making it easy for the user to choose according to their needs. Furthermore, the paper presents how machine learning, deep learning, and natural language processing, a subset of artificial intelligence, can be effectively used in digital forensic investigation. Finally, the future direction of the challenges and study scope in digital forensics and artificial intelligence is also mentioned for potential researches.

Keywords Digital forensic · Forensic tools · Artificial intelligence · Digital crime

K. Barik (✉) · K. Konar · S. Das
JIS Institute of Advanced Studies & Research, JIS University, Kolkata, India

S. Das
e-mail: saptarshi@jisiasr.org

A. Abirami
Bannari Amman Institute of Technology, Erode, India

1 Introduction

Internet users are expanding exponentially; subsequently, digital devices such as desktops, laptops, and mobile devices are overgrowing. These devices are interconnected to the form of the network and exchange a substantial amount of data. The usage of these devices and the Internet are one of the causes of cybercrime. Concerning different forensic sciences, digital forensic is a moderately modern age. Digital forensic is the technique responsible for the reorganization of digital crime after happening [1]. Digital forensics is the technique while discovering, extracting, and interpreting data from various devices, which experts interpret as legal evidence [2, 3]. One of the most encouraging technological developments in digital forensics is digital forensic tools/ software. Tools have initiated the investigation process effectively and efficiently. The data store platforms vary from IoT devices, mobile devices, desktop devices, cloud computing, interconnected devices, autonomous devices, etc. [4]. Different digital forensic tools have focused features on preserving original files or data recovered from the devices or systems [5].

The digital forensic tool protects from online financial fraud, online share trading fraud, source code theft, virus attacks, identity theft, money laundering [6], Phishing, unauthorized access, preventing sexual harassment, and stealing more sensitive information. Digital forensic tools are classified into several sub-branches: network forensic, computer forensic, cyber forensic, mobile forensic, operating system forensic, and live forensic [7]. While using forensic tools, the primary principle to be adopted is that the data should not be altered in the data collection process. In addition, while working with such tools throughout the investigation process, the processes to be well documented to access the original document and strictly restrict any alterations [8]. The investigation is generally performed in the clone of the original data to avoid alteration in the original data [9]. The duplication of the original data helps recovers the data even when there is some flaw in the investigation process.

The abundance of forensic tools will make it difficult for users to select the appropriate tool for their needs. When employing forensic technologies, it's critical to understand the user's background. There will be a variety of users with varying levels of computer knowledge [10]. The crimes also differ, and the equipment used varies depending on the sort of crime. Various forensic tools under different categories such as Desktop Forensic Tool, Live Forensic Tool, Email Forensic Tool, and Operating Forensic Tools are examined in light of these characteristics. The most popular tools from each category mentioned above are tabulated with numerous attributes to help users choose the right tool for their investigation.

There is an indispensable role of machine learning, deep learning as a subset of Artificial intelligence applications [11]. The AI technique will intensify the process by supervising a massive amount of data with less time processing, high accuracy, and excellent outcome. Therefore, investigators are prompted to practice these approaches in digital forensic analysis to investigate varieties of digital crimes.

In this context, the motivation of this work deals with, detailed study of various phases of digital forensic tools, digital forensic methods, usage of AI applications in digital forensic investigations and challenges faced during the investigation process, careful analysis, and evaluation of digital forensic tools. The remaining paper is structured as per guidelines given by Misra [12]; Sect. 2 discusses related literature surveys. Section 3 elucidated the various stages of the digital forensic method considered during this study. Section 4 comprises the related work while evaluating digital forensic tools based on a practical approach and comparative analysis. Section 5 elaborates on the challenges and future direction of research in digital forensics using artificial intelligence. Finally, the paper is concluded Sect. 6.

2 Literature Review

Garfinkel et al. [13] presented Digital forensics research work and analyzed the crisis; Due to the increase in the storage size, it takes a more extended time to examine the device image and create a disk image. In addition, it is challenging to develop a tool that supports most operating systems, increasing the expansion cost of the tools.

Furthermore, improving encryption techniques is also a field of matter to decrypt the lost data to recover back the original data is a significant concern. Moreover, the convenience of cloud computing is rising, and services are employed in various segments. As a result, it is challenging to find components throughout the investigation. In addition, different types of legal issues are dealt with while investigating cybercrime.

Abirami et al. [14] suggested a method to recognize and identify suspicious packets across a wide variety of attacks employing a machine learning technique. The empirical outcomes revealed that Naive Bayesian provided superior accuracy compared to other classifiers. Yamin et al. [15] examined modern cyberattacks that employed AI-based methods and recognized several mitigation strategies that better control the before-mentioned attacks. Kebande et al. [16] highlighted novel forensic aptness methods in the cloud context. Empirical outcomes of the extended prototype are defined, concluding proposed techniques that improve the preparation.

Xiao et al. [17] introduce a forensic video investigation structure and apply compelling video intensifying techniques for low-quality footage. Amato et al. [18] analyzed digital forensic evidence through NLP techniques. Lee et al. [19] explore the features, applications, and constraints of digital forensic tools and match them among others regarding the facility of practice through assisting investigators in utilizing combined digital forensic tools for their examination. Wu et al. [20] highlighted the recent advances, readiness of digital forensic tools in today's complex environment. Cosic et al. [21] presented an approach to construct a digital forensic intelligence repository readiness. Hemden et al. [22] proposed an effective Cloud Forensics Investigation Model (CFIM) to examine cloud crimes in a forensically reliable and convenient mode.

Jang et al. [23] recommended a methodology for a digital forensic framework for the social network. This suggested method includes effective processes, classifying digital devices, obtaining digital evidence, and analyzing. Jospesh et al. [24]

presented digital forensic in terms of the cyber world and acquainted a comparative analysis of the current state of forensic. Costantini et al. [25] highlighted building infrastructure for Artificial Intelligence (AI) application in the digital forensic area, and the significant impediment is to accumulate evidence in the analysis phase. Krivchenkov et al. [26] impersonated a review of current intelligent systems for digital forensic to promote these methods in the forensic discipline. Quick et al. [27] manifested a research gap concerning the digital forensic data volume. A data compression process can affect a variety of digital forensic platforms, such as collection, processing, analysis, and provision for intelligence, awareness, and future demands.

Mohammad et al. [28] analyzed the usability of different ML techniques in recognizing proof by tracking file system initiate to discover how various application programs can handle those files. The ML algorithms achieved satisfying outcomes. Pluskal et al. [29] proposed network traffic classification utilizing machine learning methods. The experimental result revealed that the random forest classifier produced the most encouraging outcomes.

Alhawi et al. [30] performed an analysis on discovering windows ransomware network traffic applying machine learning evaluation and achieved a true positive rate of 97.1% using the decision tree technique. Srinivasan et al. [31] suggested a process that supports the text description of NLP and outlines spam email discovery. Sachdeva et al. [32] proposed an attack Classification in the cloud environment system using machine learning techniques among a digital forensic method. Sarker et al. [33] presented a broad view on AI-driven Cybersecurity that can perform an essential purpose for intelligent cybersecurity services and management. The model uses AI techniques to achieve the cybersecurity analysis process more efficiently associated with traditional security mechanisms.

Digital forensic can be described while utilizing systematic methodologies to store, collect, verify, investigate, evaluate, record, and present digital devices for criminal inquiries to ascertain and contest cybercrime. With the speedy improvement of technologies, it is essential to choose digital forensic frameworks and methods, gripping in understanding that the cyber warning panorama will evolve. Therefore, a sketch of digital forensic methodologies from 2016 to 2021 is highlighted and presented in Table 1.

The study compared and contrasted several digital forensic tools in categories such as Live Forensic Tool, Operating System Forensic Tool, Email Forensic Tool, and Desktop Forensic Tool. As a result, the study made it easier for users to select the right tool for their needs. The study also focused on using the application of artificial intelligence in digital forensic investigation. In addition, the majority of open source tools are presented, allowing the user to have a more pleasant experience.

Table 1 Digital forensic methodology

Forensic methodology	Phases
Jadhao et al. [34]	Examine, examine the circumstances, browse unusual information, call information base rule, report, and control link
Reza et al. [35]	Conventional data acquisition method design, consistent path, and support law toward determining the reliability of digital evidence
Kigwana et al. [36]	A digital forensic examination on ISO/IEC standards
Dokko et al. [37]	Defining crime emphasizes generated based on the review of five industrial espionage cases
Singh et al. [38]	DFR mechanism assessed employing in compliance with the ISO/IEC 27,043 framework
Montasari et al. [39]	A process model to accomplish forensic investigations
Mothi et al. [40]	Verify framework by AF methods concerning different stages in a digital forensic process
Sun et al. [41]	Online NLP based proposed framework for forensic investigation

3 Phases of Digital Forensic Investigation

The digital forensic investigation consists of five steps, i.e., identification, preparation, analysis, documentation, and presentation.

3.1 Identification Phase

This phase is associated with identifying relevant data as evidence and examining the occurrence location. In addition, this phase intends to preserve the morality of the pieces of evidence and securely covers them with a log. The evidence is to be handled carefully by following proper procedures.

3.2 Preparation Phase

The data are generated using multiple tools and techniques. In this phase, the image of the data is created as a copy of the hard disk. According to law enforcement, three acquisitions are accepted: mirror image, live acquisition, and forensics duplications.

Investigator questions whether sufficient information is available to confer an explicit request is in hand. However, there is ample information to respond to it. If desiring, they coordinate with the requester; otherwise, they start preparing the investigation method. Validation of all hardware and software is the first step in any forensic process to function correctly. Next, the investigator duplicates the forensic data provided at the request. Once the forensic platform is ready, forensic data is

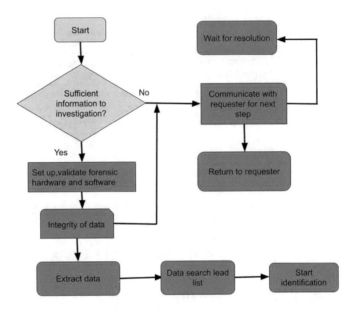

Fig. 1 Preparation phase

verified for its integrity. If the investigator receives facts, they require a persuasive sample and protect the original's series of data with care from assuring that record in their possession is intact and unchanged. After confirming the integrity of the document, they produce a method to obtain data. Figure 1 depicts process flow throughout the preparation phase.

3.3 Analysis Phase

Three types of analysis can be employed in this phase: limited, partial, and complete investigation. The short study includes solely the evidence stipulated by legal documents. The partial analysis apportions including cookies, log files, email folders. In addition, several kinds of tools can be used for forensic analysis, such as FTK (Forensic Tool Kit), Encase, etc. Fig. 2 describes process details in the analysis phase.

Forensic investigators respond to issues like who, whatever, while, wherever, and whereby to determine which user or application built, edited, collected, or transferred a specific object and when it incipiently evolved within continuation. Then, investigators reveal why all this information is essential and what it implies to the case wherever they obtained it.

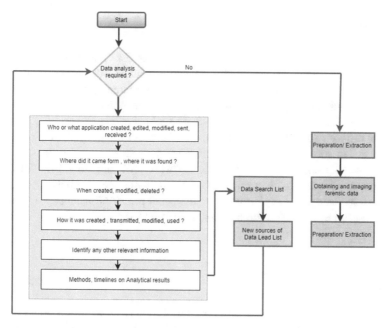

Fig. 2 Analysis phase

3.4 Documentation Phase

The presentation phase serves to find forensic processes are documented based on the evidence, with the aid of cybercrime laws, and outlines conclusion for further investigation.

After examiners have gone through all the phases, they can explore the forensic appeal and report as per the investigation. The forensic recording is outside the extent of this feature; however, it can not ignore its significance. The final report is the most authentic document of the examiners to deliver the verdicts to the requester.

3.5 Presentation Phase

It is the final step in the investigation and concerns presenting the investigator's findings and methods in the digital Crime investigation.

4   Digital Forensic Tools

Tools are the predefined application that is available for the investigation of digital Crimes. Various digital forensic tools are freely available in the market. The tools are either general or commercially licensed versions. Different types of digital forensic tools are discussed in the introduction part are presented in this section. This section will highlight a detailed analysis of tools in various categories and a comparative study of different tools in each type. The criteria for selecting this digital tool include (a) Ability to solve general issues related to management (b) Different technical considerations used in specific forensic software tool categories [42] (c) Certain technical considerations regarding two essential categories, namely disk imaging and sting searching (d) Legal issues regarding civil prosecution depending on digital evidence used by a specific forensic software tool [43].

4.1   Desktop Forensic Tool

In Desktop forensic, the investigator concentrates on recovering secondary memory such as a hard disk. Table 2. shows a comparative analysis of five Desktop forensic tools based on five key parameters, i.e., imaging, hashing, recovery, seizer, and acquire. In this study, the Pro Discover basis tool has been explored for desktop forensic analysis.

Pro Discover Basic: Pro discover tool facilitates the investigator to find whatever files they require on computer disk. It advances productivity and preserves the data as it is necessary for legal proceedings. It has an excellent searching capability that enables data to be searched efficiently and recovered. This tool can be used in incident response, corporate policy investigations, E-discovery, and computer forensic. Figure 3 shows the process of capturing an image using the Pro discover tool. Source Drive indicates the source file's location; destination path indicates 'where the outcomes are stored. The output file can be compressed with a secret password. Figure 4 represents the detailed analysis section of files during the investigation process.

Table 2 Comparative study of desktop forensic tools

Tools	Availability	Imaging	Hashing	Recovery	Seizer	Acquire
Pro discover [44]	Trial	No	Yes	Yes	Yes	Yes
Stellar [45]	Commercial	Yes	Yes	Yes	Yes	Yes
Autopsy [46]	Trial	Yes	No	Yes	Yes	Yes
Cyber check suit [47]	License	Yes	Yes	Yes	Yes	Yes
Encase [48]	Trial	No	No	Yes	Yes	No

Fig. 3 Capture image using pro discover

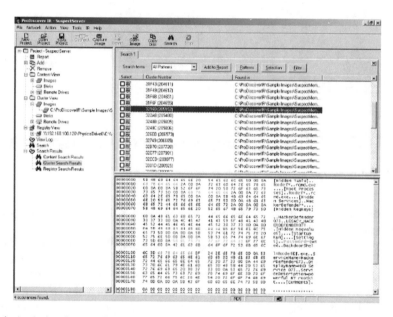

Fig. 4 Analysis using pro discover

4.2 Network Forensic Tool

Network forensic deals with controls and interprets computer networks to detect unknown malicious threats across devices and networks. This study selects five different network forensic tools based on five essential parameters, i.e., protocol, packet analyzer, packet spoofing, topology, and the open port. Table 3 represents the

Table 3 Related comparison of network forensic tools

Tools	Availability	Protocol	Packet analyzer	Packet spoofing	Topology	Open port
Wireshark [49]	Free	Yes	Yes	Yes	No	No
Nmap [50]	Yes	Yes	Yes	Yes	Yes	Yes
Nessus [51]	Yes	Yes	Yes	Yes	Yes	No
Snort [52]	Yes	Yes	Yes	Yes	No	Yes
Ettercap [53]	Yes	Yes	Yes	Yes	Yes	No

Fig. 5 Packet analysis

comparative study of network forensic tools, and the Wireshark network forensic tool has been explored for analysis in this work.

Wireshark: Wireshark is the most popular, immeasurable tool in the open-source category to perform network traffic analysis. This tool supports most operating systems, and a graphical-based user interface and a command-line interface are included to examine purposes. In addition, various filters can make customizable in this tool. Detailed packet analysis has been performed shown in Fig. 5. In contrast, Fig. 6 represents the sequence number of packet flows described graphically with time plotted in the x-axis, and the sequence no has been devised in the y-axis.

4.3 AI Application in Digital Forensic

In this section, the discussion is about how Artificial Intelligence can be used in digital forensics. During the digital inquiry, network logs are collected. The recorded

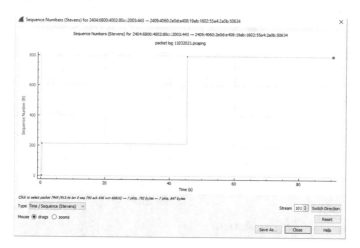

Fig. 6 Graphical representation of sequence number and packet log

Table 4 Classification of traffic

Sr. No	Category	Description
1	Normal	Normal traffic
2	Dos	An attack to produce network rescores unavailable to deliberate users
3	Probe	An object is used to learn something about the state of the network
4	r2l	To obtain unauthorized access to a victim machine
5	u2r	For illegally obtaining the root's privileges

logs are analyzed, and Artificial Intelligence is used to train the system for analysis purposes using Machine Learning techniques.

The data set contains 10,242 samples, and 18 features are selected. Traffics are classified into five categories shown in Table 4. Dataset has been split into 75–25 ratios as the training and testing phase.

In a Python-based environment, the experiment is conducted. GaussianNB, DecisionTreeClasifier, RandomForestClasifier, SVC, Linear Regression, GradientBoostingClasifier, and SVC are the Machine Learning-based techniques that were chosen to build the model for the prediction.

Figure 7 shows the number of identified cyberattacks versus protocol type, and in Fig. 8 the number of attacks identified based on IP address is displayed.

Figure 9 depicts the summary of different types of attack detected, and Fig. 10 shows the correlation matrix.

Figure 11 shows the confusion matrix of the model between the predicted label and the actual label. Figure 12 represents performance analysis of different models based on Accuracy, Precision, Recall, and F1-Score.

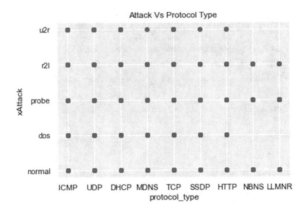

Fig. 7 Cyber attack versus protocol type

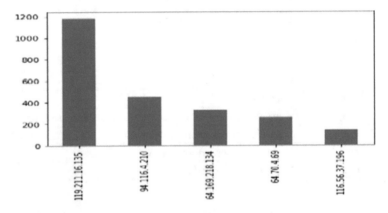

Fig. 8 IP address wise attack detected

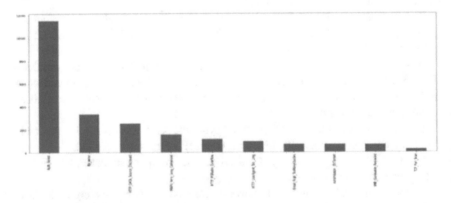

Fig. 9 Different types of attack detected

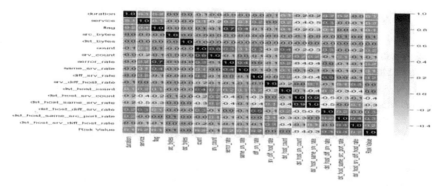

Fig. 10 Correlation matrix

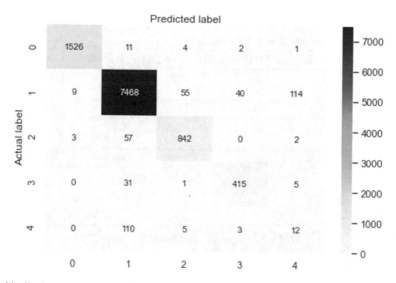

Fig. 11 Confusion matrix

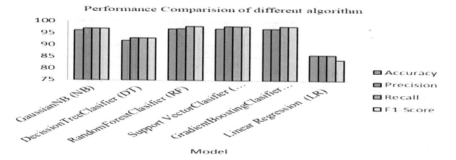

Fig. 12 Performance analysis

4.4   Live Forensic Tools

Live forensic conducts forensic analysis of active systems and focuses on RAM feature extraction, usually are sunning systems. As a result, live forensic presents accurate and consistent data for investigation than incomplete data supplied by another digital forensic process. Four different live forensic tools have been selected based on five key parameters, i.e., RAM dumping, Live logs, Live analysis, Search, Logs shown in Table 5. In this work, the Magnet RAM tool has been explored for live forensic.

Magnet RAM: The magnet RAM acquisition tool extracts live memory and examine a volatile trace. The investigator can utilize this for memory testing concerning the discovery of malware and data recovery. Figure 13 represents the creation of raw data dum with a dump extension, and Fig. 14 shows the successful outcome of the dmp file with raw extension.

Table 5 Comparative analysis of live forensic tools

Tools	Availability	RAM dumping	Live logs	Live analysis	Search	Logs
Magnet RAM [54]	Trial	Yes	Yes	Yes	No	No
OSF mount [55]	Trial	Yes	Yes	Yes	Yes	Yes
Volatility framework [56]	Trial	Yes	Yes	Yes	No	No
Belkasoft [57]	Trial	Yes	Yes	Yes	No	No

Fig. 13 Magnet RAM tool

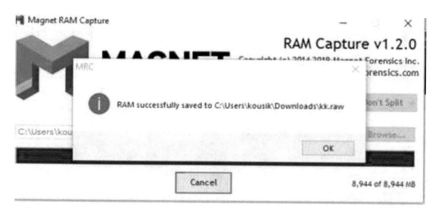

Fig. 14 Creation of DMP file using magnet RAM tool

4.5 Operation System Forensic Tools

Operating System Forensics is the method of recovering helpful information from the OS of the device. The intention of accumulating this information is to gather practical proof against the perpetrator. There are four methods for OS forensics: disk-to-image file, disk-to-disk copy, disk-to-data file & the sparse copy of a file. Table 6 shows the comparative study of four open-source Operating System tools based on popularity. In this work, OSForensic tool have been explored for analysis purposes.

OSForensic V8: It allows identifying unusual files and pursuit with hash matching, force signature connections, emails, memory. Furthermore, this quickly extracts forensic testimony from computers effectively. The overall view of the OSForensic tool is represented in Fig. 15, and version number 8 is used in this work. Data has been loaded in live acquisition mode, and Fig. 16 shows the loading process of the physical memory dump.

The loaded items are exported, and a detailed overview of the exported items is presented in Fig. 17, which includes case item ids, title, module, case item, category, and date added. Figure 18 represents program artifacts, including the application name, run count, file size, prefetch file, prefetch hash, and last run time.

Tools	Availability
OSForensic [58]	Trial
ExifTool [59]	Trial
Autospy [46]	Trial
Hashmyfiles [54]	Trial

Table 6 Different open sources of operating system tools

Fig. 15 Overall view of OSForensic V8

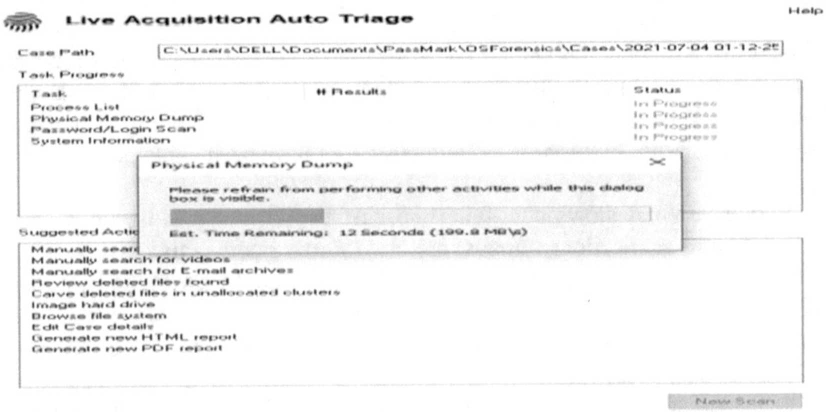

Fig. 16 Loading data in live acquisition

The evidence generated and presented in the form of a report that includes process, PID, total CPU, user time, threads, I/O bytes, I/O reads, I/O write, etc., are illustrated in Fig. 19.

4.6 Email Forensic Tool

Emails played a critical role in business communications and emerged as essential applications on the Internet. It is a helpful method for transmitting information

Fig. 17 Snap of exported items

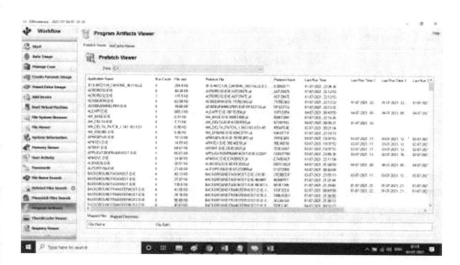

Fig. 18 Snap of program Artifacts

and records from machines and different computerized devices. However, there are various challenges in email forensics like fake emails, spoofing, and Anonymous Re-emailing. The examiner has the following goals while conducting email forensics: identifying the offender, collecting the required proof, manifesting the judgments, developing the case. Four different open source email forensic tools are included based on popularity shown in Table 7.

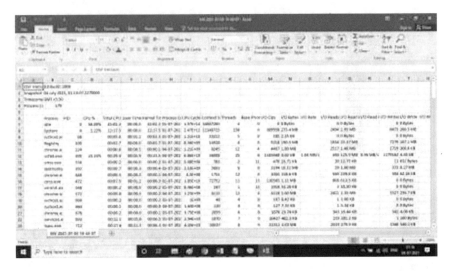

Fig. 19 Report exportation

Table 7 Open source email forensic tools	Tools	Availability
	Add4Mail [60]	Trial
	MailXaminer [61]	Trial
	eMailTrackerPro [62]	Trial
	Paraben E-mail examiner [63]	Trial

This study explores the Add4Mail Professional Trial version of Email investigation software for forensic investigation. It maintains different Email formats and can examine mail by time, header content, and message body content.

Figure 20 represents the processing stage while loading, reading the IMAP folder, including the important documents stored in the Email. Once the necessary files are exported, emails and necessary folders are displayed, the detailed overview is shown in Fig. 21.

Multiple options are available for analysis, saving, exporting like filtering based on email source, filter options, saving the outcomes in various format like .pst, .msg, .eml, .emlx, .mbox, .pdf, .html, .mht, .xps, .rtf etc. and iuulustarted in Fig. 22.

Fig. 20 Exporting emails

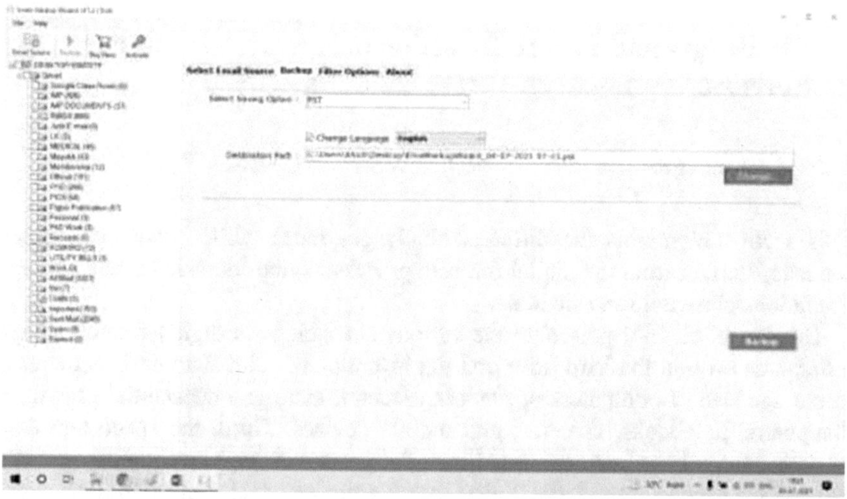

Fig. 21 Snapshot of email analysis

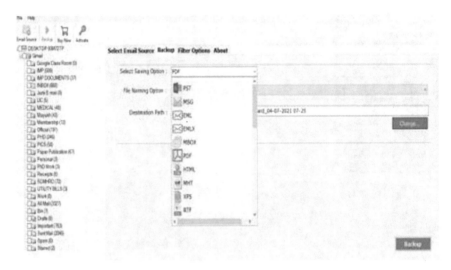

Fig. 22 Option for analysis

5 Challenges and Future Direction of Research in Digital Forensic

5.1 Challenges

This section highlights the different challenges faced while conducting digital forensic, starting from the digital forensic process, digital forensic framework, and limitations of digital forensic tools.

Guarino et al. [64] presented the various changelogs in digital forensic. First, a dramatic drop in the hard drive and the invention of solid-state devices increase in storage size of computers or devices. Second, there is a substantial growth of computers, notebooks, cameras, and mobile devices. Third, the speed and ease of connectivity have been increased over the years with rapid technology growth. Finally, we need to rethink the digital forensic process. Quick et al. [65] proposed a digital forensic data modification process by discriminating the imaging in Big data forensic to the faster massive amount of data. However, the accelerated increase of the Internet of Things (IoT) created a significant digital forensic examination method challenge. Alabdulsalam et al. [66] presented a reflection of the smartwatch in concern with the digital forensic process.

Karie et al. [67] highlighted four different types of changes faced in digital forensic. First, technological challenges are encountered: encryption, a large volume of data, volatility of digital evidence, bandwidth crunch, emerging technology, and cloud computing. Second, the challenges with legal systems or law of enforcement, namely, Jurisdiction, legal process, insufficient support for lawful criminal, ethics issues, and privacy. Third, the personal–related challenges include lack of qualified digital forensic professionals, lack of forensic knowledge, and lack of unified formal representation. Fourth, the operational difficulties include incidence detection response, lack of standardized process, significant mutual intervention, the readiness of digital forensic, and the trust of audit trails. Hraiz et al. [68] highlighted digital forensic difficulties faced in the cloud context, namely records, volatile data, creation of the forensic image, and data integrity.

5.2 Future Direction of Research

This section discusses the future direction of digital forensic research, namely IoT forensic, Big Data Forensic, Digital Forensic as a service, and new digital forensic tools. The foremost aspects of digital evidence in IoT contexts posture inherent challenges. For example, identifying specific custom data is challenging the examiner to determine where to direct exploration and fit if the location and derivation of data cannot be resolved. Accordingly, consider those above as a research opportunity to continue investigating.

Interpreting big forensic data in an appropriate and forensically reliable practice postures meaningful challenges in digital forensic. Different research opportunities to discuss big data continue to advance innovative tools and techniques or accommodate everyday things in context with digital forensics. For example, the researcher can continue with neural networks to support complicated pattern identification in different categories of digital forensics, namely cloud and network forensics. Researchers can also develop Natural Language Processing (NLP) methods for analyzing unregulated digital forensic information. Finally, the researcher community may explore the usage of artificial intelligence in the digital forensic investigation process.

The present digital forensic tools are utilized to carry out on violator's machine. Nevertheless, those tools present limited capacity for analyzing complicated cyberspace such as the cloud [69, 70]. Therefore, one of the significant difficulties that demand future research is advancing modern tools, techniques, dealing with the massive volume of data, and implementing inherent evidence to digital forensics for further investigations.

6   Conclusion

The use of the Internet and communication devices is the basis for Hackers to induce cyber-attacks. The Hackers monitor the system to find the vulnerability of the system to launch Cyber attacks. Once the cyber attack is launched, a proper investigation process is to be followed to complete the digital crime investigation.

The various phases of Digital Forensic Investigation are discussed with the relevant illustration. A comparative analysis of different Digital Forensic frameworks is tabulated. Many Digital Forensic tools are available in the market. Based on the type of attack, the tools are utilized for the investigation. The comparative analysis of various tools with some set of parameters is discussed and tabulated. AI plays a major role in Network log analysis and prediction. The network logs are analyzed based on various Machine Learning Techniques and are evaluated with various metrics to recognize the preferred techniques. The study analyzes multiple tools under Desktop, Network, Live, Operating System Forensic Tools. In Desktop forensic, Pro Discover tool, in-network forensic Wireshark tool, live forensic magnet ram tool, operating forensic OS-Forensic, and Email forensic AidMail tools are discussed in this study. It also featured the application of artificial intelligence in the digital forensic process. Further, several challenges are highlighted while accompanying the digital forensic investigation process. The entities, as mentioned earlier, will require concentrating periodically in the future and attempt to address the challenging phases of digital forensic. Furthermore, more enhanced skill sets, tools, experience, and participation are needed to build digital evidence. Therefore, it is concluded that the future research paths described can undoubtedly influence further research directions in digital forensics.

References

1. Sindhu K (2012) Digital forensics and cyber crime datamining. J Inf Secur 03:196–201. https://doi.org/10.4236/jis.2012.33024
2. Alhassan JK, Oguntoye RT, Misra S, Adewumi A, Maskeliūnas R, Damaševičius R (2018) Comparative evaluation of mobile forensic tools https://doi.org/10.1007/978-3-319-73450-7_11
3. Olajide F, Misra S (2016) Forensic investigation and analysis of user input information in business application. Indian J Sci Technol 9. https://doi.org/10.17485/ijst/2016/v9i25/95211
4. Osho O, Mohammed UL, Nimzing NN, Uduimoh AA, Misra S (2019) Forensic analysis of mobile banking apps. https://doi.org/10.1007/978-3-030-24308-1_49
5. Malin CH, Casey E, Aquilina JM (2014) Chapter 1—Malware incident response: volatile data collection and examination on a live linux system. In: Malin CH, Casey E, Aquilina JM (eds) Malware forensics field guide for linux systems, syngress, pp 1–106. https://doi.org/10.1016/B978-1-59749-470-0.00001-2. ISBN 9781597494700
6. Barker K, Askari M, Banerjee M, Ghazinour K, Mackas B, Majedi M, Pun S, Williams A (2009) A data privacy taxonomy. In: BNCOD 26: proceedings of the 26th British national conference on databases. Springer, Berlin, Heidelberg, pp 42–54
7. Patankar M, Bhandari D (2014) Forensic tools used in digital crime investigation forensic science
8. Kabir SMS (2016) Basic guidelines for research. In: An introductory approach for all disciplines, pp 168–180
9. Shimeall TJ, Spring JM (2014) Chapter 1—Motivation and security definitions. In: Shimeall TJ, Spring JM (eds) Introduction to information security, syngress, pp 1–20. ISBN 9781597499699
10. Hibshi H, Vidas T, Cranor L (2011) Usability of forensics tools: a user study. In: Proceedings—6th international conference on IT security Incident management and IT forensics, IMF, pp 81–91. https://doi.org/10.1109/IMF.2011.19

11. Qadir AM, Varol A (2020) The role of machine learning in digital forensics. In: 2020 8th international symposium on digital forensics and security (ISDFS). IEEE, pp 1–5
12. Misra S (2020) A step by step guide for choosing project topics and writing research papers in ICT related disciplines. In: International conference on information and communication technology and applications. Springer, Cham, pp 727–744
13. Garfinkel SL (2010) Digital forensics research: the next 10 years. Int J Digit Foren Incid Response 7(Supplement):S64–S73 (Naval Postgraduate School, Monterey, USA)
14. Abirami A, Palanikumar S (2021) Proactive network packet classification using artificial intelligence. In: Artificial intelligence for cyber security: methods, issues and possible horizons or opportunities. Springer, Cham, pp 169–187
15. Yamin MM, Ullah M, Ullah H, Katt B (2021) Weaponized AI for cyber attacks. J Inf Secur Appl 57:102722
16. Kebande VR, Venter HS (2018) Novel digital forensic readiness technique in the cloud environment. Aust J Forensic Sci 50(5):552–591
17. Xiao J, Li S, Xu Q (2019) Video-based evidence analysis and extraction in digital forensic investigation. IEEE Access 7:55432–55442
18. Amato F, Cozzolino G, Moscato V, Moscato F (2019) Analyse digital forensic evidences through a semantic-based methodology and NLP techniques. Futur Gener Comput Syst 98:297–307
19. Lee JU, Soh WY (2020) Comparative analysis on integrated digital forensic tools for digital forensic investigation. In: IOP conference series: materials science and engineering, vol 834, no 1. IOP Publishing, p 012034
20. Wu T, Breitinger F, O'Shaughnessy S (2020) Digital forensic tools: recent advances and enhancing the status quo. Forensic Sci Int Dig Investig 34:300999
21. Cosic J, Schlehuber C, Morog D (2021) Digital forensic investigation process in railway environment. In: 2021 11th IFIP international conference on new technologies, mobility and security (NTMS). IEEE, pp 1–6
22. Hemdan EED, Manjaiah DH (2021) An efficient digital forensic model for cybercrimes investigation in cloud computing. Multimed Tools Appl
23. Jang YJ, Kwak J (2015) Digital forensics investigation methodology applicable for social network services. Multimed Tools Appl 74(14):5029–5040
24. Joseph DP, Norman J (2019) An analysis of digital forensics in cyber security. In: First international conference on artificial intelligence and cognitive computing. Springer, Singapore, pp 701–708
25. Costantini S, De Gasperis G, Olivieri R (2019) Digital forensics and investigations meet artificial intelligence. Ann Math Artif Intell 86(1):193–229
26. Krivchenkov A, Misnevs B, Pavlyuk D (2018) Intelligent methods in digital forensics: state of the art. In: International conference on reliability and statistics in transportation and communication. Springer, Cham, pp 274–284
27. Quick D, Choo KKR (2014) Impacts of increasing volume of digital forensic data: a survey and future research challenges. Digit Investig 11(4):273–294
28. Mohammad RMA, Alqahtani M (2019) A comparison of machine learning techniques for file system forensics analysis. J Inf Secur Appl 46:53–61
29. Pluskal J, Lichtner O, Rysavy O (2018) Traffic classification and application identification in network forensics. In: IFIP international conference on digital forensics. Springer, Cham, pp 161–181
30. Alhawi OM, Baldwin J, Dehghantanha A (2018) Leveraging machine learning techniques for windows ransomware network traffic detection. In: Cyber threat intelligence. Springer, Cham, pp 93–106
31. Srinivasan S, Ravi V, Alazab M, Ketha S, Ala'M AZ, Padannayil SK (2021) Spam emails detection based on distributed word embedding with deep learning. In: Machine intelligence and big data analytics for cybersecurity applications. Springer, Cham, pp 161–189
32. Sachdeva S, Ali A (2021) Machine learning with digital forensics for attack classification in cloud network environment. Int J Syst Assur Eng Manag 1–10

33. Sarker IH, Furhad MH, Nowrozy R (2021) Ai-driven cybersecurity: an overview, security intelligence modeling and research directions. SN Comput Sci 2(3):1–18
34. Jadhao AR, Agrawal AJ (2016) A digital forensics investigation model for social networking site. In: Proceedings of the second international conference on information and communication technology for competitive strategies
35. Montasari R (2017) A standardised data acquisition process model for digital forensic investigations. Int J Inf Comput Secur
36. Kigwana I, Kebande VR, Venter HS (2017) A proposed digital forensic investigation framework for an eGovernment structure for Uganda. In: 2017 IST-Africa week conference (IST-Africa). IEEE, pp 1–8
37. Dokko J, Shin M (2018) A digital forensic investigation and verification model for industrial espionage. In: International conference on digital forensics and cyber crime. Springer, Cham, pp 128–146
38. Singh A, Ikuesan AR, Venter HS (2018) Digital forensic readiness framework for ransomware investigation. In: International conference on digital forensics and cyber crime. Springer, Cham, pp 91–105
39. Montasari R, Hill R, Carpenter V, Hosseinian-Far A (2019) The standardised digital forensic investigation process model (SDFIPM). In: Blockchain and clinical trial. Springer, Cham, pp 169–209
40. Mothi D, Janicke H, Wagner I (2020) A novel principle to validate digital forensic models. Forensic Sci Int: Dig Investig 33:200904
41. Sun D, Zhang X, Choo KKR, Hu L, Wang F (2021) NLP-based digital forensic investigation platform for online communications. Comput Secur 104:102210
42. Babiker M, Karaarslan E, Hoscan Y (2018) Web application attack detection and forensics: a survey. In: 2018 6th international symposium on digital forensic and security (ISDFS). IEEE, pp 1–6
43. Henseler H, van Loenhout S (2018) Educating judges, prosecutors and lawyers in the use of digital forensic experts. Digit Investig 24:S76–S82
44. Kamal KMA, Alfadel M, Munia MS (2016) Memory forensics tools: Comparing processing time and left artifacts on volatile memory. In: 2016 international workshop on computational intelligence (IWCI). IEEE, pp 84–90
45. Dietzel C, Wichtlhuber M, Smaragdakis G, Feldmann A (2018) Stellar: network attack mitigation using advanced blackholing. In: Proceedings of the 14th international conference on emerging networking experiments and technologies, pp 152–164
46. Dizdarevic A, Baraković S, Husic JB (2019) Examination of digital forensics software tools performance: open or not?. In: International symposium on innovative and interdisciplinary applications of advanced technologies. Springer, Cham, pp 442–451
47. Lovanshi M, Bansal P (2018) Benchmarking of digital forensic tools. In: International conference on computational vision and bio inspired computing. Springer, Cham, pp 939–947
48. Quick D, Choo KKR (2018) Digital forensic data reduction by selective imaging. In: Big digital forensic data. Springer, Singapore, pp 69–92
49. Tabuyo-Benito R, Bahsi H, Peris-Lopez P (2018) Forensics analysis of an online game over steam platform. In: International conference on digital forensics and cyber crime. Springer, Cham, pp 106–127
50. Aggarwal P, Gonzalez C, Dutt V (2020) HackIt: a real-time simulation tool for studying real-world cyberattacks in the laboratory. In: Handbook of computer networks and cyber security. Springer, Cham, pp 949–959
51. Munoz FR, Vega EAA, Villalba LJG (2018) Analyzing the traffic of penetration testing tools with an IDS. J Supercomput 74(12):6454–6469
52. Kenkre PS, Pai A, Colaco L (2015) Real time intrusion detection and prevention system. In: Proceedings of the 3rd international conference on frontiers of intelligent computing: theory and applications (FICTA) 2014. Springer, Cham, pp 405–411
53. Agrawal N, Tapaswi S (2015) Wireless rogue access point detection using shadow honeynet. Wirel Pers Commun 83(1):551–570

54. Ghafarian A, Wood C (2018) Forensics data recovery of skype communication from physical memory. In: Science and information conference. Springer, Cham, pp 995–1009
55. Hassan NA (2019) Analyzing digital evidence. In: Dig Forensics Basics. Apress, Berkeley, CA, pp 141–177
56. Seo J, Lee S, Shon T (2015) A study on memory dump analysis based on digital forensic tools. Peer-to-Peer Netw Appl 8(4):694–703
57. Hassan NA (2019) Computer forensics lab requirements. In: Digital forensics basics. Apress, Berkeley, CA, pp 69–91
58. Alqahtany S, Clarke N, Furnell S, Reich C (2016) A forensic acquisition and analysis system for IaaS. Clust Comput 19(1):439–453
59. Barton TEA, Azhar MHB (2017) Open source forensics for a multi-platform drone system. In: International conference on digital forensics and cyber crime. Springer, Cham, pp 83–96
60. Alsmadi I, Burdwell R, Aleroud A, Wahbeh A, Al-Qudah M, Al-Omari A (2018) Web forensics-chapter competencies. In: Practical information security. Springer, Cham, pp 283–296
61. Chhabra GS, Bajwa DS (2015) Review of email system, security protocols and email forensics. Int J Comput Sci Commun Netw 5(3):201–211
62. Singh V (2015) Forensic investigation of email artefacts by using various tools 2:2321–613
63. Al Fahdi M, Clarke NL, Li F, Furnell SM (2016) A suspect-oriented intelligent and automated computer forensic analysis. Digit Investig 18:65–76
64. Guarino A (2013) Digital forensics as a big data challenge. In: ISSE 2013 securing electronic business processes. Springer, Wiesbaden, pp 197–203
65. Quick D, Choo KKR (2016) Big forensic data reduction: digital forensic images and electronic evidence. Clust Comput 19(2):723–740
66. Alabdulsalam S, Schaefer K, Kechadi T, Le-Khac NA (2018) Internet of things forensics—Ch:allenges and a case study. In: IFIP international conference on digital forensics. Springer, Cham, pp 35–48
67. Karie NM, Venter HS (2015) Taxonomy of challenges for digital forensics. J Forensic Sci 60(4):885–893
68. Hraiz S (2017) Challenges of digital forensic investigation in cloud computing. In: 2017 8th international conference on information technology (ICIT)
69. Krishnan S, Zhou B, An MK (2019) Smartphone forensic challenges. Int J Comput Sci Secur (IJCSS) 13(5):183
70. Caviglione L, Wendzel S, Mazurczyk W (2017) The future of digital forensics: challenges and the road ahead. IEEE Secur Priv 6:12–17

Intelligent Authentication Framework for Internet of Medical Things (IoMT)

Timibloudi Stephen Enamamu

Abstract The rapid growth of smart wearables and body sensor networks is expected to increase over the years. The reducing cost of manufacturing, deployment and the small and unobtrusive nature of most of the wearables available have intensified the acceptability for deployment in areas such as medical devices for healthcare monitoring. This work explored the use of artificial intelligence to enhance authentication of Internet of Medical Things (IoMT) through a design of a framework. The framework is designed using wearable and or with a mobile device for extracting bioelectrical signals and context awareness data. The framework uses bioelectrical signals for authentication while artificial intelligence is applied using the contextual data to enhance the patient data integrity. The framework applied different security levels to balance between usability and security on the bases of False Acceptance Rate (FAR) and False Rejection Rate (FRR). 30 people are used for the evaluation of the different security levels and the security level 1 achieved a result based on usability vs security obtaining FAR of 5.6% and FRR of 9% but when the FAR is at 0% the FRR stood at 29%. The Intelligent Authentication Framework for Internet of Medical Things (IoMT) will be of advantage in increasing the trust of data extracted for the purpose of user authentication by reducing the FRR percentage.

Keywords Transparent authentication · Biometrics · Intelligent authentication · Internet of medical things · Artificial intelligence

1 Introduction

The ability of wearable technology to connect to the internet and other short range wireless networks like Bluetooth and near field communication has enhanced its application for Internet of Medical Things (IoMT) [1].

With the improvement of wearable devices capabilities, medical professionals are taking advantage of them to extract physiological data for improved healthcare

T. S. Enamamu (✉)
Department of Computing, Sheffield Hallam University, Sheffield S1 1WB, UK
e-mail: t.enamamu@shu.ac.uk

© The Author(s), under exclusive license to Springer Nature Switzerland AG 2022
S. Misra and C. Arumugam (eds.), *Illumination of Artificial Intelligence in Cybersecurity and Forensics*, Lecture Notes on Data Engineering and Communications Technologies 109, https://doi.org/10.1007/978-3-030-93453-8_5

management. This includes its use for interacting with the environment for the collection of data for interpreting and analyzing the individual action. The importance of the data requires secrecy and should only be accessed by an authorized personnel including the patient [2]. The data transmitted among the medical devices should be verified through data authentication [3]. Based on this, this work introduces an intelligent framework to leverage the information provided through bioelectrical signals and contextual information to improve accuracy and trust of Internet of Medical Things (IoMT) through data authentication.

There are early prior works establishing the use of bioelectrical signals like Electrocardiogram (ECG) and Electroencephalography (EEG) for authentication. The works by [4–8] used ECG for authentication, extracting the features using Morphological features extraction techniques with an acceptable accuracy. Another work by [9–12] used ECG with Wavelet Transform for feature extraction. The use of Morphological and Wavelet Transform features extraction techniques achieved a success rate of above 90%, this shows the viability of the techniques.

There are other techniques used which includes Welch Algorithm [13], Ample Maximum & Sample Minimum [14], Independent Component Analysis (ICA) [15], QRS Detection [12, 16, 17], Power Spectral Density [18], Discrete Cosine Transform [19], Power Spectral Density (PSD) [20], Mel Frequency Cepstral Coefficients (MFCC) [21] and Power spectral density (PSD) [22]. The summary of the different works with their feature extraction algorithm and their results is shown in Table 1.

The use of bioelectrical signals for authentication has a limitation because of its variability can be an issue like stress, affecting the ECG signal constancy [23], The inconsistency in the signal makes it difficult to deal [24]. The motivation for this work is to improve data trust, by proposing a framework that will apply intelligence using contextual data to enhance the authentication of the data transmitted within the Internet of Medical Things. This chapter is organized as follows with a guide [25]: Sect. 1 introduces the topic, discussing some prior work. Section 2, focused on the intelligent framework composition and configuration. Section 3 discussed the feature extraction techniques. Section 4 includes the evaluation experimentation for the feature extracted and classification. Section 5 discusses the Artificial Intelligent Decision (AIDE) Module and Sect. 6 the conclusion.

2   Intelligent Framework

The framework is designed using the standard biometric authentication process which includes data collection, signal pre-processing, feature extraction and classification of the biometric data [26, 27]. The intelligent component is incorporated in the biometric authentication process as shown in Fig. 1. The data collection component of the framework can collect different biometric data and identify them separately before pre-processing them.

The IoMT Intelligent Authentication decision is based on the biometric data and contextual information therefore, the framework is partially implemented in the data

Table 1 Showing the different prior works and the results

Work and year	Features extraction algorithm	Results (%)
Israel et al. [4]	Morphological features	97–98
Shen et al. [5]	Morphological features	100
Subasi [9]	Wavelet transform	94.5–93.2
Wang et al. [6]	Detection, morphological features (QRS)	94.47
	Principal component analysis (PCA)	97.8
Gahi et al. [7]	Morphological features	100
Chan et al. [10]	Wavelet transform	89–95
Hema et al. [13]	Welch algorithm	94.4–97.5
Kousarrizi et al. [14]	Ample maximum & sample minimum	68–100
He and Wang [15]	Independent component analysis (ICA)	5.0
Sasikala and Wahidabanu [16]	QRS detection	99
Ye et al. [11]	ICA and wavelet transform	99.6
Coutinho et al. [8]	Morphological features	100
Hema and Osman [18]	Power spectral density	78.6
Tawfik and Kamal [19]	Discrete cosine transform	99.09
Sidek and Khalil [12]	Wavelet transform	91
	QRS detection and wavelet transform	95
Hema and Elakkiya [20]	Power spectral density (PSD)	97.2–98.85
Zokaee and Faez [21]	Mel frequency cepstral coefficients (MFCC)	89
Le et al. [17]	QRS detection	85.04
Mohanchandra et al. [22]	Power spectral density (PSD)	85

Fig. 1 The intelligent authentication framework for IoMT block diagram

collection, the information from the data is used to make decisions in the intelligent component before sending the information to the classification module to enhance the data for classification. The framework includes a module for remote health-care monitoring using context awareness and bioelectrical signals. It is made up of different components which includes, wearable device, Data collection module, context awareness module and Intelligent component.

2.1 Data Collection Modules

The data collection is done using a wearable/mobile device and further processed in a data collection module which should be able to collect and process sufficient data [28]. Some of the required data for implementing the authentication framework includes the patient's heart rate, hydration, heart rate variability, respiratory rate, skin temperature, peripheral arterial oxygen saturation (SpO2), blood pressure, respiratory rate, depending on the sensor available in the data collection device. Most of these data are useful for health diagnoses and monitoring of the state of health.

2.2 Biometric Data

The data collection device could be a mobile device, smart watch or wearable device worn on any part of the body [29]. The data is collected based on the data available during the authentication process as shown in Fig. 2.

Based on the sensors available on the mobile, smartwatch or wearable, the data is collected and separated by the data collection module. The data collection module transmits the data either to the intelligent component or the pre-processing module for feature extraction. The data separation as shown in Fig. 3 depends on the data, either a signal or contextual data. The data collection module also extracts the activity information when the data is collected. The activity information includes walking, standing, lying down or in a moving car/train, or running. This helps in enhancing the application of the contextual information extracted for intelligent decisions.

Fig. 2 Showing the available data indicator

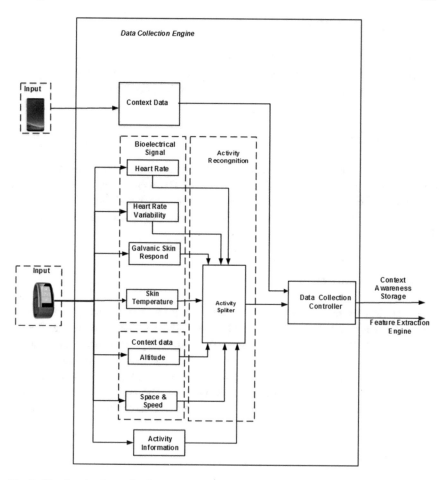

Fig. 3 Showing the data collection module

2.2.1 Heart Rate

The heart rate is the beats measured in the number of contractions of the heartbeat per minute (bpm). It varies depending on the body's physiological state. The heart rate helps in accessing the physical fitness and overall health and wellness of a person. The gradual increase of the heart rate through regular exercise will in turn increase the body's oxygen circulation [30]. This is useful for health care monitoring and accessing the wellbeing as one ages.

2.2.2   Heart Rate Viability (HRV)

The use of the Heart Rate Viability (HRV) is a tool for determining stress and other psychological events with the body including hypertension [31], Fatigue [32]. The HRV measures the change in time between successive heart beats.

2.2.3   Pulse Rate Variability (PRV)

The pulse rate is the number of heart beats in one minute which is directly related to the heart rate. The heart rate viability and Pulse Rate Variability contain different information [33] therefore, both can be useful for different health related information. The PRV has been used for investigation of mental diseases, diabetes, hypo- or hypertension, or cardiac arrhythmias [34].

2.2.4   Respiratory Rate

The respiration rate is the number of breaths per minute usually measured when at rest and can be counted by the number of breaths within a minute. This can be affected by illness, and other medical conditions. The respiratory rate is used for a variety of clinical use including obtaining a baseline for monitoring blood transfusion or drug reaction [35]. The respiration rate when abnormal can be used as an indicator for predicting cardiac arrest. The monitoring of cardiac arrest can be done using respiration rate which helps in a quick response for intensive care units (ICU) [36].

2.2.5   Skin Temperature

The Skin Temperature is the rise and fall of the human skin temperature which depends on the atmospheric temperature. Skin temperature has been used for regulating stress, headache activity and measuring the activity of the sympathetic nervous system [37].

2.2.6   Galvanic Skin Response (GSR)

The Galvanic Skin Response (GSR) is defined as the change or variation in the sweat gland activity. This is useful for monitoring emotional arousal and stress, diagnosis of diabetes [38, 39].

2.2.7 Peripheral Arterial Oxygen Saturation (SpO2)

The Peripheral Arterial Oxygen Saturation is the estimated amount of oxygen in the blood. The SpO2 can be used to monitor apnoea and hypopnoea which are types of Obstructive sleep apnoea (OSA). OSA leads to medical problems like loud snoring, disrupted sleep and excessive daytime sleepiness [40].

2.2.8 Blood Pressure

Blood pressure is the force measurement as the heart pumps blood around your body. The blood pressure can either be low, high, and ideal blood pressure. High blood pressure can lead to health conditions such as heart and kidney disease while low blood pressure can lead to heart failure and dehydration [41, 42].

2.3 Pre-processing

The signal processing involves the removal of unnecessary portions of the signal that is noisy. There can be random noise, spike that can affect the quality of the signal [43]. This is most likely in a non-stationary signals like that changes in frequency from time to time [44]. Noises in non-stationary signal can be categorized as follows [45].

Power line interference: This is a common noise in bioelectrical signals which occurs mostly when the signal is recorded from the skin surface [46].

Baseline wandering: This a low-frequency activity that interferes with the bioelectrical signals which in most cases can lead to inaccurate reading of the signal measurement. The Baseline wandering could occur due to body movements, perspiration, or respiration. The poor electrode contact with the body can cause Baseline wandering too [47].

Electronic pop: A wearable device sensor electrode could create an electronic pop noise when contact with the skin in a brief contact lost with the device.

Electrode motion artifacts: This occurs due to the physical movement of the body during recording. This type of noise is difficult to separate from the bioelectrical signal due to the overlapping nature of the motion artefact [48].

3 Feature Extraction

The feature extraction is the conversion of the bioelectrical signal into sets of vectors for discrimination discriminating between patterns of different classes. When extracting features, there should be careful consideration of the process because of the characteristic properties of the biometric including the non-stationary nature of signals.

3.1 Feature Characteristics

When processing biometric signals, some characteristics must be considered for quality for pattern recognition like distinctiveness, repeatability, efficiency, and quantity [28].

3.1.1 Repeatability

The feature characteristic of each signal vector should have closeness in estimate measurement of the properties of the signal pattern.

3.1.2 Distinctiveness

Each feature vector should be distinguishable from other feature vectors based on feature representation value. These attributes give the pattern representation robustness and enhance the accuracy of the pattern recognition.

3.1.3 Quantity

Each feature vector should be rich enough, containing useful information for representing the pattern.

3.1.4 Efficiency

The most important aspect of feature extraction is for the feature vector to distinguish different subjects based on the features extracted effectively.

3.2 Feature Extracted

The feature extraction is the reduction of data to only useful set useful information in the dataset. Most of the features used for this work have been used for similar works [3], the features are.

3.2.1 Variance (V)

The Variance is the sum of square distance of the extracted biometric signal. This is express in formula as:

$$V = \frac{\sum (X - \mu)^2}{N}$$

3.2.2 Mean of the Energy (ME)

The Mean of the Energy is the average energy in value of the extracted biometric signal. The Mean of the Energy express in formula as:

$$ME = \frac{\sum_r \exp(-\beta E_r) E_r}{\sum_r \exp(-\beta E_r)}$$

3.2.3 Minimum Energy (MinE)

This is the lowest energy value of the extracted biometric signal. This is express in formula as:

$$MinE = Minimum\ Signal\ Energy$$

3.2.4 Maximum Energy (MaxE)

The Maximum Energy is the highest energy value of the extracted biometric signal. This is expressed in formula as:

$$MaxE = Maximum\ Signal\ Energy$$

3.2.5 Mean(M)

This is the value diversity of the extracted biometric signal median. This is expressed in formula as:

$$M = \frac{1}{n} \sum_{i=1}^{n} x_i$$

3.2.6 Minimum Amplitude (MinA)

The Minimum Amplitude is the lowest point from the equilibrium point of the extracted biometric signal. This is expressed in formula as:

$$MinA = Minimum\ displacement$$

3.2.7 Standard Deviation (STD)

This is the square root of the variance of a random variation. of the extracted biometric signal. This is expressed in formula as:

$$STD = \sqrt{\frac{1}{n}\sum_{i=1}^{n}(x_i - \mu)^2}$$

3.2.8 Maximum Amplitude (MaxA)

The Maximum Amplitude is the highest point from the equilibrium point of the extracted biometric signal. This is expressed in formula as:

$$Max \cdot A = Maximum\ displacement$$

3.2.9 Range (R)

The Range is the difference between the highest extracted biometric signal value and the lowest extracted biometric signal value and expressed in formula as:

$$R = Maximum\ signal - minimum\ signal$$

3.2.10 Peak2peak (P2P)

The Peak2peak is the difference between the maximum and minimum values of the extracted biometric signal. This is expressed in formula as:

$$P2P = Signal\ Maxi\ to\ Min\ diff \cdot displacement$$

3.2.11   Root Mean Square (RMS)

The RMS is the measurement of the magnitude of a set value within the extracted biometric signal. This is expressed in formula as:

$$\text{Root Mean Square} = \sqrt{\frac{1}{n}\sum_{i=1}^{N} X_I^2}$$

3.2.12   Peak Magnitude to RMS Ratio (PMRMS)

This is the ratio of the largest absolute value of the extracted biometric signal to the root-mean-square (RMS) value of the extracted biometric signal and expressed in formula as:

$$\text{PMRMS} = \sqrt{\frac{X_\infty}{\frac{1}{N}\sum_{n=1}^{N} X_n^2}}$$

3.2.13   Average Frequency (AF)

The is the arithmetic mean the extracted biometric signal frequency and it is expressed in formula as:

$$\text{AF} = \frac{X_1 + X_2 + X_3 \ldots X_n}{N}$$

4   Evaluation Experimentation

The experimentation is carried out in two phases, the first is the evaluation of the features extracted and the second is the evaluation of the algorithm's selection for the framework.

4.1   Features Evaluation

To test the feature extraction, a sample from 12 participants is extracted for evaluating all the features. A biometric data of heart rate viability is used as the data for testing the

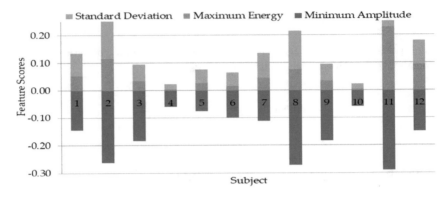

Fig. 4 Statistical features of SD, MaxE and MA

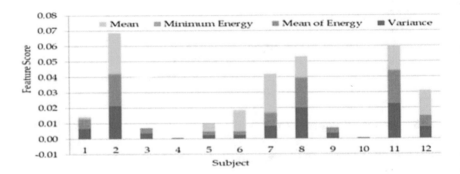

Fig. 5 Statistical features of M, MinE, ME and V

features. The result shows a good performance for its use for the intelligent authentication framework. The result of the statistical features is shown in Figs. 4, 5, 6 and 7. The different features showed it can be used to segregate patient's data based on the biometric signal extracted.

4.2 Signal Classification

To optimize the Intelligent Authentication Framework, use a bi-algorithmic approach to enhance the performance of the system. Different machine learning algorithms can be used based on the most viable one for each signal [28]. To compare the multi-algorithmic performance, the different classifiers are applied to classify different feature templates of each bioelectrical signal but applying all useful contextual information. The signal for testing the classification is the approximation of the detail and approximate coefficients of the heart rate variability. The different families of Biorthogonal wavelets are used for decomposing the signal and each signal classified

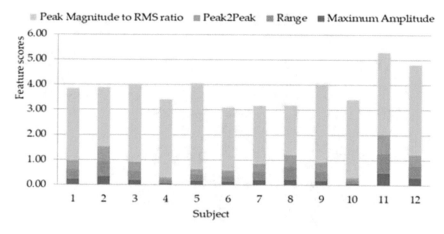

Fig. 6 Statistical features of PMRMS, P2P, R and MA

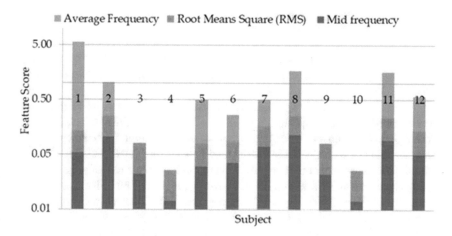

Fig. 7 Statistical features of AF, RMS and MF

for maximizing the performance. The different Biorthogonal wavelet family signal performance is shown in Fig. 8. The advantages of different Biorthogonal wavelet families are that they use different filter banks for noise removal when decomposing a signal.

4.3 Algorithm Performance

The algorithm performance evaluation used Feed-Forward Neural Network (NN-FF) and random forest for analyzing extracted data. The bioelectrical data is extracted

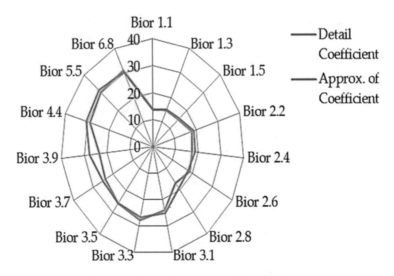

Fig. 8 Showing the different biorthogonal wavelet family signal performance

from 30 participants, the features used are the ones described earlier. From research carried out by [49–51] using random forest for classification, it improves the accuracy of the classification.

The result of the algorithm performance evaluation using either NN-FF or random forest showed a significant performance in the classification of the extracted bioelectrical signal. The comparison of the two-classification output, on few neuron sizes using the random forest classifier performed better than the NN-FF classifier. For example, the layer using the 35 neurons using random forest classifier has the best performance with an EER of 14.22% as shown in Fig. 9. The NN-FF classifier using 70 neurons achieved 3.4% at its best as shown in Fig. 10. The random forest classifier achievement below 15% EER at 14.94% and 14.31% EER for 30 and 40 neurons respectively while the NN-FF had most of the neurons performing below EER of 5%.

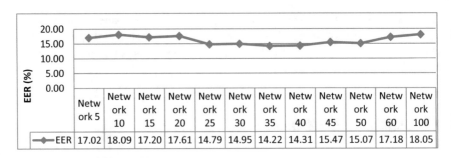

Fig. 9 The extracted data classification using random forest

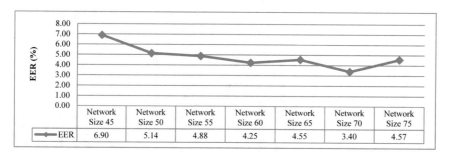

	Network Size 45	Network Size 50	Network Size 55	Network Size 60	Network Size 65	Network Size 70	Network Size 75
EER	6.90	5.14	4.88	4.25	4.55	3.40	4.57

Fig. 10 The extracted data classification using NN-FF

5   Artificial Intelligent Decision (AIDE) Module

The different contextual data can also be used by health professionals for assessing the state and activity of a patient. The IoMT intelligent authentication Framework should be able to use most of the context awareness data for enhancing authentication based on the patient activity and location. The contextual data will be selected based on the threshold set by the Intelligent Authentication Framework.

5.1   Context Awareness Data

The bioelectrical signal is classified while the context awareness data is used to support the authentication process [28]. The contextual data extraction should involve biometric data that could be able to give useful information, some of such shown in Fig. 11.

5.1.1   GPS Information

GPS sensors are integrated into most modern smartphones for location tracking and navigation. This can be utilized for other useful purposes as well as some negative intent like spying and tracking one's movement. The GPS information is useful location identification and can be used to predict the user's daily route and time the event occurred. This can also be used for managing and tracking vulnerable people for their safety. This GPS sensor could be proprietary, or a third-party application designed for improved service. The GPS information is useful for implementing the intelligent authentication framework. It is expected that at a certain time of a day, the person with a regular pattern is expected to be at a particular location depending on many factors. These factors could include working pattern, activities the person is engaged in and locations. For example, if a patient has an appointment at 11 am on a particular day and leaves the house at 11:30 am. The framework should be

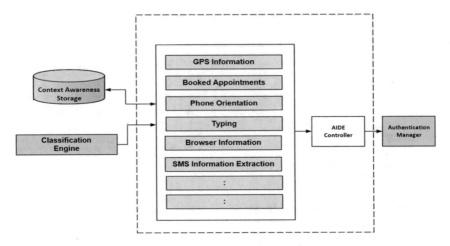

Fig. 11 Showing artificial intelligent decision (AIDE) module

able to alert both the user and General Practitioner (GP) that the patient is still at home or just leaving the house or going toward the health care facility or going to another place based on the route. For authentication purposes, the location should be a determining factor for accurate user authentication while extracting health data for the health care facility use. If the location of the patient is not expected or irregular, other context information will be used to verify the patient.

5.1.2 Online Social Network (OSN)

The use of Online Social Network (OSN) has enhanced the way we communicate. It has also allowed interaction and reuse of data, messages, and photos with friends across multiple OSNs accounts [52]. This will benefit the context awareness of the intelligent system. The subject's interaction across these OSNs will be used to identify the person's term of behavioral activities online. For example, if the subject first few daily activities include visiting particular social networks and when this is not done, the system will take note of it and look out for other context information that is it different from the pattern known to the subject. This will trigger a signal for monitoring and comparing the different intellectual information to take a definite action alerting the human agent for further action.

5.1.3 Walking Pace

A pace is the distance between the heel of a foot to the heel of the other foot. The use of gait for authentication purposes has been researched extensively. The work by Hind [53] demonstrated the use of unobstructive capturing of gait data for the user authentication. For this framework, the intelligent authentication will use the GPS and gait to determine if the person is going for appointment, running late or still at home busy based on gait information and the GPS.

5.2 Artificial Intelligent Authentication

The Intelligent Authentication Module is the "brain" of the framework by applying artificial intelligence to enhance authentication. The module uses activity information and authentication manager to decide the authentication process. This module contains data storage for storing behavioral and signal patterns for comparing with newly collected data.

Health care data transmitted within an IoMT is expected to be trusted because of the sensitivity of the data. The intelligent authentication framework for IoMT uses the extracted data for authentication to increase trust as the data is extracted and transmitted. The data after classification is passed for authentication using majority voting to either accept or deny the patient's data. A voting system can enhance the output of the classification. The use of majority voting has its advantage, this includes low space usage which makes it easy for use in IoT devices because power and storage space. The acceptance or rejection is represented as '1' or '0' respectively. The result from the majority voting with a score of either '1' or '0' is for final decision to either accept or reject the patient's data. The authentication process using the different biometric data should be flexible to accommodate different biometric databases on the most viable data at any time. For example, the GPS data can be used to analyze the patient's location but if the location is not the expected, another context awareness data like pace, gait is combined with the biometric data like heart rate, heart rate variability. If the location is known, the framework authentication will go through, or data accessed by the patient.

The context awareness data from the different sensor's data contain different contextual information therefore, scores are given to it as the data are captured based on the quality of the data capture. These scores are used to validate the classification output after cross checking the context information. If the GPS verifies it, it will score it '−1' but if it is the wrong location or not recognized, it will score '0'. This is done for different context awareness data but where the context awareness data is not available, it scores it '0' because the data cannot be verified as shown in Table 2. The total scores it tabulated for final decision.

As the table indicates, the scores total are used to decide the authentication of the process but where the score is below a threshold determined by the framework, the authentication will fail to go through. The classification of the biometric data and the contextual awareness must be above the set threshold for the final authentication to take place.

Table 2 Showing the data scoring table

Context awareness	Data 1	Data 2	Data 3	Data 4	Data 5
GPS	1	1	1	1	0
SMS	1	1	1	1	1
Device orientation	1	1	1	0	−1
OSN	0	1	1	−1	1
Steps	1	1	1	1	1
:	:	:	:	:	:
Score	5	4	6	3	2

5.2.1 Activity Information

The activity information is important because with it, the context awareness data can be with either good to use or not. For example, if the patent is lying down, the pace/steps is not useful there; the pace information will not be used for context awareness information, but the GPS can be used. If the patient is using a smartphone the phone's orientation could be used together with the gyroscope and accelerometer can be used. The activity splitter as shown in Fig. 4 is used to differentiate and segment the data with the activity type and time indicated before sending it to the pro-processing and the intelligent component. To authenticate the patient, framework identities the most suitable data for authenticating the patient based on different parameters set to accept the data as meeting a threshold. The context information is also defined before passing it to the authentication manager.

5.2.2 Data Template

The data templates are generated based on the different activity and contextual information. The templates are generated for both biometric data after features have been created and the contextual template only when the information will be useful for enhancing the authentication process and when it will not be created but discarded. The templates are created based on time intervals, this could be 3, 5 or 7 seconds based on individual testing to use the best time interval that will give maximum security without affecting usability.

5.2.3 Storage

The storage is very important for biometric authentication system because it use for storing the template. The template is renewed periodically base of the frequency of extraction. Both the contextual and biometric data are both store as a template for classification. The data stored includes a date and time stamp, Activity identity, the data type and the data name, temporary storage location as illustrated in Table 3.

Table 3 Showing data storage table

Date/Time		Biometric information				Contextual information					Temporary storage
		HR	GRS	Skin temp	Resp	Pace	GPS	Acce	Gyro		
4/09/2021/12:40	1	0.91256	66	0.91256	..	952	..	..	..	\\dataprofile\Motion
4/09/2021/12:40	1	0.91256	66	0.91256	..	966	..	..	..	\\dataprofile\Motion
4/09/2021/12:40	2	0.91256	66	0.91256	..	976	..	..	..	\\dataprofile\Motion
4/09/2021/12:40	2	0.91367	66	0.91256	..	979	..	..	..	\\dataprofile\Motion
4/09/2021/12:41	3	0.91367	66	0.91256	..	1001	..	..	..	\\dataprofile\Non-motion
4/09/2021/12:41	2	0.91367	66	0.91256	..	1002	..	..	..	\\dataprofile\Non-motion
4/09/2021/12:41	1	0.91274	66	0.91256	..	1003	..	..	..	\\dataprofile\Non-motion

The table stores the information as templates and discards the templates as new ones are generated. The generated templates go through the comparison to make sure they meet the requirement and are then used for authentication before being stored. This authentication manager process and determines the templates to be stored. The date and time stamps are in microseconds and are expected to be extracted daily. The activity identity involves the type of activity the patient is involved in, this includes if the person is sitting, walking, lying down, running. The data type includes the types of either contextual or biometric data while the data name is stored the name of the signal like heart rate, heart rate viability, pace, GPS etc. The temporary storage is the location the data templates are store. These different templates are stored in different locations which are either motion or non-motion storage. The storage is divided into motion and non-motion because some data captured when not in motion is not useful like the pace of the patient therefore, the data of the non-motion is not processed as it contains no useful information.

5.2.4 Authentication Manager

The Authentication Manager applies the information from the contextual information score and voting output to make the final decision. This process at the initial start will need the patient's knowledge base authentication to start doing the initial user authentication as shown in Table 4. This knowledge-based user authentication is useful just like every biometric authentication system for verification/revalidate the patients in case of unforeseen issue. The knowledge base user authentication can also be used to reset the intelligent authentication framework. This could also include fingerprint authentication for patient that can't use knowledge based due to memory issues like someone with dementia.

Another important factor the authentication manager use is the uses of threshold to set the security level. This aspect is used to further enhance the security but if a high threshold is set, this will also affect the usability therefore, in setting the threshold usability and security must be considered. The security level will factor in a process that will reduce false rejection leveraging on the contextual information provided. To improve the usability vs security, a feedback mechanism is put in place to access the

Table 4 Showing knowledge base authentication

Patient security question	Answer
1. Email	flourish@favour.com
2. What is your closest brother's name	Ebiketomon
3. What is your closest friend's name	Edwin
4. Which town is your favourite park located	Yenagoa city
5. Which city is your wife living	Abuja
6. PIN	4323

Table 5 Showing the security levels

Level in EER (%)	Authentication level
0–0.2	5
0.2–0.4	4
0.4–0.6	3
0.6–0.8	2
0.8–1.0	1
1.0>	0

mechanism from time to time. The authentication using the security level depends on the how often the user is rightly authenticated and wrongly reject. That is the error rate and success rate of the system. The security level is in the range of 0 – 5. 0 indicates the worst and 5 is the strongest security level while EER of 0 – 0.2% is the best performance of the authentication and 1.0% EER is the worst performance. The security levels are shown in Table 5.

5.3 The Framework Evaluation

The evaluation of the framework is analyzed using only the bioelectrical data for the evaluation based on the False Acceptance Rate (FAR) and the False Rejection Rate (FRR). This is done using the testing data extracted from the 30 participants used for the algorithm evaluation. The evaluation output is shown in Table 6 and Fig. 12.

Figure 12 shows the performance based on false identification and false rejection of the data. Level 4 and 5 show a false acceptance rate of 0% which is ideal for health care data, but the false rejection rate is high at 29% at level 4. When usability is considered, Level 1 can be considered but for health care management, the false acceptance will be a concern.

Table 6 Showing the framework evaluation output

Performance of the evaluation		
Security level (%)	FRR (%)	FAR (%)
0.8–1.0	9	5.6
0.6–0.8	18.3	1.6
0.4–0.6	22.6	0.2
0.2–0.4	29	0
0–0.2	77.7	0

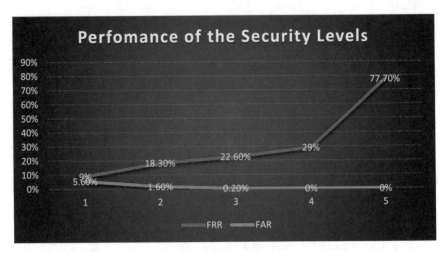

Fig. 12 Showing the framework evaluation output in graph

6 Conclusion

The intelligent authentication framework for IoMT is developed to increase the biometric data trust while transmitted wirelessly. This will also improve the use of multiple IoT devices that can extract biometric information for the purpose of health care monitoring. This framework considered the convenience of the patient by removing process that involve the direct participation of the patient in activating the authentication and transmission of health-related data. This is done not minding the environmental condition like if the person is active or non-active. The result shows the two levels of 4 and 5 achieving false acceptance of the 0% respectively however, using these levels will equally affect the usability of the framework therefore with the application of artificial intelligence using the contextual data, the rate of false rejection will reduce if not attaining 0%.

The research has proven to be useful for patients that require human assistance by making the patient non-active participant in the procedure. The framework is also considered a not-intrusive collection of Biometric data. This approach will improve the overall user experience of smart mobile devices usage for healthcare because mobile users are always with their device most of the time.

References

1. Ghubaish A, Salman T, Zolanvari M, Unal D, Al-Ali AK, Jain R (2020) Recent advances in the internet of medical things (IoMT) systems security. IEEE Internet Things J
2. Yachongka V, Yagi H (2020) Biometric identification systems with both chosen and generated secrecy. In: 2020 International symposium on information theory and its applications (ISITA).

IEEE, pp 417–421
3. Enamamu T, Otebolaku A, Marchang J, Dany J (2020) Continuous m-health data authentication using wavelet decomposition for feature extraction. Sensors 20(19):5690
4. Israel SA, Irvine JM, Cheng A, Wiederhold MD, Wiederhold BK (2005) ECG to identify individuals. Pattern Recogn 38:133–142
5. Shen C, Cai Z, Guan X, Du Y, Maxion RA (2013) User authentication through mouse dynamics. IEEE Trans Inf Forensics Secur 8:16–30
6. Wang Y, Agrafioti F, Hatzinakos D, Plataniotis KN (2008) Analysis of human electrocardiogram for biometric recognition. EURASIP J Adv Signal Process 2008:19
7. Gahi Y, Lamrani M, Zoglat A, Guennoun M, Kapralos B, El-Khatib K (2008) Biometric identification system based on electrocardiogram data. In: New technologies, mobility and security. NTMS'08. IEEE, pp 1–5
8. Coutinho DP, Fred AL, Figueiredo MA (2010) One-lead ECG-based personal identification using Ziv-Merhav cross parsing. In: 2010 20th international conference on pattern recognition (ICPR). IEEE, pp 3858–3861
9. Subasi A (2007) EEG signal classification using wavelet feature extraction and a mixture of expert model. Expert Syst Appl 32:1084–1093
10. Chan AD, Hamdy MM, Badre A, Badee V (2008) Wavelet distance measure for person identification using electrocardiograms. IEEE Trans Instrum Meas 57(2):248–253
11. Ye C, Coimbra MT, Kumar B (2010) Investigation of human identification using two-lead electrocardiogram (ECG) signals. 2010 fourth IEEE international conference on biometrics: theory applications and systems (BTAS). IEEE, pp 1–8
12. Sidek KA, Khalil I (2011) Automobile driver recognition under different physiological conditions using the electrocardiogram. In: 2011 computing in cardiology. IEEE, pp 753–756
13. Hema CR, Paulraj M, Kaur H (2008) Brain signatures: a modality for biometric authentication. In: ICED 2008. International conference on electronic design, 2008. IEEE, pp 1–4
14. Kousarrizi MN, Teshnehlab M, Aliyari M, Gharaviri A (2009) Feature extraction and classification of EEG signals using wavelet transform, SVM and artificial neural networks for brain computer interfaces. In: 2009. International joint conference on bioinformatics, systems biology and intelligent computing. IJCBS'09. IEEE, pp 352–355
15. He C, Wang ZJ (2009) An independent component analysis (ICA) based approach for EEG person authentication. In: 3rd international conference on bioinformatics and biomedical engineering. ICBBE 2009. IEEE, pp 1–4
16. Sasikala P, Wahidabanu R (2010) Identification of individuals using electrocardiogram. Int J Comput Sci Netw Secur 10:147–153
17. Lee J, Chee Y, Kim I (2012) Personal identification based on vectorcardiogram derived from limb leads electrocardiogram. J Appl Math
18. Hema C, Osman A (2010) Single trial analysis on EEG signatures to identify individuals. In: 2010 6th international colloquium on signal processing and its applications (CSPA). IEEE, pp 1–3
19. Tawfik MM, Kamal HST (2011) Human identification using QT signal and QRS complex of the ECG. Online J Electron Elect Eng 3:383–387
20. Hema C, Elakkiya A (2012) Recurrent neural network based recognition of EEG biographs
21. Zokaee S, Faez K (2012) Human identification based on ECG and palmprint. Int J Electr Comput Eng 2:261
22. Mohanchandra K, Lingaraju G, Kambli P, Krishnamurthy V (2013) Using brain waves as new biometric feature for authenticating a computer user in real-time. Int J Biometr Bioinform (IJBB) 7:49
23. Revett K (2012) Cognitive biometrics: a novel approach to person authentication. Int J Cogn Biometr 1(1):1–9
24. Lee SW, Woo DK, Son YK, Mah PS (2019) Wearable bio-signal (PPG)-based personal authentication method using random forest and period setting considering the feature of ppg signals. J Comput 14(4):283–294

25. Misra S (2020) A step by step guide for choosing project topics and writing research papers in ICT related disciplines. In: International conference on information and communication technology and applications. Springer, Cham, pp 727–744
26. Poh N, Bengio S, Korczak J (2002) A multi-sample multi-source model for biometric authentication. In: Proceedings of the 12th IEEE workshop on neural networks for signal processing. IEEE, pp 375–384
27. Tamil EBM, Kamarudin N, Salleh R, Tamil A (2008) A review on feature extraction & classification techniques for biosignal processing (Part I: Electrocardiogram). In: 4th Kuala Lumpur international conference on biomedical engineering 2008. Springer, pp 107–112
28. Enamamu TS (2019) Bioelectrical user authentication. University of Plymouth
29. Murphy C (2011) Cellular phone evidence data extraction and documentation
30. Myers J (2003) Exercise and cardiovascular health. Circulation 107(1):e2–e5
31. Kemp AH, Quintana DS (2013) The relationship between mental and physical health: insights from the study of heart rate variability. Int J Psychophysiol 89(3):288–296
32. Tran Y, Wijesuriya N, Tarvainen M, Karjalainen P, Craig A (2009) The relationship between spectral changes in heart rate variability and fatigue. J Psychophysiol 23(3):143–151
33. Mejía-Mejía E, May JM, Torres R, Kyriacou PA (2020) Pulse rate variability in cardiovascular health: a review on its applications and relationship with heart rate variability. Physiol Meas 41(7):07TR01
34. Mejía-Mejía E, May JM, Elgendi M, Kyriacou PA (2021) Differential effects of the blood pressure state on pulse rate variability and heart rate variability in critically ill patients. NPJ Dig Med 4(1):1–11
35. Elliott M (2016) Why is respiratory rate the neglected vital sign? A narrative review. Int Arch Nurs Health Care 2(3):050
36. Cretikos MA, Bellomo R, Hillman K, Chen J, Finfer S, Flabouris A (2008) Respiratory rate: the neglected vital sign. Med J Aust 188(11):657–659
37. Andrasik F, Rime C (2007) Chapter 121—Biofeedback. In: Waldman SD, Bloch JI (eds) Pain management, W.B. Saunders, pp 1010–1020. https://doi.org/10.1016/B978-0-7216-0334-6.50125-4. ISBN 9780721603346
38. Kurniawan H, Maslov AV, Pechenizkiy M (2013) Stress detection from speech and galvanic skin response signals. In: Proceedings of the 26th IEEE international symposium on computer-based medical systems. IEEE, pp 209–214
39. Goshvarpour A, Goshvarpour A (2020) The potential of photoplethysmogram and galvanic skin response in emotion recognition using nonlinear features. Phys Eng Sci Med 43(1):119–134
40. Strollo J Jr, Rogers RM (1996) Obstructive sleep apnea. N Engl J Med 334(2):99–104
41. Voroneanu L, Cusai C, Hogas S, Ardeleanu S, Onofriescu M, Nistor I et al (2010) The relationship between chronic volume overload and elevated blood pressure in hemodialysis patients: use of bioimpedance provides a different perspective from echocardiography and biomarker methodologies. Int Urol Nephrol 42(3):789–797
42. Gheorghiade M, Vaduganathan M, Ambrosy A, Böhm M, Campia U, Cleland JG et al (2013) Current management and future directions for the treatment of patients hospitalized for heart failure with low blood pressure. Heart Fail Rev 18(2):107–122
43. Moukadem A, Abdeslam DO, Dieterlen A (2014) Time-frequency domain for segmentation and classification of non-stationary signals: the stockwell transform applied on bio-signals and electric signals. Wiley
44. Hammond J, White P (1996) The analysis of non-stationary signals using time-frequency methods. J Sound Vib 190:419–447
45. Zokaee S, Faez K (2012) Human identification based on ECG and palmprint. Int J Electr Comput Eng (IJECE) 2:261–266
46. do Vale Madeiro JP, Cortez PC, da Silva Monteiro Filho JM, Rodrigues PRF (2019) Techniques for noise suppression for ECG signal processing. In: Developments and applications for ECG signal processing. Academic Press, pp 53–87
47. Sörnmo L, Laguna P (2005) The electrocardiogram—A brief background. In: Bioelectrical signal processing in cardiac and neurological applications, pp 411–452

48. Ghaleb FA, Kamat MB, Salleh M, Rohani MF, Abd Razak S (2018) Two-stage motion arte-
    fact reduction algorithm for electrocardiogram using weighted adaptive noise cancelling and
    recursive Hampel filter. PloS one 13(11):e0207176
49. Liaw A, Wiener M (2002) Classification and regression by random forest. R News 2(3):18–22
    (R package version 4.6. 10)
50. Breiman L (1999) Random forests. UC Berkeley TR567
51. Breiman L (2001) Random forests. Mach Learn 45:5–32
52. Youssef BE (2014) Online social network internetworking analysis. Int J Next-Gener Netw
    6(2):1
53. Al-Obaidi H, Li F, Clarke N, Ghita B, Ketab S (2018) A multi-algorithmic approach for gait
    recognition. In: ECCWS 2018 17th European conference on cyber warfare and security, p 20

Parallel Faces Recognition Attendance System with Anti-Spoofing Using Convolutional Neural Network

Stephen Bassi Joseph⬭, Emmanuel Gbenga Dada⬭, Sanjay Misra⬭, and Samuel Ajoka⬭

Abstract Face recognition which is a sub-discipline of computer vision is gaining a lot of attraction from large audience around the world. Some application areas include forensics, cyber security and intelligent monitoring. Face recognition attendance system serves as a perfect substitute for the conventional attendance system in organizations and classrooms. The challenges associated with most face recognition techniques is inability to detect faces in situations such as noise, pose, facial expression, illumination, obstruction and low performance accuracy. This necessitated the development of more robust and efficient face recognition systems that will overcome the drawbacks associated with conventional techniques. This paper proposed a parallel faces recognition attendance system based on Convolutional Neural Network a branch of artificial intelligence and OpenCV. Experimental results proved the effectiveness of the proposed technique having shown good performance with recognition accuracy of about 98%, precision of 96% and a recall of 0.96. This demonstrates that the proposed method is a promising facial recognition technology.

Keywords Face recognition · Artificial intelligence · Attendance system · Deep learning · Forensics

S. B. Joseph (✉) · S. Ajoka
Department of Computer Engineering, Faculty of Engineering, University of Maiduguri, Maiduguri, Nigeria
e-mail: sjbassi74@unimaid.edu.ng

E. G. Dada
Department of Mathematical Sciences, Faculty of Science, University of Maiduguri, Maiduguri, Nigeria
e-mail: gbengadada@unimaid.edu.ng

S. Misra
Department of Computer Science and Communication, ï¿½stfold University College, Halden, Norway

1 Introduction

Facial recognition has become a valuable tool for forensics [40], cyber security [3, 42] and surveillance monitoring [34]. Attendance tracking an aspect of surveillance monitoring is a crucial factor for ensuring performance of employees and or students in various organizations and institutions. Conventionally, most attendance records of organizations and institutions are manually taken using attendance sheets or registers provided by management as part of the regulation [46]. Attendance tracking is used in offices/organizations to keep records of staffs attendance and punctuality, and it has also been implemented in academia such as institutions of higher learning to capture/record students' attendance to track students' performance and activities on campus. The conventional technique of taking attendance is inconvenient, time-consuming, inefficient and can be easily compromised [10]. The conventional approach also makes it cumbersome to track the attendance of individuals either in organizations or schools. Some shortcomings of the conventional method of attendance systems includes scalability, reliability, efficiency in data collection and filtering, anti-spoofing and impossibility to remotely check attendance status [44]. Recently, the use of human bio-metric vitals for face recognition. The human face is one of the vital biometric example that has attracted the interest of several researchers from different fields of endevour for recognition [1, 18, 32].

Recently, several works have been proposed on attendance like the use of RFID tags inserted in identification cards [33]. Any attempt to read multiple tags at once may result in signal collision and, as a result, data loss, as well as tag damage [25]. Voice verification often has a problem of decoding accents.To avoid making mistakes, one must learn to speak consistently and clearly at all times. The software will not execute your instruction if you mumble, talk too quickly, or run words into one other. A siren environment is very germane to the optimal performance of voice recognition software as such software will perform abysmally in the presence of background noise [15]. Fingerprint scanners technology although an environmentally improved device compared to the voice recognition software, is designed for a single user at a time only [24]. Hence, fingerprint scanners do not address the problem of students/employees carrying multiple cards, identity theft and individuals having to queue for a long time to register their attendance [4]. Automating work and tracking participants' positions in real time are two popular incentives offered by attendance systems. The use of such systems promotes organizational efficiency and performance, as well as individual pleasure. In terms of effort, time, and cost, manual paperwork is inefficient [30]. Furthermore, there may be a number of complications with manual paperwork, including the possibility of major errors owing to large amounts of data [4]. Therefore, there is a need to proffer solution to challenges associated with conventional systems by designing and implementing an automated real-time parallel faces recognition attendance system with anti-spoofing.

Convolutional Neural Networks are a type of deep neural network that can take in an image as input, assign importance (learnable weights and biases) to various aspects/objects in the image, and distinguish between them [35, 45]. Convolutional

Neural Network has made incredible progress in tasks involving images [1]. CNN has become one of the most well-known neural networks in the field of deep learning. Face recognition, driverless vehicles, and intelligent medical treatment have all been made possible thanks to computer vision based on Convolutional neural networks, which was previously thought to be impossible [27]. Details of CNN and it numerous application areas in research can be seen in [17, 27].

In this paper a method to identify and classify human faces in real time for the purpose of attendance monitoring with anti-spoofing capability is proposed. Human faces are are captured and identified as they arrive. The proposed system uses multiple cameras to capture faces and classify them using CNN. The contribution of this work can be summarized as follows:

i. An up to date related literature on face recognition approaches
ii. Convolutional Neural Networks with anti-spoofing capacity were used to design and develop a multi-face recognition attendance system.

The remainder of the paper is laid out as follows. Section 2 discussed several studies that are relevant to face recognition attendance system. Section 3 presents the methods used for data collection, preprocessing operations, system design, modeling, implementation and performance evaluation metrics. Section 4 presents the results and discussions of the results of the experiments conducted. Section 5 is the conclusion of the paper and also presents future direction.

2 Related Works

Numerous works on facial recognition and attendance tracking system have been proposed using various face detection and recognition algorithms such as Haar Cascade [43], Principle component analysis (PCA) for face recognition [9], Linear Binary Pattern (LBP) [11], Voronoi decomposition-based random region erasing (VDRRE) for face palsy recognition [2] and Linear Discriminant Analysis (LDA) [13]. However, these systems did not make use of multiple cameras to capture data from different angles and could not recognize more than two faces at a time. For instance, the work of [11] the system couldn't recognize faces where there were accidental changes in a person's appearance, such as tonsuring, wearing a scarf, or having a beard.

Chin et al. in [10] suggested a face recognition-based integrated attendance management system as part of an in-house created learning management suite (libri). The system was unable to identify each student present in class, leading to the conclusion that such an attendance system could be improved. Their approach aims to solve the issues by integrating face recognition in the attendance process and to improve processing time by using algorithms such as Haar Cascade (Viola Jones) for face detection and Principal Component Analysis (PCA/Eigenface) for face recognition technique. However, the system lacks the ability to identify each student present in class, so there is still much more room for improvement.

Islam et al. in [22] in their work on face recognition system based on Real Time Classroom Attendance Management by using Kinect camera had issues with recognition if someone's picture is stored in the database or not. The authors also did not make use of any sensor such as IR sensor in the system. The Viola-Jones method is used to detect human faces in video streams. The feature extraction techniques of Speeded Up Robust Features (SURF), Histogram of Oriented Gradients (HOG), and Linear Binary Pattern (LBP) were evaluated and compared on the created dataset for recognition purposes. The suggested system employs the Kernel Based Filtering approach to normalize the images. When a student's face matches one in the dataset, the student is recorded as present. However, the system was unable to handle the following issues: it was unable to determine whether or not one person's dataset photographs were recorded in the database, and it lacked the implementation of sensors such as an infrared sensor into the system.

Chintalapati and Raghunadh in [11] employed face recognition techniques to construct an automated attendance management system The frontal photographs of the students are captured using a camera situated at the classroom's entrance. The photos are then analyzed using Viola Jones (face detection algorithm). The image is enlarged to 100×100 gray scale and stored in a database after applying histogram equalization and resizing. PCA, LDA, LBPH, PCA + SVM, and PCA + Bayesian are used in this order. SVM and Bayesian pattern classifiers are utilized for feature extraction, whereas PCA, LDA, and LBPH are used for pattern classification. When the recognition is finished, an Excel Sheet is generated. LBPH beats other algorithms, according to the authors, with a higher recognition rate and a lower false positive rate. However, the system lacks many cameras to record images from various angles and can only detect two faces at a time. Additionally, the recognition does not work when a person's appearance changes unintentionally, such as tonsuring, wearing a scarf, or having a beard.

Sharanabasappa and Sanjaypande [38], proposed a strategy for face detection. In their method, faces are extracted from each image using a complex algorithm. Face detection is done with the Haar cascade, while face segmentation is done with the Gaussian mixture model (GMM). A test of 1000 photos yielded an accuracy of 83%, where accuracy is defined as the number of genuine faces detected vs the total number of faces in the scene. The test is performed at various angles and with varying levels of light intensity. The algorithm, according to the authors, achieves a good mix of speed and efficiency. The proposed system, on the other hand, does not have anti-spoofing features.

Fuzail et al. [16] presented a facial detection system for tracking student attendance in class. Real-time face detection algorithms are integrated into an existing learning management system in this system (LMS). It can detect and register students who are attending a lecture automatically. In the front of the classroom, a rotating camera was placed in the center. The system employed the Haar classifier (Viola-jones) for face detection. The classifier works by using positive and negative facial photos to train a model. A positive image is one that contains the intended object to be detected (a face), while a negative image is one that does not. After the model has been trained, it can recognize facial traits, which are then saved in an XML file. The Eigenface

(PCA) benefit is used to recognize a face. Furthermore, the method is unable to identify each student in the class. As a result, the teacher correlates the unidentified faces to the collected faces manually.

Pande et al. [31] suggested a robust method for real-time parallel processing face recognition using live video pictures. The authors integrate the Eigenface technique with Haar-like features to detect both eyes and face, which improves the algorithmic efficiency of face detection. It also employs a Robert cross edge detector to determine the human face's position, as well as a 2-D spatial gradient measurement on an image, which results in gray scale edge segmentation by detecting regions of high spatial frequency that closely correlate to edges. A simple Euclidean distance measure between the projected vector and each face image feature vector is used to classify each face image. The authors employ Principal Component Analysis (PCA) to minimize the dimensionality of the training set, keeping only the features that are important for face recognition. This allows the system to respond quickly. The proposed approach greatly enhances recognition performance, according to the results of the experiments. The system outperforms other strategies in general. Over a wide number of databases of faces, the algorithm had a success rate of 92%.

Balcoh et al. [7] proposed a face recognition-based system for efficient attendance management. The paper outlined an effective method for automatically marking attendance without the need for human interaction. This attendance is tracked using a camera mounted in front of the classroom that captures photographs of pupils on a continual basis. It then uses Haar cascade (Viola-jones) to detect faces in photos, and then uses Principal Component Analysis to compare the discovered faces to a database for facial recognition (Eigenface). The attendance is then marked after the image noise is removed using a median filter. The system's flaw is that it is incapable of recognizing faces hidden behind a veil.

By integrating Eigenfaces with fisher faces and employing wavelets, a novel face recognition method was proposed by Devi et al. in 2010 [13]. The suggested system employs the 2FNN facial recognition approach, which is based on appearance (Two-Feature Neural Network). Face features are extracted using two separate feature extraction algorithms: PCA and LDA, and then these extracted features are integrated using wavelet fusion. To boost classification accuracy, wavelet fusion and neural networks are applied. The suggested system outperforms existing approaches while having a low computing complexity. The technology can also detect faces with lighting issues, facial expressions, and aging, according to the findings. Preliminary trial results showed that the system was very accurate in terms of correct identification rate 98.50% and equal error rate 1.50%.

Deep learning-based facial recognition attendance system was proposed by Arsenovic et al. [6]. Face detection and image preprocessing that deals with locating face landmarks and face positioning uses an affine transformation to center these landmarks as much as feasible. Per face, the system generates 128-byte embeddings. The Deep CNN was trained using the OpenFace library, which included a pre-trained FaceNet network, and classification was done using SVM. On a limited dataset, the proposed approach had a 95.02% accuracy. However, because it was exposed to daylight while the door was open, some of the photos were forecasted inaccurately.

This indicates that the proposed model is light-sensitive. The model was manually trained with a variety of photos of the same person in various poses (Blurred, Noised, Frontal, Sideways, from Above, Flip, Shifted, Sheared, Scaled, Translated, and Rotated) (offline augmentation). The automatic augmentation (online augmentation) process that was employed by the machine learning toolkit made the training procedure to be very tedious and time consuming.

Lukas et al. [28] suggested a technique for students attendance systen in a class room. The suggested method was created utilizing a combination of Discrete Wavelet Transforms (DWT) and Discrete Cosine Transform (DCT) to extract facial features before classifying the facial objects using the Radial Basis Function (RBF). The system was able to identify students with a performance accuracy of about 82%. This shows an averagely good strategy, however the system is not equipped with anti-spoofing capability.

Other attempts for automated attendance systems includes: Android based system proposed by [41], local binary pattern by [14], Haar Cascade classifier [37] and Convolutional neural network [20, 36]. Most suggested attendance monitoring systems uses a single camera and also lacks the ability to prevent spoofing. A good reference guide to manuscript structuring and layout presentation can be found in [29]. A summary of related works is presented in Table 1.

3   Attendance System Using CNN

In order to enhance the attendance system scheme this work proposes a real time attendance scheme with anti-spoofing ability. The architectural design of proposed scheme is depicted in Fig. 1. Implementation of the proposed real time parallel face detection attendance system was conducted using Convolutional neural network developed in Python programming language with some machine learning libraries like; OpenCV, Dlib, Face-Recognition and NumPy. Furthermore, a software program for the motion sensing, panning the camera was written in a sketch programmer kit for the Arduino, it was compiled and checked for errors. Then the programmer kit was used to transfer the source code onto the Arduino micro controller board, before embedding it onto the construction platform.

The system architecture is composed of four distinct phases: preprocessing phase, face detection phase, anti-spoofing phase and decision phase. Each of these stages is segmented and adaptable. This is done to make the system flexible, so that changes to one layer do not effect the others.

3.1   Preprocessing Phase

The first stage, the preprocessing stage acts as the capturing stage of real time images. Faces were detected by the use of Histogram of Oriented Gradient (HOG) algorithm

Table 1 Summary of some previous related works

S/N	References	Proposed method	Strength	Weakness
1.	Abayomi-Alli et al. [2]	Viola-Jones method for the face detection algorithm and Eigenfaces method for the face recognition algorithm	Tracking of students and promising for detection	Low accuracy, use of single camera and No antispoofing
2.	Islam et al. [21]	Viola-Jones algorithm (MATLAB)	Good recognition	Does not identify new students No antispoofing
3.	Hernandez-Ortega et al. [18]	Face recognition algorithms (PCA, LDA, e.tc)	Fast recognition	Only recognizes faces at 30° angle variation. Cannot identify changes on faces like scarf and beard. No anti spoof
4.	Islam et al. [22]	OpenCV and Gaussian mixture model	Good recognition accuracy	Lacks antispoofing capability
5.	Jha et al. [23]	Face detection algorithms integrated on an existing Learning Management System	Time saving	Inability to recognize all students and low identification rate
6.	Johnson [24]	Eigenfaces method using Haar-like features	Performance efficiency	Lacks anti-spoofing capability
7.	Kaur et al. [25]	Viola-Jones algorithm and EigenFace method	Reliability and performance	Single camera and no antispoofing.
8.	Hung and Khang [20]	Two -Feature Neural Network	Low computational complexity illumination problems, facial expressions. Good performance efficiency	Single camera and no antispoofing.
9.	Khan et al. [26]	Convolutional neural network (CCNN), OpenCV and SVM	Good performance efficiency	Single camera and no antispoofing
10.	Li et al. [27]	Discrete Wavelet Transforms (DWT) and Discrete Cosine Transform (DCT)	Good performance accuracy	No anti spoofing
11.	Lukas et al. [28]	Haar Cascade classifier	Good performance accuracy	Noise intolerant

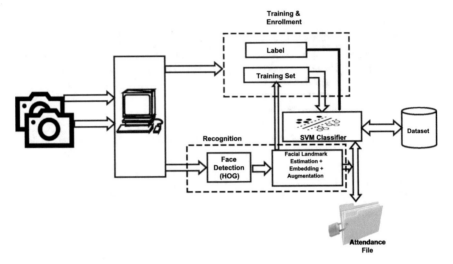

Fig. 1 Proposed system architecture

[12] for face detection in video frames with a multiple web camera. The HOG algorithm runs feature descriptor in every frames of the video frame in order to detect new faces entering the field-of-view (FOV) of the cameras. The HOG algorithm also reduces the impact of lighting effects, tiny changes in position or appearance, and improves illumination, shadowing, and edge contrast invariance [12].

3.2 Face Detection Phase

For real-time photos, the face detection stage serves as a screening point. Here, captured face images are buffered, examined and filtered to identify new members or defaulters based on registered members. Identified valid members are sent to the second stage while new members are either sent to egress or registration point. This stage involves the use of detection algorithms. The face landmark Estimation algorithm [23, 26] was applied to localize the important 68 facial landmarks that exist on a face regions which including eyes, eyebrows, nose, ears, and mouth: so that the face is centered and project the centered face image for generation of face embedding (face image Vector) with the help of an algorithm called face landmark estimation available in Dlib which is a machine learning library toolkit and then project it to Convolutional neural network for classification in this case, the eyebrows, eyes, nose, lips, and chin line are located. The Face image embedding is a pre-trained Convolutional Neural Network model for feeding the face images that generate 128 dimensions or embedding or vector. The vector of faces is generated from the Fully-Connected layer. In the database, each person has only one image (dataset), which is passed through the neural network to generate 128 dimensions vector. The embedding vector extracted is compared with the vector generated for the person (test set) who has to be recognized by the Support Vector Machine (SVM). Details of facial landmarks used for this work can be seen in [8].

3.3 Anti-Spoofing Phase

A metric termed the eye aspect ratio (EAR) was computed to develop the eye blink detector, which was introduced by Soukupová and Cech in 2016 [39]. Unlike traditional image processing methods for computing blinks, which often entail a combination of eye localization, thresholding to detect the whites of the eyes, and determining if the "white" region of the eyes vanishes for a length of time, this method does not require any of these steps (indicating a blink). Instead, the eye aspect ratio is a far more elegant method that incorporates a very basic computation based on the ratio of distances between the eyes' facial landmarks. The eye aspect ratio (EAR) and how it can be used to assess whether or not a person in a video frame is blinking. Facial landmark detection was used to locate key facial features such as the eyes, brows, nose, ears, and mouth. More information on blink detection and equations can be found in [19, 21].

3.4 Decision Phase

The resolutions of anti-spoofing stage are forwarded to the decision phase. The decision phase is composed of databases for attendance register and decision list. The data base stores the information/ policies that describes how to compute the attendance based on every outcome. The record system is generated into a comma-separated values(CSV) file format. Appropriate decisions are either printed for further actions or they are buffered for future use.

3.5 Proposed Detection Flowchart/Algorithm

The proposed system flowchart is presented in Fig. 2.

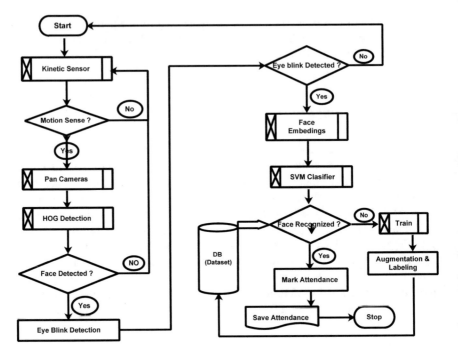

Fig. 2 Proposed system flowchart

Table 2 Evaluation metrics

		Predicted class	
Actual class		Positive	Negative
	Present	True positive (TP)	False negative (FN)
	Absent	False positive (FP)	True negative (TN)

4   Results and Discussion

4.1   Evaluation Metrics

The performance metrics employed to evaluate the various system performance is the confusion matrix (Error Matrix). It visualizes the performance of the system like Accuracy, Precision and Recall. True Positive (TP), False Positive (FP), True Negative (TN), and False Negative (FN) are the four components of the confusion matrix (FN). Table 2 shows the evaluation metrics.

Where;

- **True Positive (TP)**: when the dataset is present, TP is the test result that properly detected the subject.
- **True Positive (TN)**: when the dataset is missing, TN is the test result that fails to recognize the subject.
- **False Positive (FP)**: when the dataset is missing, the test result FP recognizes the subject as unknown.
- **False Negative (FN)**: when the dataset is present, FN is the test result that does not recognize the subject as known.

1. **Accuracy**: The fraction of correctly identified members represented by Eq. 1.

$$\text{Accuracy} = \frac{TP + TN}{TP + TN + FP + FN} \tag{1}$$

2. **Precision**: Eq. 2 represents the proportion of accurately identified members to all positive findings.

$$\text{Precision} = \frac{TP}{TP + FP} \tag{2}$$

3. **Recall**: Eq. 3 represents the proportion of accurately identified members to the sum of correct and incorrectly classed outcomes.

$$\text{Recall} = \frac{TP}{TP + FN} \tag{3}$$

4.2 Datasets

The proposed system was implemented utilizing real captured datasets in order to test its performance. The dataset was captured from department of Computer Engineering, University of Maiduguri, Borno state Nigeria. Details of datasets can be found in [5]. For training, a supervised learning technique termed back propagation neural network was used. The model was trained using deep neural network learning technique, where by the faces are labeled, Dlib library toolkit for machine learning was used for data Augmentation to enlarge the dataset into multiple fold that was used to train the model.

4.3 Performance Evaluation

The performance measures provided in Eqs. 1, 2 and 3 were used to analyze the proposed system's performance. Table 3 presents the summary of results. One hundred

Table 3 Summary of results

Subject type	Faces	TP	TN	FP	FN	Detected	Recognised
Faces with eyeglass	9	5	2	1	1	9	6
Faces with cap	10	5	5	0	0	10	5
Faces with without any	56	4	52	0	0	56	4
Faces with scarf/veil	11	8	3	0	0	11	8
Faces with beard	14	8	6	0	0	14	8
Total	100	30	68	1	1	100	31

faces were tested, thirty faces that were actually present in the dataset were correctly identified corresponding to recognition accuracy of about 98%. The main reason for false recognition was due reflection sun glasses of the test-image, lighting conditions present during image capture also affects the recognition results as in Table 3. The overall performance obtained are: Accuracy = 98%, Precision = 0.96% and Recall = 0.96%.

To test the performance of the proposed system two different experiments were conducted. the first experiment was conducted without anti-spoofing capability while the second experiment to demonstrate fraud (spoofing) using mobile phones. In order to address the issue of spoofing which is the main objective of this work, a eye blink detection algorithm was incorporated into the proposed detection system. The proposed system was able to identify the spoofed images.

5 Conclusion and Future Works

This paper proposed the design and implement a parallel faces recognition attendance system using Convolutional Neural Network with anti-spoofing capabilities. The system monitors and identifies faces for the purpose of attendance monitoring. The system exploited the capabilities of CNN and eye blink algorithm for face identification. The proposed system's performance was calculated using a real-world dataset. Experimental results shows that the proposed system was able to capture and recognize faces with performance accuracy of 98%, precision of 96% and a recall of 0.96%.

References

1. Abayomi-Alli A, Atinuke O, Onashoga et al (2020) Facial image quality assessment using an ensemble of pre-trained deep learning models (EFQnet). In: 2020 20th international conference on computational science and its applications (ICCSA). IEEE, pp 1–8. https://doi.org/10.1109/ICCSA50381.2020.00013

2. Abayomi-Alli OO et al (2021) Few-shot learning with a novel voronoi tessellation-based image augmentation method for facial palsy detection. Electronics 10(8):978. https://doi.org/10.3390/electronics10080978

3. Abbas NN, Ahmed T, Shah SHU, Omar M, Park HW (2019) Investigating the applications of artificial intelligence in cyber security. Scientometrics 121(2):1189–1211. https://doi.org/10.1007/s11192-019-03222-9

4. Agulla EG, Rúa EA, Castro JLA, Jiménez DG, Rifón LA (2009) Multimodal biometrics-based student attendance measurement in learning management systems. In: 2009 11th IEEE international symposium on multimedia. IEEE, pp 699–704. https://doi.org/10.1109/ISM.2009.25

5. Ajoka S, Joseph SB (2019) Design and implementation of parallel face recognition system using CNN. Tech. rep., Department of computer engineering. University of Maiduguri, Borno State, unpublished Final year project report

6. Arsenovic M, Sladojevic S, Anderla A, Stefanovic D (2017) Facetime-deep learning based face recognition attendance system. In: 2017 IEEE 15th international symposium on intelligent systems and informatics (SISY). IEEE, pp 000053–000058

7. Balcoh NK, Yousaf MH, Ahmad W, Baig MI (2012) Algorithm for efficient attendance management: face recognition based approach. Int J Comput Sci Issues (IJCSI) 9(4):146

8. Burgos-Artizzu XP, Perona P, Dollár P (2013) Robust face landmark estimation under occlusion. In: Proceedings of the IEEE international conference on computer vision, pp 1513–1520. https://doi.org/10.1007/s11263-018-1097-z

9. Chavan R, Phad B, Sawant S, Futak V, Rawat A (2015) Attendance management system using face recognition. IJIRST-Int J Innov Res Sci Technol

10. Chin ET, Chew WJ, Choong F (2015) Automated attendance capture and tracking system. J Eng Sci Technol 10:45–59

11. Chintalapati S, Raghunadh M (2013) Automated attendance management system based on face recognition algorithms. In: 2013 IEEE international conference on computational intelligence and computing research. IEEE, pp 1–5

12. Dalal N, Triggs B (2005) Histograms of oriented gradients for human detection. In: 2005 IEEE computer society conference on computer vision and pattern recognition (CVPR'05), vol 1. IEEE, pp 886–893

13. Devi BJ, Veeranjaneyulu N, Kishore K (2010) A novel face recognition system based on combining eigenfaces with fisher faces using wavelets. Procedia Comput Sci 2:44–51

14. Elias SJ, Hatim SM, Hassan NA, Abd Latif LM, Ahmad RB, Darus MY, Shahuddin AZ (2019) Face recognition attendance system using local binary pattern (LBP). Bull Electr Eng Inform 8(1):239–245. https://doi.org/10.11591/eei.v8i1.1439

15. Finch C (2009) The disadvantages of voice recognition software. https://www.techwalla.com/articles/the-disadvantages-of-voice-recognition-software

16. Fuzail M, Nouman HMF, Mushtaq MO, Raza B, Tayyab A, Talib MW (2014) Face detection system for attendance of class' students. Int J Multidiscip Sci Eng 5(4)

17. Guo G, Zhang N (2019) A survey on deep learning based face recognition. Comput Vis Image Understand 189:102805. https://doi.org/10.1016/j.cviu.2019.102805

18. Hernandez-Ortega J, Galbally J, Fierrez J, Haraksim R, Beslay L (2019) Faceqnet: quality assessment for face recognition based on deep learning. In: 2019 international conference on biometrics (ICB). IEEE, pp 1–8. https://doi.org/10.1109/ICB45273.2019.8987255

19. Houssaini AS, Sabri MA, Qjidaa H, Aarab A (2019) Real-time driver's hypovigilance detection using facial landmarks. In: 2019 international conference on wireless technologies, embedded and intelligent systems (WITS). IEEE, pp 1–4. https://doi.org/10.1109/WITS.2019.8723768

20. Hung BT, Khang NN (2021) Student attendance system using face recognition. In: proceedings of integrated intelligence enable networks and computing. Springer, pp 967–977. https://doi.org/10.1007/978.981.33.6307.698

21. Islam A, Rahaman N, Ahad MAR (2019) A study on tiredness assessment by using eye blink detection. Jurnal Kejuruteraan 31(2):209–214. https://doi.org/10.17576/jkukm-2019-31(2)-04

22. Islam M, Mahmud A, Papeya AA, Onny IS et al (2017) Real time classroom attendance management system. Ph.D. thesis. BRAC University
23. Jha P, Pradhan S, Thakur SK, Singh R (2021) Face detection and recognition using open cv. Ann Roman Soc Cell Biol 25(6):11799–11804. https://www.annalsofrscb.ro/index.php/journal/article/view/7769
24. Johnson S (2019) Biometric fingerprint scanners advantages and disadvantages. https://www.techwalla.com/articles/biometric-fingerprint-scanners-advantages-disadvantages
25. Kaur M, Sandhu M, Mohan N, Sandhu PS (2011) RFID technology principles, advantages, limitations & its applications. Int J Comput Electr Eng 3(1):151
26. Khan M, Chakraborty S, Astya R, Khepra S (2019) Face detection and recognition using opencv. In: 2019 international conference on computing, communication, and intelligent systems (ICCCIS). IEEE, pp 116–119. https://doi.org/10.1109/ICCCIS48478.2019.8974493
27. Li Z, Liu F, Yang W, Peng S, Zhou J (2021) A survey of convolutional neural networks: analysis, applications, and prospects. IEEE Trans Neural Netw Learn Syst. https://doi.org/10.1109/TNNLS.2021.3084827
28. Lukas S, Mitra AR, Desanti RI, Krisnadi D (2016) Student attendance system in classroom using face recognition technique. In: 2016 international conference on information and communication technology convergence (ICTC). IEEE, pp 1032–1035
29. Misra S (2020) A step by step guide for choosing project topics and writing research papers in ict related disciplines. In: International conference on information and communication technology and applications. Springer, pp 727–744. https://doi.org/10.1007/978.3.030.69143.1.55
30. bin Mohd Nasir MAH, bin Asmuni MH, Salleh N, Misra S (2015) A review of student attendance system using near-field communication (NFC) technology. In: International conference on computational science and its applications. Springer, pp 738–749. DoI:https://doi.org/10.1007/978.3.319.21410.8.56
31. Pande V, Elleithy KM, Almazaydeh L (2012) Parallel processing for multi face detection and recognition
32. Patil M, Dhawale C, Misra S (2016) Analytical study of combined approaches to content based image retrieval systems. Int J Pharm Technol 8(4):22982–22995
33. Pss S, Bhaskar M (2016) RFID and pose invariant face verification based automated classroom attendance system. In: 2016 international conference on microelectronics, computing and communications (MicroCom). IEEE, pp 1–6. https://doi.org/10.1109/MicroCom.2016.7522434
34. Rasti P, Uiboupin T, Escalera S, Anbarjafari G (2016) Convolutional neural network super resolution for face recognition in surveillance monitoring. In: International conference on articulated motion and deformable objects. Springer, pp 175–184. https://doi.org/10.1007/978.3.319.41778.3.18
35. Saha S, Saha S (2018) A comprehensive guide to convolutional neural networks—The eli5 way, vol 17
36. Sanivarapu PV (2021) Multi-face recognition using CNN for attendance system. In: Machine learning for predictive analysis. Springer, pp 313–320. https://doi.org/10.1007/978.981.15.7106.031
37. Shah K, Bhandare D, Bhirud S (2021) Face recognition-based automated attendance system. In: International conference on innovative computing and communications. Springer, pp 945–952. https://doi.org/10.1007/978.981.15.5113.079
38. Sharanabasappa R, Sanjaypande M (2012) Real time multiple face detection from live camera, a step towards automatic attendance system. Int J Comput Appl 975:8887
39. Soukupová T, Cech J (2016) Eye blink detection using facial landmarks. In: 21st computer vision winter workshop, Rimske Toplice, Slovenia
40. Spaun NA (2011) Face recognition in forensic science. In: Handbook of face recognition. Springer, pp 655–670. https://doi.org/10.1007/978.0.85729.932.126
41. Sunaryono D, Siswantoro J, Anggoro R (2019) An android based course attendance system using face recognition. J King Saud Univ Comput Inf Sci. https://doi.org/10.1016/j.jksuci.2019.01.00

42. Thomas T, Vijayaraghavan AP, Emmanuel S (2020) Neural networks and face recognition. In: Machine learning approaches in cyber security analytics. Springer, pp 143–155. https://doi.org/10.1007/978.981.15.1706.8.8
43. Viola P, Jones MJ (2004) Robust real-time face detection. Int J Comput Vis 57(2):137–154
44. Younis MI, Al-Tameemi ZFA, Ismail W, Zamli KZ (2013) Design and implementation of a scalable rfid-based attendance system with an intelligent scheduling technique. Wirel Person Commun 71(3):2161–2179. https://doi.org/10.1007/s11277-012-0929-3
45. Yuan L, Qu Z, Zhao Y, Zhang H, Nian Q (2017) A convolutional neural network based on tensorflow for face recognition. In: 2017 IEEE 2nd advanced information technology, electronic and automation control conference (IAEAC). IEEE, pp 525–529. https://doi.org/10.1109/IAEAC.2017.8054070
46. Zeng W, Meng Q, Li R (2019) Design of intelligent classroom attendance system based on face recognition. In: 2019 IEEE 3rd information technology, networking, electronic and automation control conference (ITNEC). IEEE, pp 611–615. https://doi.org/10.1109/ITNEC.2019.8729496

A Systematic Literature Review on Face Morphing Attack Detection (MAD)

Mary Ogbuka Kenneth, **Bashir Adebayo Sulaimon**,
Shafii Muhammad Abdulhamid, and **Laud Charles Ochei**

Abstract Morphing attacks involve generating a single artificial facial photograph that represents two distinct qualities and utilizing it as a reference photograph on a document. The high quality of the morph raises the question of how vulnerable facial recognition systems are to morph attacks. Morphing Attack Detection (MAD) systems have aroused a lot of interest in recent years, owing to the freely available digital alteration tools that criminals can employ to perform face morphing attacks. There is, however, little research that critically reviews the methodology and performance metrics used to evaluate MAD systems. The goal of this study is to find MAD methodologies, feature extraction techniques, and performance assessment metrics that can help MAD systems become more robust. To fulfill this study's goal, a Systematic Literature Review was done. A manual search of 9 well-known databases yielded 2089 papers. Based on the study topic, 33 primary studies were eventually considered. A novel taxonomy of the strategies utilized in MAD for feature extraction is one of the research's contributions. The study also discovered that (1) single and differential image-based approaches are the commonly used approaches for MAD; (2) texture and keypoint feature extraction methods are more widely used than other feature extraction techniques; and (3) Bona-fide Presentation Classification Error Rate and Attack Presentation Classification Error Rate are the commonly used performance metrics for evaluating MAD systems. This paper addresses open issues and includes additional pertinent information on MAD, making it a valuable resource for researchers developing and evaluating MAD systems.

M. O. Kenneth (✉) · B. A. Sulaimon
Department of Computer Science, Federal University of Technology, Minna, Nigeria
e-mail: kenneth.pg918157@st.futminna.edu.ng

B. A. Sulaimon
e-mail: bashirsulaimon@futminna.edu.ng

S. M. Abdulhamid
Department of Cyber Security Science, Federal University of Technology, Minna, Nigeria
e-mail: shafii.abdulhamid@futminna.edu.ng

L. C. Ochei
Department of Computer Science, University of Port Harcourt, Port Harcourt, Nigeria
e-mail: laud.ochei@uniport.edu.ng

© The Author(s), under exclusive license to Springer Nature Switzerland AG 2022
S. Misra and C. Arumugam (eds.), *Illumination of Artificial Intelligence in Cybersecurity and Forensics*, Lecture Notes on Data Engineering and Communications
Technologies 109, https://doi.org/10.1007/978-3-030-93453-8_7

Keywords Face morphing · Morphing attack detection · Systematic literature review · Feature extraction techniques · Performance metrics

1 Introduction

Biometric characteristics such as face, iris, voice and fingerprint are natural tool in carrying out identification task such as in border control, e-Government application, law enforcement, surveillance, e-commerce applications, user verification in mobile phones and many more [1–3]. Face as a biometric characteristics are regularly used as a means of identification because of the noninvasive nature of its capture process and consumer usability [4]. Face as a means of identification are presented for many forms of documentation worldwide, including, voters card, national identity card, international passports and driving licenses. Face recognition systems are commonly used for automatic recognition of individuals by observing their facial biometric characteristics [5–7].

The deployment of face recognition systems are on the rise due to its accurate and reliable face recognition algorithms, hence the attacks on these systems become more creative [8–10]. Examples of attacks faced by face recognition systems includes the presentation attack [11] such as spoofing that presents a copy of an individual characteristics in order to impersonate that individual [12], and concealed face attacks that aim to disable face recognition using physical objects. Another form of attack identified by Seibold [13] is the face morphing attack. This attack aims to present one face comparison picture which is automatically matched successfully to more than one individual and by human experts [14, 15].

Face morphing can present a serious security threat when these morphed photographs are used in identification or passports, enabling multiple individuals (subjects) to verify their identity with that linked to the presented paper [15–17]. This defective connection of multiple subjects with the document could result in a variety of illegal activities such as human trafficking, financial transaction, and illegal immigration [8]. A targeted offender would morph his face photograph with another of the lookalike partners in a real-life situation of a face-morphing attack. If the partner requests an e-passport with the transformed face photograph, he/she will obtain a legitimate e-passport configured with document security features that match. Both the partner (accomplice) and the criminal could be authenticated against the morphed image stored in the e-passport with success. This means that the offender can use the e-passport granted to the accomplice to pass through the Automatic Border Control gates or maybe even pass through the human inspections at the gate [18]. Hence automatic detection of this face morphing attack is of great importance.

In the previous years, there have been few authors who have worked on detection of face morphing attacks. In 2014 Ferrara [19] introduced the face morphing attack which was called the magic passport. The viability of attacks on Automated Border Control (ABC) systems using morphed face images was examined and it was concluded that when the morphed passport is presented; if the passport is not substantially different from the applicant's face, the officer will recognize the photograph and release the document. And thus the released document passes all authenticity checks carried out at the gates. Raghavendra [20] carried out a novel research on how this face morphing attack can be detected. The research was conducted using facial micro-textures retrieved via statistically independent filters which are trained on natural photographs. This variation in micro-texture was extracted using Binarized Statistical Image Features (BSIF) and classification had been made via Support Vector Machine (SVM). This was the first research done towards detection of face morphing attacks.

Later in 2017 Seibold [21] aimed to detect face morphing attack using deep neural network. Three Convolutional neural network architecture were trained from scratch and using already trained networks for the initialization of the weights. Pretrained networks was noticed to outperform the networks trained from scratch for each of the three architecture. Hence it has been concluded that the features acquired for classification tasks are also useful for MAD. In 2018 and 2019 researchers such as Singh [22] and Wandzik [23] proposed MAD using deep decomposed 3D form and diffuse Reflectance and a General-Purpose Face Recognition System, respectively. Peng [22] did not just stop at detecting face morphing attack but went further to de-morph the morphed face image using generative adversarial network to rebuild facial image of the accomplice.

Other researchers were able to perform review of image morphing and face morphing attacks in a general scope but no related works were found that conducted a SLR of face morphing attack. A gap in the domain of biometric systems that needs filling is the lack of existing literature that provides systematic knowledge regarding MAD with the ability to further research, given vital information. The aim of this paper is thus to review the current literatures on MAD techniques in a systematic way.

The paper's primary contributions are to:

1. Present a novel taxonomy of feature extraction techniques used in face morphing attack detection (MAD).
2. Present information on commonly used approaches for morph attack detection, feature extraction techniques, and performance evaluation measures for evaluating morphing attack detection systems.
3. Present open issues and challenges of face morphing attack detection.

The remainder of this paper is structured according to: a summary of previous works on MAD was presented in Sect. 2. The Review method used in carrying out the study is presented in Sect. 3. Section 4 shows the results obtained after review and the presented results were discussed. Section 5 presents the Parametric used in MAD. Taxonomy of MAD techniques are presented in Sect. 6. Section 7 presents open issues and future directions in the field of MAD. In Sect. 6 conclusions were drawn and Appendix A presents a list of the primary studies identified.

2 Previous Related Surveys

Face Recognition (FR) systems were found vulnerable to morphing attacks. Based on this vulnerability, Korshunov [24] focused on assessing the vulnerability of FR systems to deep fake videos where actual faces are replaced by an adversarial generative network that generates images trained on two subjects/people's faces. Two existent FR algorithm based on Facenet and VGG neural networks were evaluated and it suggests that both algorithms were susceptible to deep morphed video as they do not differentiate morphed videos from the actual videos with a Detection Equal Error Rate (D-EER) of up to 95.00%. It was also observed that baseline detection algorithms based on the image quality measurements with SVM classifier could identify high quality deep morph videos with a D-EER of 8.97%.

Scherhag [25] conducted review of the currently proposed morphed facial image detectors regarding their robustness across various databases. The aim of this survey was to identify reliable algorithms for detection. It was concluded that the majority of current detection techniques do not appear to have great performance across various databases showing that morph detectors on a single database could cloud the overall appearance of the real detection results.

Kramer [26] conducted four different experiments to investigate the performance of humans and computers with high quality facial morphs. These four experiments include morph detection using computer simulation, using live-face matching, induced morph detection and tips, and finally research based on Robertson [17] replication using morphs of higher quality. Based on these tests, it was discovered that humans were extremely susceptible to error when detecting morph and also human training on MAD did not yield change. In a live matching experimentation, morphs were also acknowledged as bona fide images; and poses a major concern for security agencies, therefore this demonstrated that identification was again prone to error. Finally, it was established that a simple computer model outperformed the human participants. Ultimately it was established that the human participants were outperformed by a simple computer model.

Makrushin [27] conducted a survey on recent developments in the assessment and mitigation of face morphing attack. It was discovered that the identification of morphed facial images at the human and automated facial recognition systems level was needed to mitigate morph assault It was also found that existing MAD algorithms still have significant high error rates and that the performance of these MAD algorithms severely degrades with images that are re-digitalized and manipulated anti-forensically. It was also proposed that extensive work on the limitations of the MAD techniques should be carried out.

Scherhag [28] conducted a study on facial recognition systems under morphing attacks. This survey was based on conceptual categorization and metrics for an assessment of MAD techniques and a rigorous survey of related literature, in addition open issues and challenges based on Face morph attacks in face recognition systems was carried out. In this survey three steps of morphing process of face images was identified. The first step was determination of the correspondence between the

Table 1 Related survey overview

S/No	References	Number of cited references	Scope of time covered
1	Scherhag [28]	124	1986–2018
2	Makrushin [27]	46	1998–2018
3	Scherhag [25]	22	2004–2018
4	Kramer [26]	39	1993–2019
5	Korshunov [24]	22	2014–2019

contributing samples. Secondly warping which entails distortion of both images to achieve geometrically alignment between sample images and the third step called blending which deals with merging the color values of the warped images. Based on their survey the quality of the created face morphed images can be accessed based on the image quality, morphing artifacts, plausibility of face morph and human insight of morphed images. Lastly drawbacks of studies related to face morphing and MAD were identified which are lack of automatic creation of high-quality face morphs, no available measures for susceptibility of FR systems with respect to morphing attacks and lastly MAD were prone to over fitting. Overview of the related survey is depicted in Table 1.

3   Review Method

Performing SLR in a specific field is necessary to define research issues, as well as to explain potential work in that area [29, 30]. SLR was selected as the tool of inquiry. This study uses SLR guidance, which is a method of secondary analysis that uses consistent and well-defined measures to classify, analyze and interpret all existing evidence relevant to a specific research question [31]. The SLR procedure aims to be as fair as possible by being auditable and repeatable [29]. The SLR process aims at being verifiable and repeatable as equally as possible [29]. The aim of SLR, based on Soledad [32], is to have a potential list of all research that are relevant to a particular subject area. Whereas, general reviews attempt to sum up findings from a variety of studies. The SLR cycle consists of three consecutive phases of preparation, executing, and reporting. The preparation process also known as the planning phase is conducted in this section which involves identifying the research questions as well as how the analysis is conducted [32].

3.1   Review Design

The review design outlines the basis of this analysis by identifying the research questions for the SLR and keywords for search.

3.1.1 SLR Research Questions

Very few researchers have performed face morphing attack identification over the years. Hence the SLR research questions that this study aims to address are:

1. Which approaches are used for detection of face morphing attacks?
2. What feature extraction techniques are used for detection of face morphing attack?
3. Which performance metric are used to evaluate face morph attack detection algorithms?

3.1.2 Search Strategy

The SLR focuses on looking for relevant books or technical papers in the academic repositories. This paper used nine (9) repositories to do the search process for SLR. The following are the repositories used:

1. Scopus (www.scopus.com)
2. Google Scholar (www.scholar.google.com)
3. IEEE Xplore (www.ieeexplore.com)
4. Semantic Scholar (www.semanticscholar.org)
5. ScienceDirect (www.sciencedirect.com)
6. ACM Digital Library (dl.acm.org)
7. Springer link (link.springer.com)
8. Taylor & Francis (taylorandfrancis.com)
9. International Scientific Indexing (isindexing.com).

The repositories were selected because they provide important and maximum impact full-text articles and conference papers, typically covering the areas of MAD.

The search keywords used to locate specific studies in the title of the document, keywords and abstract are as follows: "face morphing" OR "face image modification" AND "face morphing attack" OR "face alteration attack" AND "face morphing attack detection".

3.2 Review Conduction

This segment on review conduction specifies the evaluation process for performing the SLR. The SLR evaluation protocol refers to the review structure and rules [33].

Table 2 Inclusion and exclusion criteria

Inclusion criteria	Abstract and title are written in English Full-text article Study that concentrated on face morphing attack detection
Exclusion criteria	Study that does not address face morphing attack detection Duplicate article Short paper Study not written in English

3.2.1 Inclusion and Exclusion Criteria

This SLR used the parameters for inclusion and exclusion set out in Table 2. On the basis of Table 2, papers which do not focus on the detection of face morphing attacks have been omitted. SLR also removes duplicate papers of the same report.

3.2.2 Study Selection

Study selection was achieved using the below processes [34]:

1. Database search using the search keywords to find relevant studies.
2. Exclude studies based upon the criteria for exclusion.
3. Exclude any insignificant study based on the examination of their titles and abstracts.
4. Assessing the selected studies based complete reading and quality checklist.
5. Extracting responses relating to study issues.
6. Obtain Primary studies.

3.2.3 Quality Assessment

As per Okoli [35] SLR guidelines Quality Assessment (QA) questions must be well-defined to evaluate the credibility of the research paper and also provide a quantitative measure among them. The rating methods are Y (Yes), P (Partially) and N (No). The checklist/questions for quality assessments described in this SLR are shown in Table 3.

3.2.4 Data Extraction

The data collection tools that were used to perform an in-depth analysis for all selected primary studies is presented in Table 4.

Table 3 Quality assessment: checklist

S/No	Question	Score
1	Are the data collection methods adequately described	a. Yes: it explicitly describes the methods used to collect the face morph images b. Partially: it only mentioned the data collection method without further explanation c. No: Data collection method was not listed or clarified
2	The used techniques are they clearly described and their selection justified?	a. Yes: it either clearly describes the techniques used to detect face morphing attack b. Partially: it only gave a peripheral explanation of the techniques c. No: it neither described nor mentioned the techniques for face morphing attack detection
3	How precisely were the limitations to the research documented?	a. Yes: It clearly clarified the proposed algorithm's restriction b. Partially: The restrictions were stated but it did not clarify why c. No: The restriction was not stated
4	Were the discoveries credible?	a. Yes: The analysis has been clarified methodologically so the result can be trusted b. Partially: the analysis has been clarified in methodological terms but not in depth c. No: The research wasn't clarified methodologically

Table 4 Extracted data form

S/No	Data extracted	Description	Type
1	Bibliographic references	Authors, publication year, title and the publishing source	General
2	Study type	conference paper, Text, journal, lecture paper, workshop paper	General
3	Approaches for detection of face morphing attacks	Description of the approaches of face morphing attack detection	Research question (RQ)
4	Feature extraction techniques	Description of the feature extraction techniques for MAD	RQ
5	The performance metric	Performance metric are used to evaluate face morph attack detection algorithms	RQ
6	Findings/contribution	Displaying research results and feedback	General

3.2.5   Synthesis

Analytical results via SLR showed 102 studies for further deliberation. The selected 102 studies were thoroughly reviewed but only 33 publications were left which could address the study question of this SLR. Those 33 publications have therefore been selected as primary studies. Figure 1 Displays the number of studies per systematic procedure.

Figure 2 Displays the number of primary studies by published year. All 33 publications selected have been published from 2016, 2017, 2018, 2019 and 2020. It can be seen that 2018 has the highest selected papers with 15 articles compared to the other years

4   Results and Discussion

This segment reports the findings and discussion after conducting the SLR for answering the defined question of SLR research. Furthermore, the responses to the SLR questions which were obtained from selected primary studies based on specified forms of data extract are discussed.

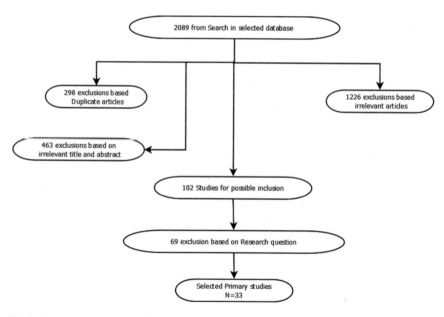

Fig. 1 Procedure for finding primary study

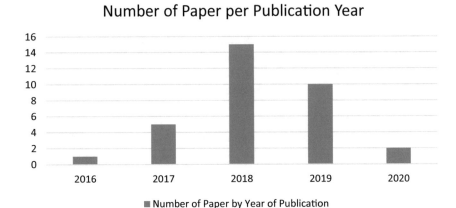

Fig. 2 Number of paper per publication year

4.1   Result

This section includes the results of the SLR research questions.

4.1.1   Finding Research Questions

RQ1: Which approaches are used for detection of face morphing attacks?

Studies were analyzed to answer this question, and as a result, two approaches to MAD was identified. This two approaches are reflected in Table 5. Table 5 provides the list of primary studies which address the MAD approaches described. Approach (1) has 16 papers dealing with approach, and two papers address approach (2).

RQ2: What feature extraction techniques are used for detection of face morphing attack?

The feature extraction techniques identified for face morphing attack detection are displayed in Table 6. In Table 6 technique (1), (5), (7), (8), (9), (10), (17) and (18) are used by one paper respectively, technique (2), (6), (11), (15), (16) and (19) are used by two papers respectively, technique (3) and (14) are used by five papers,

Table 5 Number of primary study addressing the identified approaches

S/N	Approaches	Number of papers	Study identifiers
1	Single image-based	30	P32, P33, P28, P12, P25, P1, P13, P2, P5, P3, P16, P23, P27, P11, P18, P9, P8, P30, P7, P19, P20, P21, P14, P17, P10, P15, P22, P24, P26, P29, P31
2	Differential image-based	4	P4, P6, P22, P20

Table 6 Number of preliminary survey addressing the identified feature extraction techniques

S/N	Feature extraction techniques	Number of papers	Study identifiers
1	Steerable pyramid	1	P14
2	Binarized statistical image features (BSIF)	2	P2 and P18
3	Local binary pattern (LBP)	5	P2, P15, P5, P13, P8
4	Deep neural networks	9	P2, P17, P12, P3, P16, P11, P1, P8, P7
5	Histogram of oriented gradients (HOG)	1	P2
6	Photo response non-uniformity (PRNU)	2	P10, P9
7	Scale invariant feature transform (SIFT)	1	P2
8	Regressing local binary features (LBF)	1	P6
9	Distance based	1	P4
10	Local phase quantization (LPQ)	1	P8
11	Benford features	2	P19, P32
12	FAST	4	P29, P26, P24, P29
13	AGAST	3	P24, P26, P29
14	SURF	5	P22, P23, P24, P27, P29
15	Sobel & canny	2	P29, P24
16	Discrete fourier transform	2	P29, P31
17	Shi Tomasi	1	P26
18	Laplacian pyramid	1	P28
19	Oriented BRIEF (ORB)	2	P29, P24

technique (4) has the highest number of papers with nine papers, technique (12) is used by four paper and finally technique (13) is used by three papers.

RQ3: Which performance metric are used to evaluate face morph attack detection algorithms?

Table 7 depict the identified performance metric and the number of papers that addresses them. In Table 7, 17 papers evaluate using (1), 29 papers evaluate using (2), 30 papers evaluate using (3), 1 paper evaluates using (4) and lastly 2 papers evaluate using (5).

Table 7 Number of primary study addressing the identified performance metrics

S/N	Performance metrics	Number of papers	Study identifiers
1	Detection-equal error rate (D-EER)	17	P23, P25, P27, P13, P2, P3, P10, P9, P12, P30, P33, P1, P8, P17, P7, P22, P28
2	Bona fide presentation classification error rate (BPCER)/false rejection rate (FRR)	29	P13, P2, P19, P20, P14, P5, P3, P9, P12, P1, P8, P7, P6, P16, P11, P27, P29, P30, P18, P17, P4, P22, P23, P24, P25, P28, P31, P33, P21
3	Attack presentation classification error rate (APCER)/false acceptance rate (FAR)	30	P32, P19, P20, P21, P13, P2, P27, P14, P5, P23, P24, P25, P3, P9, P12, P29, P1, P31, P8, P7, P6, P16, P11, P18, P17, P4, P22, P28, P30, P33
4	True positive rate	1	P32
5	Accuracy (ACC)	3	P11, P15, P26

4.2   Discussion

Discussion about this SLR is provided in this section. The discussion is based on the research problem set out in Sect. 4.1.1.

4.2.1   Approaches to Face Morphing Attack Detection

Single Image-Based

The existence of morph alterations is identified here on a single image, like the identity passport provided to the officer at the time of registration or the face picture read from an e-document during authentication at the gate [36, 37]. This means that the single given image is processed by the detector and classified as either morph or bona-fide without any reference image [4]. There are two types of single image-based approach. These are the print-scan attack detection and digital attack detection.

1. **Print-Scan Attack Detection**

    The original images be it morphed or bone-fide are first printed a printer and then scanned with a scanner. This method of printing/scanning alters the image content, eliminating most of the fine details (that is, digital artefacts) which might help identify morphing [38]. Literatures that addressed morphing attack detection based on print-scan images are P14, P17, P10, P15, P12, P25, P28, P33 and P1.

2. **Digital Attack Detection**

    Here the digital copy of the bona fide and morphed images captured by the camera is being used without having to undergo any kind of post-processing like print-scan. This is the most commonly used approach in the literatures. Literatures that adopted this approach includes P30, P13, P24, P27, P2, P5, P3, P16, P11, P21, P10, P23, P32, P18, P9, P33, P20, P8, P19, P31, P29, P26, P25, P22 and P7.

Differential Image-Based (Image-Pair Based Approach)

This approach deals with the contrast between a live image (for example, the image obtained at the gate) and the one stored on the electronic record in order to perform MAD [38]. Recently, some works have explored differential image-based detection approach. These researches has indicated that introducing a bone fide reference image permits for a whole new set of techniques [39].

In P4 research the angles and distances between the passport's facial landmarks and the bona fide picture are compared. The angle comparison gives the best results, but the classification error rates are not yet small enough for real-world use. Hence in future work the technique can be combined with a texture based technique to achieve small classification error rate.

To examine MAD involving a bona fide probe image, P6 created a repository of paired images between the reference images investigated (whether morphed or bona fide) and the probe images (assumed to be live captures). Each reference image is matched with similar images. The probe photographs, however, are the frontal face photographs captured by the very same users in different sessions than the reference images. Consequently P6, P4, P20 and P22 used differential image based technique.

4.2.2   Face Morphing Attack Detection Feature Extraction Techniques

Steerable Pyramid

Steerable Pyramid developed by Simoncelli & Freeman's [40] is a sequential multiple scales, multiple oriented image decomposition that offers a good front-end for computer vision and image processing applications [40, 41]. This could be seen as a selective alignment variant of the Laplacian pyramid, during which a steerable filter bank is used at each pyramid level instead of a single Laplacian or Gaussian filter. P14 used these techniques to remove scale-space features (morphed or bona-fide) from pictures. The scale-space is essentially a collection of directed filters synthesized as a linear combination of the fundamental functions. This technique has been described as being successful in morphing attack section since the extracted texture features can easily reflect visual distortion throughout the image.

Binarized Statistical Image Features (BSIF)

It is a local feature descriptor built by binarizing the reactions to linear filters but, in contrast to previous binary descriptors, the filters are learned from natural images utilizing independent component analysis [42]. This feature extraction technique was used by P2, P18, P22, P23 and P27.

Local Binary Pattern (LBP)

Local binary patterns (LBP) is a form of visual descriptor which is used in computer vision classification [7, 43]. LBP is indeed a type of gray scale that supports the local contrast estimate of an image within the reach of the texture measure. LBP is originally specified in an eight-pixel neighborhood, and the center pixel gray value is set as a threshold. All neighbors having values greater or equal to the center pixel value are given a value of 1, otherwise they are set as 0 [1]. The values upon thresholding (notably 0 or 1) will increase with the corresponding pixel weight, respectively, and their multiplicative result will be the LBP value [44]. One drawback of LBP found by [44] is its vulnerability to changes in noise and lighting. The following literatures used this extraction technique for MAD: P22, P23, P27, P28, P15, P2 and P5. P13 made use of the LBP pyramid, an extension of LBP. The features of the Pyramid-Local Binary Component (P-LBP) were being used to efficiently measure residual noise, as it has been proven by literatures to be effective in modeling the residual noise. P20 and P21 used LBP histogram.

Deep Neural Networks (Convolutional Neural Network)

The convolutional neural network (CNN) [45] is among the most popular feature extraction techniques used for MADs. This technique is also used for extracting profound features from images. CNN is a multilayer network of neurons; each layer consist of multiple 2D surfaces, and each plane consist of multiple independent neurons [46, 47]. CNNs comprises of many connections, and the architecture is composed of different types of layers, including pooling, convolution and fully-connected layers, and realize form of regularization [48]. CNN makes use of deep architecture to learn complicated features and functions that can represent high-level abstractions. P20, P25 and P21used CNN architecture for feature extraction, P16 used scratched and pre-trained alextNet, googleNet and VGG19 architecture for extraction and analysis. It was noted that pre-trained VGG19 achieved the best result. P3, P11, P12 and P33 used pre-trained AlexNet architecture, P33, P8 and P1 used VGG architecture, faceNet a popular model for face recognition and verification [49] was used by P17 for MAD while P2, P7 and P22 used the OpenFace Model.

Histogram of Oriented Gradients (HOG)

The histogram of oriented gradients (HOG) is a feature descriptor centered on the gradient approximation used for object recognition purposes in computer vision and image processing. HOG descriptor method counts gradient orientation instances in localized portions of an image identification window, or region of interest (ROI). Because HOG is invariant to photometric and geometric transformations, it is very well adapted for human detection [50]. HOG was implemented for MAD by P2, P22, P23 and P27 given the fact that the morphing process decreases the variations in high frequencies and thus the steepness of the gradients is reduced which improves MAD.

Photo Response Non-uniformity (PRNU)

The PRNU is a distortion-like pattern, which originates from small differences between individual pixels during the digital photo sensor transformation of photons into electrons. It forms an intrinsic part of those sensors, while this weak signal is incorporated in any image they capture [51–53]. P9, P10 and P30 used PRNU for MAD as all image sensors exhibit PRNU. PRNU exists in each image irrespective of the scene content, except fully dark or over-exposed images, and PRNU survives, gamma correction, filtering, loss compression and several other processing method [54, 55].

Scale Invariant Feature Transform (SIFT)

SIFT is a computer vision Keypoint extractor algorithm used for the identification and definition of local features in images [56]. By using a staggered filtering technique, the scales-invariant features are well defined. In the early stage, key locations are identified by looking for positions with a difference of Gaussian function that are maximum or minimum [57]. A character vector that defines the sampled region of the local image relative to its scale space frame is then used for each point [58]. SIFT Keypoint extractors were used for P2, P22, P23, P24, P27 and P29 as morphed photos are assumed to comprise of fewer key locations, which are described as maxima and minima resulting from the difference in function of Gaussians. This keypoint identification is used as descriptive function for MAD.

Regressing Local Binary Features (RLBF)

The detector proposed by Ren [59] is a landmark detector. Where a variety of local binary characteristics and a local theory have been used to learn these features. Lowe [58] proposes that each facial landmark learn independently of a set of highly discriminatory local binary features, then use them to learn a linear regression together for the final mark detection result. P6 used a characteristic extractor for MAD based on an approach focused on differential images.

Distance Based

In a differential image scenario, this method was used by P4 for MAD. The landmarks of both the image of bona fide and the morphed images were evaluated by the predictor of the facial landmark of dlib [60]. The landmarks are standardized to a range between 0 and 1 to achieve a scalable-robust method. Next the Euclidean distance between bona fide and morphed images, which results in a 2278 long vector known as the distance characteristics, is determined for each landmark's relative location. The results obtained by P4 are not appropriate for operational deployment but is an initial step to MAD based on reference.

Local Phase Quantization (LPQ)

Initially, Ojansivu [61] proposed the LPQ operator which is a texture descriptor. LPQ focuses on the Fourier component spectrum's blurring invariance property. It uses the local phase data extracted using the transformation of 2-D Short Term Fourier (STFT), measured on each pixel of the image over a rectangular district [61]. Due to the robustness of the image, LPQ was adopted by P8.

Benford Feature

The rule of Benford states: in a set of natural numbers, the first digits distribution is a logarithmic [62]. That is it is a likelihood distribution for the probability of the very first digit in a set of numbers. Benford characteristics can be used in natural data sets for pattern or pattern loss detection [63]. The use of Benford's features for pattern detection has led Fu [64] to suggest it in JPEG format compressed images for tamper detection. Hypothesis behind applying Benford's characteristics by P19 for MAD is that the naturally produced data are in accordance with Benford law and the altered data infringes the law. P32, also used Benford's rule for morphed face image detection.

Features from Accelerated Segment Test (FAST)

FAST is an existing algorithm for the identification of interest points in an image originally introduced by Rosten and Drummond [65]. FAST uses one variable which is the threshold of intensity between the middle pixel and the ones in a circular ring around the middle [66]. FAST is measured easily and quickly to match. The precision is pretty good, too. FAST does not represent a scale-space detector, so the detection of the edges at the particular scale can produce much more than a scale-space technique like SIFT [67]. P24, P26 and P29 used FAST descriptor for MAD.

Adaptive and Generic Accelerated Segment Test (AGAST) Features

AGAST has been designed to address the limitation of the FAST algorithm that includes: FAST needs to learn from an image dataset in the context in which it operates, and then generate a decision tree to identify each center pixel as a function or not. However, this approach cannot guarantee that every pixel combination will be discovered, and this can yield inaccurate results [68, 69]. Additionally, each time the working context shifts, the FAST feature detector must be trained from scratch [68]. AGAST is founded on the same criteria of Accelerated Segment Test feature as FAST, but utilizes another decision tree. AGAST is trained on the basis of a set of data with all possible 16 pixel combinations on the circle included. This guarantees that the decision tree is working in any setting. AGAST performance increases for random scenes, and AGAST operates with no training steps in any arbitrary environment [65]. P24, P26 and P29 used AGAST descriptor for MAD.

Shi Tomasi Features

The Shi Tomasi is an angle/corner detector entirely based on the detection of Harris corner [70]. A small change in a selection criterion, however, has enabled this detector to perform even better than the initial. Also, the Shi Tomasi can be characterized as an enhancement on the Harris technique, using only the lowest eigenvalues for discrimination, thus significantly streamlining the computation [71]. Shi Tomasi was used for MAD by P26.

Oriented FAST and Rotated BRIEF (ORB) Features

ORB is a mixture of the famous FAST key point descriptor with some modification of the Binary Robust Independent Elementary Feature (BRIEF) descriptor [72]. ORB is a simple binary descriptor founded on the BRIEF, which is noise tolerant and rotation invariant [66]. These techniques provide good performance and have low cost [72]. Firstly, ORB utilizes FAST to identify the key points. A Harris corner formula for locating top N points is then added. FAST is not used for orientation calculation, and is a variant to rotation. Hence it used to measure the intensity weighted centroid with center corners located [73]. The rotation matrix is calculated by utilizing the patch orientation, and the orientation of the BRIEF descriptors is steered [67]. ORB was used by P24 and P29 for MAD.

Discrete Fourier Transform (DFT)

DFT is a technique for signal processing [74]. It is a transform dealing with a countable discrete-time signal and a discrete amount of frequency [75]. DFT translates a signal for the time domain to its relative frequency domain. This frequency domain depiction of the time domain signal is named the signal frequency spectrum [74]. Hence the spectrum of the signal shows the range of frequencies and their amount that are present in the time domain signal. Sensor Pattern Noise (SPN) is a deterministic factor which remains almost the same if multiple images are taken from the exact same location. Because of this property, the SPN is present in any image a sensor captures, and can therefore be used to classify the source of the image [76, 77]. P31 used the differences in the Fourier frequency spectrum of the SPN of the images to differentiate between morphed and bona fide images [78]. P29 used the frequency domain representation of the time domain signal as feature for MAD [79].

Speed up Robust Features (SURF)

SURF is a powerful algorithm for image registration and object recognition. SURF describes the local texture features of key points in different directions and scales of the image and it remains invariance to rotation, brightness, and scaling changes [80]. SURF uses the Hessian Blob Detector (HBD) to identify interest points on an image [81, 82]. HBD is based on the scale-space depiction of the Hessian matrix, computed in box filters, so that Hessian matrix elements can be properly measured using integral images at really low computational expense [72, 83]. The SURF descriptor was used by P22, P23, P24, P27 and P29 for MAD.

Canny and Sobel Edge Detection

Edge Detection is an operation finding boundaries that limit two homogeneous image regions that have different brightness levels [84]. The aim of edge detection algorithms is to generate a line drawing of the loaded image. The extracted characteristics could be used to recognize and track objects. The Sobel operator is a discrete differential operator that uses two kernels measuring 3×3 pixels to calculate the gradient [85]. One kernel evaluates the gradient in the x-direction, and the other one evaluates the gradient in the y-direction [86, 87]. The gradient is determined using the formula of Eq. 1:

$$G = \sqrt{S_x{}^2 + S_y{}^2} \tag{1}$$

where G: Sobel gradient operator value, S_x: Horizontal gradient and S_y: Vertical sobel gradient.

The Canny Edge Detector is commonly referred to as the optimal detector, developed by John F. Canny in 1986 [88]. The steps involved in canny operator are: firstly, to process the images, a Gaussian filter is introduced to eliminate noise in an image. Secondly the magnitude of the gradient is calculated. Thirdly, non-max suppression is implemented by the algorithm to omit pixels that are not part of an edge. Finally the thresholding of hysteresis is used across the edges [86]. The features from the sobel and canny edge detector were used by P24 and P29 for MAD.

Laplacian Pyramid

The Laplacian pyramid, is a band-pass image decomposition originating from Gaussian Pyramid which is a multi-scale image depiction produced by a recursive reduction of the image set [89]. Laplacian pyramid was used by P28 to remove details from the spatial information by decomposing color space pictures into various scales.

4.2.3   MAD Performance Metric

Five performance metric were adopted by the primary papers in evaluation of face morphing attack detection systems. This five performance measure are as follows.

Bona Fide Presentation Classification Error Rate (BPCER) OR False Rejection Rate (FRR)

This is to be described as the percentage of genuine presentations wrongly classified as presentation attacks in a particular scenario or as the relative quantity of genuine images categorized as morphing attacks [90]. BPCER can also be characterized as the expected percentage of transactions incorrectly rejected with truthful claims of identity (in a positive identity system) [91]. P29, P13, P23, P2, P14, P5, P11, P3, P20, P9, P28, P12, P1, P27, P16, P8, P31, P25, P7, P33, P6, P19, P22, P30, P24, and P18 made use of this performance metric for MAD performance evaluation.

Attack Presentation Classification Error Rate (APCER) OR False Acceptance Rate (FAR)

This is described as the percentage of attacks that use the same presentation attack device species incorrectly classified as true (bone fide) presentations in a particular scenario or it can be described as a relative number of morphing attacks classified as true images [90, 92]. P29, P13, P23, P2, P14, P5, P11, P3, P20, P9, P28, P12, P1, P27, P16, P8, P31, P25, P7, P33, P6, P19, P22, P30, P24, and P18 made use of this performance metric for MAD performance evaluation.

Detection-Equal Error Rate (D-EER)

D-EER is an algorithm used to describe the BPCER Threshold values and it's APCER. The common value obtained when the rates are same/equal is called the equal error [90]. The common value indicate that the APCER percentage is the same as the BPCER percentage. This is the position at which BPCER = APCER. It is used as the optimal point during training. The lesser the D-EER, the greater the biometric system's precision. On the basis of the assessed decision threshold (θ), (APCER (θ)

+ BPCER (θ))/2) is used as the detection error. This performance metric was used by P13, P2, P3, P10, P9, P12, P1, P8, P22, P23, P25, P27, P28, P30, P33 and P7.

Accuracy (ACC)

This is described as the percentage of correctly categorized images in relative to all categorized images [93–95]. Accuracy was used by P11 and P26 as a performance measure. The formula for calculating ACC is presented in Eq. 2

$$ACC = \text{correctly classified images/all classified images} \qquad (2)$$

True Positive Rate (TPR)

TPR also called Sensitivity or Recall estimates the percentage of actual positive categorized as such (for example, the amount of morphed pictures recognized as an attack [93, 94, 96]. This can be calculated using the formula in Eq. 3:

$$TPR = \text{TruePostiive/(TruePositive + FalseNegative)} \qquad (3)$$

5 Parametric Discussion

This section presents a tabular discussion of parameters used in Morphing Attack Detection (MAD). The parameters used in MAD are discussed in Table 8.

6 Taxonomy of MAD Techniques

Based on the SLR performed the techniques used for MAD can be grouped into 6 broad taxonomies founded on the detected and extracted image features. This taxonomies is as shown in Fig. 3:

1. **Texture Descriptor**: Texture is an attribute used to separate images into regions of interest and to categorize those regions. Texture includes information regarding the spatial configuration of colors or intensities in an image or selected region of image. It is anticipated that the image morphing process will lead to a change in the textual properties of morphed images which will make it a useful function for differentiating between morph and bona fide images. LBP, LPQ, BSIF and RLBF are the descriptors which fall into this category.

Table 8 Parameters used in MAD

S/No	Parameters	Discussion
1	Training dataset	This are morphed and bona fide images used to train a MAD algorithm. The better the training dataset the better the MAD algorithm. It is mostly 70% of the overall dataset
2	Testing dataset	This are morphed and bona fide images used to test the efficiency of a MAD algorithm after been trained with the training dataset. With the testing dataset an algorithm accuracy can be tested. it is mostly 30% of the overall dataset
3	Landmark-detection	One important parameter used for MAD is landmark detection. This is preprocessing stage used to detect and normalize morphed and bona fide images according to important face features such as the mouth, eyes, and nose. With landmark detection the facial image can be cropped to focus on just the facial features for better MAD
4	Feature extraction	It is a sort of dimension reduction that effectively represents a compact characteristic vector for interesting sections of the images. Features extracted are used to determine whether an image is morphed and bona fide. Example of feature extractors are local binary pattern, steerable pyramid etc.
5	Classification	This is about determining which of a set of groups to which the individual testing data set belongs, based on the training data set whose membership in the category is identified. In MAD there are two category of classification which are bone fide image or morphed image
6	Scenario	Deals with approaches used in MAD. And there are only two scenario which are reference (differential) based scenario and no-reference (single-image) based scenario
7	Post-processing	Deals with parameters that can alter the natural characteristics of a morphed imaged to prevent attack detection. Example of this parameters are image sharpening, print-scan operation and image compression

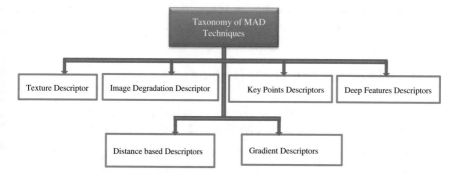

Fig. 3 Taxonomy of feature extraction techniques in MAD

2. **Image Degradation Descriptor**: The descriptors in this category takes advantage of degradations present in images. Image morphing leads to several image degradation due to the artefacts created by morphing process, hence making these degradations important features for MAD. The descriptors in this group includes: PRNU, Laplacian Pyramid, Benford features, DFT and Steerable Pyramids.

3. **Key Points Descriptors**: These descriptors does not just deal with merely 2D locations on the image but with 3D locations on the image scale space. This locations are the x, y and scale coordinates. Key point descriptors are used for MAD, as morphed images are supposed to comprise of fewer key point locations that are described as the maximum and minimal result of Gaussian function difference. Hence the quantity of extracted key points can be used as a useful feature for MAD. The descriptors in this category include: SURF, SIFT, FAST, AGAST, ORB and Shi Tomasi.

4. **Deep Features Descriptors**: A deep attribute is the coherent layer response within a hierarchical structure to an input that gives an answer relative to the final output of the model. Recent researches on face recognition has shown that the use of deep features for object recognition and classification have achieved good performance and easy adaptability. This advantages makes deep feature descriptors suitable for MAD. The descriptors that fall into this category are: VGG, AlexNet, OpenFace and FaceNet,

5. **Distance based Descriptors**: This deals with detecting the landmarks on both the bona fide and morphed image. And the distance of the relative position of the landmark between the bona-fide and morphed images is computed, resulting in a feature vector. The calculated feature vectors are referred to as the distance features. Hence this Distance based technique are used for differential image-based scenario.

6. **Gradient Descriptors**: Image gradient is a change of direction in the color or intensity of the image. These descriptors are used because the morphing process reduces the changes in high frequency of the image and thus decreases the gradient steepness which enhances MAD. The descriptors that fall into this category are: HOG, Canny and Sobel edge detectors.

7 Open Issues and Future Directions

Just like every other field MAD as a research field is not left without existing issue and challenges. The most significant issues and challenges are described follows:

1. *Lack of robust publicly available database*: it is was found that in most research work for MAD, researchers had to create/generate morphed images using morphing software as there are no extensive publicly accessible databases of morphed and bona fide database and some of the initially available databases does not exist anymore. Researches have been conducted on different in-house databases. This prevents creation of useful and robust comparative benchmarks

for existing MAD algorithms. Thus this can be a great limitation to researchers in creating a standardized and reliable MAD.

2. ***Lack of publicly available MAD algorithm***: another issue faced in MAD is lack of publicly available MAD algorithm which can be used by researchers for comprehensive experimental evaluation of new and existing MAD algorithms. This situation brings about questions such as how reliable are the current state-of-the-art MAD algorithms.

3. ***Diversity of experimental database***: most research works train and test proposed MAD algorithms on a single database generated by a single morphing software. But in reality image morphing is carried out with various morphing software which can give different effect and characteristics. This has made current MAD algorithms not to be robust or effective in detecting morph images created with various morph software.

4. ***Image post-processing***: it has been identified that post-processing of morphed images can alter it features. With this alteration it is become impossible for the current MAD algorithms to detect morphed images successfully. Image post-processing task such as image sharpening and image compression has been ignored in most works. Hence it is important to consider different or possible image post-processing task that can be performed on a morphed image in order to improve performance of MAD algorithms.

In summary a SLR which is a formal way of synthesizing the information existing from existing primary studies significant to the research questions on MAD [92]. From this SLR review issues and challenges in MAD was identified.

8 Conclusion

This systematic literature review (SLR) provides researchers and industry practitioners with a current synthesis of feature extraction techniques in face morphing attack detection, approaches of MAD and the performance metric to assess the performance of the MAD systems. This research revealed that texture descriptors, key point extractors, Gradient descriptors, image degradation descriptors, deep learning based methods and Distance-based descriptors can be used as feature descriptors in MAD.

This study illustrates that MAD is an active research area especially differential image based approach of MAD. The differential image-based approach has be adopted by only two literatures which got low detection accuracy thus making the system not suited for operational deployment. Hence it is recommended that more research should be done on differential image-based approach in order to enhance performance. Also this SLR is useful to the scholarly community in understanding of researches regarding to face morphing attack detection and to gain insight of the gaps that remain in the literature.

Appendix A

Primary Study in Review

See Table 9.

Table 9 Primary studies in review

#	Authors	Topic	Approaches To MAD	Feature extraction techniques	Performance metric
P1	Ferrara [38]	"Face morphing detection in the presence of printing/scanning and heterogeneous image sources"	Print-scan attack	CNN (VGG, AlexNet)	D-EER, BPCER and APCER
P2	Scherhag [97]	"Detection of morphed faces from single images: a multi-algorithm fusion approach"	Digital attack	LBF, BSIF, SIFT, HOG and CNN	D-EER, BPCER and APCER
P3	Venkatesh [98]	"Detecting morphed face attacks using residual noise from deep multi-scale context aggregation network"	Digital attack	CNN (AlexNet)	D-EER, BPCER and APCER
P4	Scherhag [99]	"Detecting morphed face images using facial landmarks"	Differential image-based	Distance based	D-EER, BPCER and APCER
P5	Spreeuwers [100]	"Towards robust evaluation of face morphing detection"	Digital attack	LBP	D-EER, BPCER and APCER
P6	Damer [8]	"Detecting face morphing attacks by analyzing the directed distances of facial landmarks shifts"	Differential image-based	Regressing local binary features (LBF)	D-EER, BPCER and APCER

(continued)

Table 9 (continued)

#	Authors	Topic	Approaches To MAD	Feature extraction techniques	Performance metric
P7	Damer [101]	"A multi-detector solution towards an accurate and generalized detection of face morphing attacks"	Digital attack	CNN (Openface)	BPCER and APCER
P8	Damer [102]	"On the generalization of detecting face morphing attacks as anomalies: novelty versus outlier detection"	Digital attack	Local Phase Quantization (LPQ) and CNN	D-EER, BPCER and APCER
P9	Debiasi [51]	"PRNU-based detection of morphed face images"	Digital attack	PRNU	D-EER, BPCER and APCER
P10	Scherhag [103]	"Detection of face morphing attacks based on prnu analysis"	Digital attack & print-scan attack	PRNU	D-EER
P11	Wandzik [23]	"Morphing detection using a general-purpose face recognition system"	Digital attack	CNN (AlexNet)	ACC, FAR and FRR
P12	Singh [22]	"Robust morph-detection at automated border control gate using deep decomposed 3D shape & diffuse reflectance"	Print-scan attack	CNN (pre-trained AlexNet)	D-EER, BPCER and APCER
P13	Venkatesh [104]	"Morphed face detection based on deep color residual"	Digital attack	Pyramid local binary pattern (P-LBP)	D-EER, BPCER and APCER
P14	Ramachandra [4]	"Detecting face morphing attacks with collaborative representation of steerable features"	Print-scan attack	Steerable pyramid	BPCER and APCER

(continued)

Table 9 (continued)

#	Authors	Topic	Approaches To MAD	Feature extraction techniques	Performance metric
P15	Jassim [1]	"Automatic detection of image morphing by topology-based analysis"	Print-scan attack	LBP	ACC
P16	Seibold [21]	"Detection of face morphing attacks by deep learning"	Digital attack	CNN (Pre-trained AlexNet)	FAR and FRR
P17	Scherhag [105]	"Deep face representations for differential morphing attack detection"	Print-scan	CNN (FaceNet)	D-EER, BPCER and APCER
P18	Raghavendra [20]	"Detecting Morphed Face Images"	Digital attack	BSIF	FAR and FRR
P19	Makrushin [62]	"Automatic generation and detection of visually faultless facial morphs"	Digital attack	Benford features	FAR and FRR
P20	Damer [106]	"To detect or not to detect: the right faces to morph"	Digital attack & differential image-based	LBP histogram & CNN	BPCER and APCER
P21	Damer [107]	"MorGAN: recognition vulnerability and attack detectability of face morphing attacks created by generative adversarial network"	digital attack	LBP histogram & CNN	BPCER and APCER
P22	Scherhag [18]	"Towards detection of morphed face images in electronic travel documents"	Digital attack & differential image-based	LBP, BSIF, SIFT, SURF, HOG, Deep neural network (OpenFace)	D-EER, BPCER and APCER
P23	Scherhag [25]	"Performance variation of morphed face image detection algorithms across different datasets"	Digital attack	LBP, BSIF, SIFT, SURF, HOG	D-EER, BPCER and APCER

(continued)

Table 9 (continued)

#	Authors	Topic	Approaches To MAD	Feature extraction techniques	Performance metric
P24	Kraetzer [108]	"Modeling attacks on photo-ID documents and applying media forensics for the detection of facial morphing"	Digital attack	SIFT, SURF, ORB, FAST, AGAST, sobel & canny edge detector	FAR and FRR
P25	Ortega-Delcampo [48]	"Border control morphing attack detection with a convolutional neural network de-morphing approach"	Digital attack & print-scan attack	CNN (Autoencoder)	D-EER, BPCER and APCER
P26	Neubert [109]	"Face morphing detection: an approach based on image degradation analysis"	Digital attack	FAST, AGAST, shiTomasi	ACC
P27	Scherhag [97]	"Morph detection from single face image: a multi-algorithm fusion approach"	Digital attack	LBP, BSIF, SIFT, SURF, HOG, deep neural network (OpenFace)	D-EER, BPCER and APCER
P28	Ramachandra [89]	"Towards making morphing attack detection robust using hybrid scale-space colour texture features"	Print-scan attack	Laplacian pyramid & LBP	D-EER, BPCER and APCER
P29	Neubert [79]	"A face morphing detection concept with a frequency and a spatial domain feature space for images on eMRTD"	Digital attack	ORB, Discrete fourier transformation (DFT), SURF, SIFT, AGAST, Sobel & Canny, FAST,	FAR and FRR
P30	Debiasi [52]	"PRNU variance analysis for morphed face image detection"	Digital attack	PRNU	D-EER, BPCER and APCER
P31	Zhang [78]	"Face morphing detection using fourier spectrum of sensor pattern noise"	Digital attack	Discrete fourier transformation (DFT)	BPCER and APCER

(continued)

Table 9 (continued)

#	Authors	Topic	Approaches To MAD	Feature extraction techniques	Performance metric
P32	Makrushin [63]	"Generalized Benford's Law for blind detection of morphed face images"	Digital attack	Benford features	FPR and TPR
P33	Raghavendra [110]	"Transferable deep-CNN features for detecting digital and print-scanned morphed face images"	Digital & print-scan attack	CNN (AlexNet and VGG19)	D-EER, BPCER and APCER

References

1. Jassim S, Asaad A (2018) Automatic detection of image morphing by topology-based analysis. In: 2018 26th European signal processing conference (EUSIPCO), Rome, pp 1007–1011. https://doi.org/10.23919/EUSIPCO.2018.8553317

2. Wandzik L, Garcia RV, Kaeding G, Chen X (2017) CNNs under attack: on the vulnerability of deep neural networks based face recognition to image morphing. In: Kraetzer C, Shi Y-Q, Dittmann J, Kim HJ (eds) Digital forensics and watermarking, vol 10431. Springer International Publishing, Cham, pp 121–135. https://doi.org/10.1007/978-3-319-64185-0_10

3. Olanrewaju L, Oyebiyi O, Misra S, Maskeliunas R, Damasevicius R (2020) Secure ear biometrics using circular kernel principal component analysis, Chebyshev transform hashing and Bose–Chaudhuri–Hocquenghem error-correcting codes. Signal Image Video Process 14(5):847–855. https://doi.org/10.1007/s11760-019-01609-y

4. Ramachandra R, Venkatesh S, Raja K, Busch C (2020) Detecting face morphing attacks with collaborative representation of steerable features. In: Chaudhuri BB, Nakagawa M, Khanna P, Kumar S (eds) Proceedings of 3rd international conference on computer vision and image processing, vol 1022. Springer, Singapore, pp 255–265. https://doi.org/10.1007/978-981-32-9088-4_22

5. Wu J (2011) Face recognition jammer using image morphing. Boston University, Saint Mary's Street, Boston, ECE-2011-03

6. Rathgeb C, Dantcheva A, Busch C (2019) Impact and detection of facial beautification in face recognition: an overview. IEEE Access 7:152667–152678. https://doi.org/10.1109/ACCESS.2019.2948526

7. Kenneth OM, Bashir SA, Abisoye OA, Mohammed AD (2021) Face morphing attack detection in the presence of post-processed image sources using neighborhood component analysis and decision tree classifier. In: Misra S, Muhammad-Bello B (eds) Information and communication technology and applications, vol 1350. Springer International Publishing, Cham, pp 340–354. https://doi.org/10.1007/978-3-030-69143-1_27

8. Damer N et al (2019) Detecting face morphing attacks by analyzing the directed distances of facial landmarks shifts. In: Brox T, Bruhn A, Fritz M (eds) Pattern recognition, vol 11269. Springer International Publishing, Cham, pp 518–534. https://doi.org/10.1007/978-3-030-12939-2_36

9. Tolosana R, Gomez-Barrero M, Busch C, Ortega-Garcia J (2020) Biometric presentation attack detection: beyond the visible spectrum. IEEE Trans Inf Forensics Secur 15:1261–1275. https://doi.org/10.1109/TIFS.2019.2934867

10. Ferrara M, Franco A, Maltoni D (2016) On the effects of image alterations on face recognition accuracy. In: Bourlai T (ed) Face recognition across the imaging spectrum. Springer International Publishing, Cham, pp 195–222. https://doi.org/10.1007/978-3-319-28501-6_9

11. Mohammadi A, Bhattacharjee S, Marcel S (2018) Deeply vulnerable: a study of the robustness of face recognition to presentation attacks. IET Biom 7(1):15–26. https://doi.org/10.1049/iet-bmt.2017.0079

12. Jayashalini R, Priyadharshini S (2017) Face anti-spoofing using robust features and fisher vector encoding based innovative real time security system for automobile applications. Inf Commun Technol 11

13. Seibold C, Hilsmann A, Eisert P (2018) Reflection analysis for face morphing attack detection. In: 2018 26th European signal processing conference (EUSIPCO), Rome, pp 1022–1026. https://doi.org/10.23919/EUSIPCO.2018.8553116

14. Zanella V, Fuentes O (2004) An approach to automatic morphing of face images in frontal view. In: Monroy R, Arroyo-Figueroa G, Sucar LE, Sossa H (eds) MICAI 2004: advances in artificial intelligence, vol 2972. Springer, Berlin, Heidelberg, pp 679–687. https://doi.org/10.1007/978-3-540-24694-7_70

15. Robertson DJ, Kramer RSS, Burton AM (2017) Fraudulent ID using face morphs: experiments on human and automatic recognition. PLOS ONE 12(3):e0173319. https://doi.org/10.1371/journal.pone.0173319

16. Seibold C, Samek W, Hilsmann A, Eisert P (2021) Accurate and robust neural networks for security related applications exampled by face morphing attacks

17. Robertson DJ, Mungall A, Watson DG, Wade KA, Nightingale SJ, Butler S (2018) Detecting morphed passport photos: a training and individual differences approach. Cogn Res Princ Implic 3(1):27. https://doi.org/10.1186/s41235-018-0113-8

18. Scherhag U, Rathgeb C, Busch C (2018) Towards detection of morphed face images in electronic travel documents. In: 2018 13th IAPR international workshop on document analysis systems (DAS), Vienna, pp 187–192. https://doi.org/10.1109/DAS.2018.11

19. Ferrara M, Franco A, Maltoni D (2014) The magic passport. In: IEEE international joint conference on biometrics, Clearwater, FL, USA, pp 1–7. https://doi.org/10.1109/BTAS.2014.6996240

20. Raghavendra R, Raja KB, Busch C (2016) Detecting morphed face images. In: 2016 IEEE 8th international conference on biometrics theory, applications and systems (BTAS), Niagara Falls, NY, USA, pp 1–7. https://doi.org/10.1109/BTAS.2016.7791169

21. Seibold C, Samek W, Hilsmann A, Eisert P (2017) Detection of face morphing attacks by deep learning. In: Kraetzer C, Shi Y-Q, Dittmann J, Kim HJ (eds) Digital forensics and watermarking, vol 10431. Springer International Publishing, Cham, pp 107–120. https://doi.org/10.1007/978-3-319-64185-0_9

22. Singh JM, Ramachandra R, Raja KB, Busch C (2021) Robust morph-detection at automated border control gate using deep decomposed 3D shape and diffuse reflectance. http://arxiv.org/abs/1912.01372

23. Wandzik L, Kaeding G, Garcia RV (2018) Morphing detection using a general- purpose face recognition system. In: 2018 26th European signal processing conference (EUSIPCO), Rome, pp 1012–1016. https://doi.org/10.23919/EUSIPCO.2018.8553375

24. Korshunov P, Marcel S (2018) Vulnerability of face recognition to deep morphing. In: International conference on biomedical, p 5

25. Scherhag U, Rathgeb C, Busch C (2018) Performance variation of morphed face image detection algorithms across different datasets. In: 2018 international workshop on biometrics and forensics (IWBF), Sassari, pp 1–6. https://doi.org/10.1109/IWBF.2018.8401562

26. Kramer RSS, Mireku MO, Flack TR, Ritchie KL (2019) Face morphing attacks: investigating detection with humans and computers. Cogn Res Princ Implic 4(1):28. https://doi.org/10.1186/s41235-019-0181-4

27. Makrushin A, Wolf A (2018) An overview of recent advances in assessing and mitigating the face morphing attack. In: 2018 26th European signal processing conference (EUSIPCO), Rome, pp 1017–1021. https://doi.org/10.23919/EUSIPCO.2018.8553599

28. Scherhag U, Rathgeb C, Merkle J, Breithaupt R, Busch C (2019) Face recognition systems under morphing attacks: a survey. IEEE Access 7:23012–23026. https://doi.org/10.1109/ACCESS.2019.2899367

29. Swartz MK (2011) The PRISMA statement: a guideline for systematic reviews and meta-analyses. J Pediatr Health Care 25(1):1–2. https://doi.org/10.1016/j.pedhc.2010.09.006

30. Torres-Carrion PV, Gonzalez-Gonzalez CS, Aciar S, Rodriguez-Morales G (2018) Methodology for systematic literature review applied to engineering and education. In: 2018 IEEE global engineering education conference (EDUCON), Tenerife, pp 1364–1373. https://doi.org/10.1109/EDUCON.2018.8363388

31. Misra S (2021) A step by step guide for choosing project topics and writing research papers in ICT related disciplines. In: Misra S, Muhammad-Bello B (eds) Information and communication technology and applications, vol 1350. Springer International Publishing, Cham, pp 727–744. https://doi.org/10.1007/978-3-030-69143-1_55

32. Hordri NF, Yuhaniz SS, Shamsuddin SM (2017) A systematic literature review on features of deep learning in big data analytics. Int J Adv Soft Comput Appl 9(1):33–49

33. Dang DD, Pekkola S (2017) Systematic literature review on enterprise architecture in the public sector, vol 15, no 2, p 25

34. Yannascoli SM, Schenker ML, Carey JL, Ahn J, Baldwin KD (2013) How to write a systematic review: a step-by-step guide, vol 23, p 6

35. Okoli C (2015) A guide to conducting a standalone systematic literature review. Commun Assoc Inf Syst 37. https://doi.org/10.17705/1CAIS.03743

36. Lin EOW, Pan F, Moscheni F (2003) A no-reference quality metric for measuring image Blur. In: Seventh international symposium on signal processing and its applications, vol 1. IEEE, p 4

37. Wang Z, Sheikh HR, Bovik AC (2002) No-reference perceptual quality assessment of JPEG compressed images. In: Proceedings. International conference on image processing, rochester, vol 1, NY, USA, pp I-477–I-480. https://doi.org/10.1109/ICIP.2002.1038064

38. Ferrara M, Franco A, Maltoni D (2019) Face morphing detection in the presence of printing/scanning and heterogeneous image sources. IET Biom 10(3):290–303. https://doi.org/10.1049/bme2.12021

39. Witlox K (2019) Face unmorphing. In: 31th Twenty student conference on IT, Netherlands, pp 1–7

40. Simoncelli EP, Freeman WT (1995) The steerable pyramid: a flexible architecture for multi-scale derivative computation. In: Proceedings, international conference on image processing, vol 3, Washington, DC, USA, pp 444–447. https://doi.org/10.1109/ICIP.1995.537667

41. Ehsaeyan E (2016) An improvement of steerable pyramid denoising method. Electron Eng 12(1):7

42. Kannala J, Rahtu E (2012) BSIF: binarized statistical image features. In: 21st international conference on pattern recognition, vol 1, Tsukuba, Japan, pp 1363–1366

43. Huang D, Shan C, Ardabilian M, Wang Y, Chen L (2011) Local binary patterns and its application to facial image analysis: a survey. IEEE Trans Syst Man Cybern Part C Appl Rev 41(6):765–781. https://doi.org/10.1109/TSMCC.2011.2118750

44. Song K-C, Yan Y-H, Chen W-H, Zhang X (2013) Research and perspective on local binary pattern. Acta Autom Sin 39(6):730–744. https://doi.org/10.1016/S1874-1029(13)60051-8

45. Liu Z, Luo P, Wang X, Tang X (2015) Deep learning face attributes in the wild. In: 2015 IEEE international conference on computer vision (ICCV), Santiago, Chile, pp 3730–3738. https://doi.org/10.1109/ICCV.2015.425

46. Winiarti S, Prahara AM, Pramudi D (2018) Pre-trained convolutional neural network for classification of tanning leather image. Int J Adv Comput Sci Appl 9(1). https://doi.org/10.14569/IJACSA.2018.090129

47. Korshunova I, Shi W, Dambre J, Theis L (2017) Fast face-swap using convolutional neural networks. http://arxiv.org/abs/1611.09577
48. Ortega-Delcampo D, Conde C, Palacios-Alonso D, Cabello, E (2020) Border control morphing attack detection with a convolutional neural network de-morphing approach. IEEE Access 1–1. https://doi.org/10.1109/ACCESS.2020.2994112
49. Schroff F, Kalenichenko D, Philbin J (2015) FaceNet: a unified embedding for face recognition and clustering. In: 2015 IEEE conference on computer vision and pattern recognition (CVPR), pp 815–823. https://doi.org/10.1109/CVPR.2015.7298682
50. Surasak T, Takahiro I, Cheng C, Wang C, Sheng P (2018) Histogram of oriented gradients for human detection in video. In: 2018 5th international conference on business and industrial research (ICBIR), Bangkok, pp 172–176. https://doi.org/10.1109/ICBIR.2018.8391187
51. Debiasi L, Scherhag U, Rathgeb C, Uhl A, Busch C (2018) PRNU-based detection of morphed face images. In: 2018 international workshop on biometrics and forensics (IWBF), Sassari, pp 1–7. https://doi.org/10.1109/IWBF.2018.8401555
52. Debiasi L, Rathgeb C, Scherhag U, Uhl A, Busch C (2018) PRNU variance analysis for morphed face image detection. In: 2018 IEEE 9th international conference on biometrics theory, applications and systems (BTAS), Redondo Beach, CA, USA, pp 1–9. https://doi.org/10.1109/BTAS.2018.8698576
53. Bonettini N et al (2018) Fooling PRNU-based detectors through convolutional neural networks. In: 2018 26th European signal processing conference (EUSIPCO), Rome, pp 957–961. https://doi.org/10.23919/EUSIPCO.2018.8553596
54. Chierchia G, Parrilli S, Poggi G, Verdoliva L, Sansone C (2011) PRNU-based detection of small-size image forgeries. In: 2011 17th international conference on digital signal processing (DSP), Corfu, Greece, pp 1–6. https://doi.org/10.1109/ICDSP.2011.6004957
55. Chierchia G, Cozzolino D, Poggi G, Sansone C, Verdoliva L (2014) Guided filtering for PRNU-based localization of small-size image forgeries. In: 2014 IEEE international conference on acoustics, speech and signal processing (ICASSP), Florence, Italy, pp 6231–6235. https://doi.org/10.1109/ICASSP.2014.6854802
56. Verma SB, Sravanan C (2016) Analysis of SIFT and SURF feature extraction in palmprint verification system. In: IEEE international conference on computing, communication and control technology
57. Lowe DG (2004) Distinctive image features from scale-invariant keypoints. Int J Comput Vis 60(2):91–110. https://doi.org/10.1023/B:VISI.0000029664.99615.94
58. Lowe DG (1999) Object recognition from local scale-invariant features. In: Proceedings of the seventh IEEE international conference on computer vision, vol 2, Kerkyra, Greece, pp 1150–1157. https://doi.org/10.1109/ICCV.1999.790410
59. Ren S, Cao X, Wei Y, Sun J (2014) Face alignment at 3000 FPS via regressing local binary features. In: 2014 IEEE conference on computer vision and pattern recognition, Columbus, OH, USA, pp 1685–1692. https://doi.org/10.1109/CVPR.2014.218
60. King DE (2009) Dlib-ml: a machine learning toolkit. J Mach Learn Res 10:1755–1758
61. Ojansivu V, Heikkilä J (2008) Blur insensitive texture classification using local phase quantization. In: Elmoataz A, Lezoray O, Nouboud F, Mammass D (eds) Image and signal processing, vol 5099. Springer, Berlin, Heidelberg, pp 236–243. https://doi.org/10.1007/978-3-540-69905-7_27
62. Makrushin A, Neubert T, Dittmann J (2017) Automatic generation and detection of visually faultless facial morphs. In: Proceedings of the 12th international joint conference on computer vision, imaging and computer graphics theory and applications, Porto, Portugal, pp 39–50. https://doi.org/10.5220/0006131100390050
63. Makrushin A, Kraetzer C, Neubert T, Dittmann J (2018) Generalized Benford's Law for blind detection of morphed face images. In: Proceedings of the 6th ACM workshop on information hiding and multimedia security, Innsbruck Austria, pp 49–54. https://doi.org/10.1145/3206004.3206018
64. Fu D, Shi YQ, Su W (2007) A generalized Benford's law for JPEG coefficients and its applications in image forensics. San Jose, CA, United States, p 65051L. https://doi.org/10.1117/12.704723

65. Zhang H, Wohlfeil J, Grießbach D (2016) Extension and evaluation of the AGAST feature detector. In: ISPRS annals of the photogrammetry, remote sensing and spatial information sciences, vol III–4, pp 133–137. https://doi.org/10.5194/isprsannals-III-4-133-2016
66. Kulkarni AV, Jagtap JS, Harpale VK (2013) Object recognition with ORB and its implementation on FPGA. Int J Adv Comput Res 3(3):6
67. Karami E., Prasad S, Shehata M (2015) Image matching using SIFT, SURF, BRIEF and ORB: performance comparison for distorted images. In: 2015 newfoundland electrical and computer engineering, Canada, p 5
68. Hutchison D et al (2010) Adaptive and generic corner detection based on the accelerated segment test. In: Daniilidis K, Maragos P, Paragios N (eds) Computer vision—ECCV 2010, vol 6312. Springer, Berlin, Heidelberg, pp 183–196. https://doi.org/10.1007/978-3-642-15552-9_14
69. Biadgie Y, Sohn K-A (2014) Feature detector using adaptive accelerated segment test. In: 2014 international conference on information science & applications (ICISA), Seoul, South Korea, pp 1–4. https://doi.org/10.1109/ICISA.2014.6847403
70. Cooke T, Whatmough R (2005) Detection and tracking of corner points for structure from motion. In: Defence science and technology organisation, Australia, Technical DSTO-TR-1759
71. Juranek L, Stastny J, Skorpil V (2018) Effect of low-pass filters as a shi-tomasi corner detector's window functions. In: 2018 41st international conference on telecommunications and signal processing (TSP), Athens, pp 1–5. https://doi.org/10.1109/TSP.2018.8441178
72. Urban S, Weinmann M (2015) Finding a good feature detector-descriptor combination for the 2d keypoint-based registration of TIS point clouds. In: ISPRS annals of the photogrammetry, remote sensing and spatial information sciences, vol. II-3/W5, pp 121–128. https://doi.org/10.5194/isprsannals-II-3-W5-121-2015
73. Rublee E, Rabaud V, Konolige K, Bradski G (2011) ORB: an efficient alternative to SIFT or SURF. In: 2011 international conference on computer vision, Barcelona, Spain, pp 2564–2571. https://doi.org/10.1109/ICCV.2011.6126544
74. Tchagang AB, Valdes JJ (2019) Discrete fourier transform improves the prediction of the electronic properties of molecules in quantum machine learning. In: 2019 IEEE Canadian conference of electrical and computer engineering (CCECE), Edmonton, AB, Canada, pp 1–4. https://doi.org/10.1109/CCECE.2019.8861895
75. Mironovova M, Bíla J (2015) Fast fourier transform for feature extraction and neural network for classification of electrocardiogram signals. In: 2015 fourth international conference on future generation communication technology (FGCT), Luton, United Kingdom, pp 1–6. https://doi.org/10.1109/FGCT.2015.7300244
76. Luka J, Fridrich J, Goljan M (2006) Digital camera identification from sensor pattern noise. IEEE Trans Inf Forensics Secur 1(2):205–214. https://doi.org/10.1109/TIFS.2006.873602
77. Liu B, Wei X, Yan J (2015) Enhancing sensor pattern noise for source camera identification: an empirical evaluation. In: Proceedings of the 3rd ACM workshop on information hiding and multimedia security, Portland Oregon USA, pp 85–90. https://doi.org/10.1145/2756601.2756614
78. Zhang L-B, Peng F, Long M (2018) Face morphing detection using fourier spectrum of sensor pattern noise. In: 2018 IEEE international conference on multimedia and expo (ICME), San Diego, CA, pp 1–6. https://doi.org/10.1109/ICME.2018.8486607
79. Neubert T, Kraetzer C, Dittmann J (2019) A face morphing detection concept with a frequency and a spatial domain feature space for images on eMRTD. In: Proceedings of the ACM workshop on information hiding and multimedia security, Paris, France, pp 95–100. https://doi.org/10.1145/3335203.3335721
80. Zhu Z, Zhang G, Li H (2018) SURF feature extraction algorithm based on visual saliency improvement, vol 5, no 3, p 5
81. Anjana MV, Sandhya L (2017) Implementation and comparison of feature detection methods in image mosaicing. IOSR J Electron Commun Eng 2(3):7–11

82. Bhosale SB, Kayastha VS, Harpale (2014) Feature extraction using surf algorithm for object recognition. Int J Tech Res Appl 2(4):3
83. Oyallon E, Rabin J (2015) An analysis of the SURF method. Image Process Line 5:176–218. https://doi.org/10.5201/ipol.2015.69
84. Asmaidi A, Putra DS, Risky MM, FUR (2019) Implementation of sobel method based edge detection for flower image segmentation. SinkrOn 3(2):161. https://doi.org/10.33395/sinkron. v3i2.10050
85. Gao W, Zhang X, Yang L, Liu H (2010) An improved Sobel edge detection. In: 2010 3rd international conference on computer science and information technology, Chengdu, China, pp 67–71. https://doi.org/10.1109/ICCSIT.2010.5563693
86. Sumeyya I, Fatma SH, Merve T, Suhap S (2017) The enhancement of canny edge detection algorithm using Prewitt, Robert, Sobel Kernels. In: Conference: international conference on engineering technologies, Turkey
87. Vincent O, Folorunso O (2009) A descriptive algorithm for Sobel image edge detection. In: SITE 2009: informing science + IT education conference. https://doi.org/10.28945/3351
88. Canny J (1986) A computational approach to edge detection. IEEE Trans Pattern Anal Mach Intell 8(6):679–698. https://doi.org/10.1109/TPAMI.1986.4767851
89. Ramachandra R, Venkatesh S, Raja K, Busch C (2019) Towards making morphing attack detection robust using hybrid scale-space colour texture features. In: 2019 IEEE 5th international conference on identity, security, and behavior analysis (ISBA), Hyderabad, India, pp 1–8. https://doi.org/10.1109/ISBA.2019.8778488
90. El-Abed M, Charrier C, Rosenberger C (2012) Evaluation of biometric systems. In: Yang J (ed) New trends and developments in biometrics, InTech. https://doi.org/10.5772/52084
91. Mansfield AJ (200) Best practices in testing and reporting performance of biometric devices. National Physical Laboratory, USA, Technical NPL Report CMSC
92. Vaidya AG, Dhawale AC, Misra A (2016) Comparative analysis of multimodal biometrics. Int J Pharm Technol 8(4):22969–22981
93. Sokolova M, Lapalme G (2009) A systematic analysis of performance measures for classification tasks. Inf Process Manag 45(4):427–437. https://doi.org/10.1016/j.ipm.2009. 03.002
94. Flach P (2019) Performance evaluation in machine learning: the good, the bad, the ugly, and the way forward. In: Proceedings of the AAAI conference on artificial intelligence, vol 33, pp 9808–9814. https://doi.org/10.1609/aaai.v33i01.33019808
95. Olaleye T, Arogundade O, Adenusi C, Misra S, Bello A (2021) Evaluation of image filtering parameters for plant biometrics improvement using machine learning. In: Patel KK, Garg D, Patel A, Lingras P (eds) Soft computing and its engineering applications, vol 1374. Springer, Singapore, pp 301–315. https://doi.org/10.1007/978-981-16-0708-0_25
96. Sharma D, Yadav UB, Sharma P (2009) The concept of sensitivity and specificity in relation to two types of errors and its application in medical research, vol 2, p 7
97. Scherhag U, Rathgeb C, Busch C (2018) Detection of morphed faces from single images: a multi-algorithm fusion approach, p 7
98. Venkatesh S, Ramachandra R, Raja K, Spreeuwers L, Veldhuis R, Busch C (2020) Detecting morphed face attacks using residual noise from deep multi-scale context aggregation network. In: 2020 IEEE winter conference on applications of computer vision, pp 269–278. https://doi. org/10.1109/WACV45572.2020.9093488
99. Scherhag U, Budhrani D, Gomez-Barrero M, Busch C (2018) Detecting morphed face images using facial landmarks. In: Mansouri A, El Moataz A, Nouboud F, Mammass D (eds) Image and signal processing, vol 10884. Springer International Publishing, Cham, pp 444–452. https://doi.org/10.1007/978-3-319-94211-7_48
100. Spreeuwers L, Schils M, Veldhuis R (2018) Towards robust evaluation of face morphing detection. In: 2018 26th European signal processing conference (EUSIPCO), Rome, pp 1027–1031. https://doi.org/10.23919/EUSIPCO.2018.8553018
101. Damer N, Zienert S, Wainakh Y, Saladie AM, Kirchbuchner F, Kuijper A (2019) A multi-detector solution towards an accurate and generalized detection of face morphing attacks. In: 2019 22th International conference on Information Fusion, pp 1–8

102. Damer N, Grebe JH, Zienert S, Kirchbuchner F, Kuijper A (2019) On the generalization of detecting face morphing attacks as anomalies: novelty versus outlier detection. In: 2019 IEEE 10th international conference on biometrics theory, applications and systems (BTAS), Tampa, FL, USA, pp 1–5. https://doi.org/10.1109/BTAS46853.2019.9185995
103. Scherhag U, Debiasi L, Rathgeb C, Busch C, Uhl A (2019) Detection of face morphing attacks based on PRNU analysis. IEEE Trans Biom Behav Identity Sci 1(4):302–317. https://doi.org/10.1109/TBIOM.2019.2942395
104. Venkatesh S, Ramachandra R, Raja K, Spreeuwers L, Veldhuis R, Busch C (2019) Morphed face detection based on deep color residual noise. In: 2019 ninth international conference on image processing theory, tools and applications (IPTA), Istanbul, Turkey, pp 1–6. https://doi.org/10.1109/IPTA.2019.8936088
105. Scherhag U, Rathgeb C, Merkle J, Busch C (2020) Deep face representations for differential morphing attack detection. http://arxiv.org/abs/2001.01202
106. Damer N et al (2019) To detect or not to detect: the right faces to morph. In: 2019 international conference on biometrics (ICB), Crete, Greece, pp 1–8. https://doi.org/10.1109/ICB45273.2019.8987316
107. Damer N, Saladie AM, Braun A, Kuijper A (2018) MorGAN: Recognition vulnerability and attack detectability of face morphing attacks created by generative adversarial network. In: 2018 IEEE 9th international conference on biometrics theory, applications and systems (BTAS), Redondo Beach, CA, USA, pp 1–10. https://doi.org/10.1109/BTAS.2018.8698563
108. Kraetzer C, Makrushin A, Neubert T, Hildebrandt M, Dittmann J (2017) Modeling attacks on photo-ID documents and applying media forensics for the detection of facial morphing. In: Proceedings of the 5th ACM workshop on information hiding and multimedia security, Philadelphia Pennsylvania USA, pp 21–32. https://doi.org/10.1145/3082031.3083244
109. Neubert T (2017) Face morphing detection: an approach based on image degradation analysis. In: Kraetzer C, Shi Y-Q, Dittmann J, Kim HJ (eds) Digital forensics and watermarking, vol 10431. Springer International Publishing, Cham, pp 93–106. https://doi.org/10.1007/978-3-319-64185-0_8
110. Raghavendra R, Raja KB, Venkatesh S, Busch C (2017) Transferable deep-CNN features for detecting digital and print-scanned morphed face images. In: 2017 IEEE conference on computer vision and pattern recognition workshops (CVPRW), Honolulu, HI, USA, pp 1822–1830. https://doi.org/10.1109/CVPRW.2017.228

Averaging Dimensionality Reduction and Feature Level Fusion for Post-Processed Morphed Face Image Attack Detection

Mary Ogbuka Kenneth and Bashir Adebayo Sulaimon

Abstract Facial morphing detection is critical when applying for a new passport and using the passport for identity verification due to the limited ability of face recognition algorithms and people to detect morphed photographs. As a result of face recognition systems' vulnerability to morphing attacks, the value of detecting fake passports at the ABC gate is undeniable. Nonetheless, identifying morphed images after they have been altered using image operations like sharpening, compression, blurring, print-scan and resizing is a significant concern in Morphing Attack Detection (MAD). These image operations can be used to conceal the morphing artefacts, which makes MAD difficult. Several researchers have carried out MAD for print-scan images; few researchers have done MAD for compressed images; however, just one paper has considered image sharpening operation. Hence, this paper proposes a MAD technique to perform MAD even after image sharpening operation using averaging dimensionality reduction and feature level fusion of Histogram of Oriented Gradient (HOG) 8×8 and 16×16 cell size. The 8×8 pixels cell size was used to capture small-scale spatial information from the images, while 16×16 pixels cell size was used to capture large-scale spatial details from the pictures. The proposed technique achieved a better accuracy of 95.71% compared with the previous work, which reached an accuracy of 85% when used for MAD on sharpened image sources. This result showed that the proposed technique is effective for MAD on sharpened post-processed images.

Keywords Face morphing attack · Bona-fide images · Sharpening · Morphed images · Machine learning

M. O. Kenneth (✉) · B. A. Sulaimon
Department of Computer Science, Federal University of Technology, Minna, Nigeria
e-mail: kenneth.pg918157@st.futminna.edu.ng

B. A. Sulaimon
e-mail: bashirsulaimon@futminna.edu.ng

1 Introduction

Face Recognition Systems (FRS) automatically recognize people based on their facial features [1]. FRS is built on data acquired over the last forty years from signals and patterns processing algorithms, resulting in accurate and trustworthy facial recognition algorithms. Biometrics allows people to be identified based on their physiological traits [2]. Face biometrics are now utilized in forensics, criminal identification in airports and train stations, surveillance, credit card authentication, and logical access control to electronic commerce and electronic government services, among other uses. In addition, biometric facial photographs are an essential component of electronic passports [3], which have now been used to create almost 800 million passport instances after ten years of development. As a result, face recognition using these passports has become popular in border checks [4]. Face recognition was chosen for border enforcement because, in the event of a false negative device judgment, the border enforcement officer would conduct a visual comparison, which is a clear benefit over all other multimodal biometric such as fingerprint identification [5]. Face recognition's usefulness in Automatic Border Control (ABC) e-gates is justified by these factors [1]. By matching the live collected face photographs with the face reference picture contained in the electronic Machine Readable Travel Document (eMRTD) passport, a conventional ABC system examines the connection between both the eMRTD and the passport holder (the individual who submits the eMRTD to the border agent). The value of ABC systems, which are based on highly efficient and precise border control operations, has increased as a result [1].

FRS, as a critical component of an ABC system, are vulnerable to a variety of attacks. These attacks can be classified into two kinds. The initial attack targets the ABC system itself is commonly accomplished by introducing a facial artefact into the capture unit. Face spoofing or presentation attacks [6, 7] are examples of this type of attack. On the other hand, these attacks necessitate a significant amount of effort in both creating a face artefact and submitting it to the ABC e-gate. Aside from that, this form of attack will only succeed if the adversary is able to have in possession a lost eMRTD passport and create a facial artefact that matches the eMRTD passport's face photograph [8]. The assault against the eMRTD biometric reference is the second type of attack: The biometric data recorded in the (stolen) passport's logical information structure is changed here to replace the reference image. Because most passport applications allow for a printed face picture as part of the application process, this assault is straightforward to carry out. In addition, for passport renewal and VISA applications, many nations will permit digital photo uploads to a web gateway. This provides intruders with numerous opportunities to submit a false face picture to the passport's issuing body and receive a legitimate eMRTD passport that includes both physical and digital security elements as well as the bogus photo [1]. Simple changes can be made using freely available software to attack the EMRTD biometric reference picture [9].

Among the different face picture adjustments, face morphing is recognized as the most severe attack on the ABC border protection mechanism [10]. Face morphing is a method of constructing a new face image by combining the precise details from two or more input face photographs belonging to different people. As a result, the morphed face image would eventually reflect the facial appearance elements of many data subjects, contributing to the morphed face [11]. As a result, any invader can morph their face into another data subject and seek an eMRTD passport that both subjects can use. This defective connection of multiple subjects with the document could result in illegal activities such as human trafficking, financial transactions, and illegal immigration [9].

Ferrara [3], Damer [3], Kramer [11] and Scherhag [11] recently proved that humans are unable to discern altered facial photos. Additionally, because eMRTD passports are widely utilized with ABC border control systems, this morphing attack may be carried out without fabricating a passport paper. As a result, to ensure the dependability of border control activities, these types of threats must be prevented.

In the previous years, there have been few authors who have worked on MAD. In 2014 Ferrara [3] introduced a face morphing attack which was called the magic passport. The viability of attacks on Automated Border Control (ABC) systems using morphed face images was examined. It was concluded that when the morphed passport is presented, the officer will recognize the photograph and release the document if the passport is not substantially dissimilar from the candidate's face. And thus, the released document passes all authenticity checks carried out at the gates.

Raghavendra [1] carried out novel research on how this face morphing attack can be detected. The study was conducted using facial micro-textures retrieved via statistically autonomous filters trained on natural photographs. This micro-texture dissimilarity was extracted using Binarised Statistical Image Features (BSIF), and classification had been made via Support Vector Machine (SVM). This was the first research done towards the MAD. Later in 2017, Seibold [12] aimed to perform MAD using a deep neural network. Three Convolutional Neural Network (CNN) architectures were trained from scratch and using already trained networks to initialize the weights. Pretrained networks were noticed to outperform the networks trained from scratch for each of the three architecture. Hence it has been concluded that the attributes acquired for classification tasks are also beneficial for MAD. In 2018 Wandzik [13] suggested a method for MAD based on a general-purpose FRS. This work combined a general-purpose FRS with a simple linear classifier to detect morph images successfully.

In 2019 Venkatesh [14] presented a novel approach for MAD focused on quantifying residual noise caused by the morphing phase Venkatesh [14] used an aggregation of several denoising methods estimated using a deep Multi-Scale Context Aggregation Network (MSCAN) to quantify the morphing noise. In 2020 Ortega-Delcampo [14] did not just stop at detecting face morphing attacks but went further to de-morph the morphed face images. Finally, in 2021 Kenneth [15] proposed a method for MAD in the presence of post-processed image sources.

The necessity of detecting false passports at the ABC gate, as a result, is undeniable. Nonetheless, recognizing altered images after they have been processed is a significant challenge in MAD. For example, after creating the morphing facial image, the image could be further treated with image compression, image sharpening, print-scan, or blurring operations to improve or diminish image quality purposely. This morph picture alteration technique could be used to hide morph artefacts. The automated production of morphed face pictures, in particular, can result in morphing artefacts. Shadow or phantom artefacts may be caused by missing or misplaced landmarks. The facial region can be substituted by an adapted outer space of one of the people to alleviate the issue of the morphed face image.

Weng [16] offers an interpolation of the hair region to hide artefacts in the hair region. On the other hand, interpolating the hair region can disguise morphing objects in the hair region [16]. However, unnatural colour gradients and edges may occur due to insufficient interpolation methods, which can be mitigated by sharpening or blurring. Furthermore, this morphing artefact can be made to obtain realistic histogram forms by adjusting the colour histogram during the sharpening process, which prevents Morphing Attack Detection (MAD) systems from identifying such transformed images.

Hence, in line with these identified challenges, this research proposes a MAD after image sharpening operation using averaging feature dimensionality reduction and summation feature level fusion of 8×8 and 16×16 scale Histogram of Oriented Gradient (HOG) descriptor. As a result, the following are the paper's main contributions:

1. Development of a dimensionality reduction and feature-level fusion technique for MAD even after image sharpening post-processing operation.
2. Evaluation of the technique's performance in (1) using False Acceptance Rate, False Acceptance Rate and Accuracy performance metric.

The following is how the rest of the paper is organized: A survey of relevant studies is included in Sect. 2. Section 3 describes the approach utilized to conduct the research. The findings of the experiment are described in Sect. 4. Section 5 draws conclusions, while Sect. 6 discusses future work.

2 Related Works

MAD's relevance cannot be emphasized, especially after picture post-processing operations such as image sharpening, print-scan, and image compression.

Singh [17] used a deconstructed 3D geometry and diffuse reflectance to accomplish MAD. This approach was recommended because it can detect morphing attacks through print-scan, posture, and illumination abnormalities.In this investigation, actual picture was captured at the ABC gate, and these components are then used to train a MAD linear SVM that compares the real picture taken at the ABC gate to the eMTRD face image. The problem with this paper is that the suggested algorithm ignores image post-processing tasks including print-scan, contrast enhancement, image compression, blurring, and sharpening.

Makrushin [18] performed MAD using Benford features. In this work, a splicing based approach was used to produce blurred facial images which are visually faultless automatically. A spread of Benford highlights extricated from quantised Discrete Cosine Change (DCT) coefficients of JPEG-compacted transformed pictures were utilized as feature vectors, and the Support Vector Machine (SVM) was applied for grouping. The upside of the suggested system is that it could perform well even on JPEG-compacted transformed pictures. Anyway, the strategy could not distinguish morphed images in the wake of performing print and scan operation on the images.

A de-morphing configuration based on a convolutional neural network (CNN) model was proposed by Ortega-Delcampo [19]. This method is characterized by two images: the passport's potentially altered image and the person's photo in the ABC system. The de-morphing procedure aims to reveal the chip picture. Assume the chip image has been altered. The exposing procedure between the in vivo image and the morphing chip image in that case will give the person in the ABC system a different facial identity, disclosing the impostor. If the chip photo is a non-morphing image, the end image will be comparable to a genuine traveler. The significant contribution of this work is the enhancement of picture quality and graphic aspects accomplished following the de-morphing procedure, as well as the discovery of the impostor's concealed identity. The CNN model extracted a large number of features that made training or learning of these features slow. Feature selection could increase learning speed and enhance the result as more relevant features would be used for training.

Kenneth [15] proposed a method for MAD in the presence of post-processed Images sources based on feature selection using Neighborhood Component Analysis (NCA) and classification using decision tree. The Local Binary Pattern (LBP) descriptor was used to extract morphed and bona-fide image features. The image sharpening post-processing operation was considered. In this paper it was identified that image sharpening operation on morph images could alter the morph artefacts making the altered morphed images difficult for a MAD system to detect. Two experiments were conducted in this study. In the first experiment the decision tree classifier was trained with the original LBP without feature selection. In the second experiment the decision tree classifier was trained using just the NCA selected LBP feature sets. The classification using the NCA selected features produced a higher accuracy than classification using all the LBP features. For non-post-processed image sources the proposed system attained and accuracy of 94% while an accuracy of 85% was achieved for the sharpened image sources. A drawback of this study is that the accuracy of the sharpening image sources is low in comparison with the normal images. This result shows that more accurate methods are needed for improved MAD in the presence of image sharpening operation.

Raghavendra [1] used Binarized Statistical Image Features (BSIF) to perform MAD. The suggested technique used BSIF to extract a micro-texture variation from a facial image, and the classification was done with a linear Support Vector Machine (SVM). The image's BSIF characteristics are extracted, and each pixel's response to a filter trained on statistical properties of natural images is computed to represent it as a binary code. With an Attack Presentation Classification Error Rate (APCER) of

1.73%, the system performed well, demonstrating its relevance to real-world circumstances. The work's drawback is its lack of robustness when it comes to the datasets employed. The dataset was generated using a single morphing tool (GNU Image Manipulation Program), limiting its performance. However, in real-world different morphing tools are used to carry out morphing attack.

CNN was used by Raghavendra [20] as a feasible feature extractor and classifier for MAD. To perform MAD for print-scan and digital morphed photos, the proposed method used transferrable features obtained from a pre-trained CNN. VGG19 and AlexNet were two CNN algorithms employed in the feature mining process. The picture features were retrieved independently from the AlexNet and VGG19 models' fully-connected layers. The feature level fusion approach was used to combine these qualities into a single feature vector. In both cases, the proposed method gave better outcomes for digital photos with an Equal Error Rate (EER) of 8.22% than print-scan images with an EER of 12.47%. However, the print-scan post-processing technique was the only one investigated. The effects of compression, resizing, and sharpening were not considered.

Face recognition algorithms based on CNN and hand-crafted characteristics are used by Wandzik [21] to tackle the MAD problem. The face characteristics were mined using four feature extractors: Dlib, FaceNet, High-Dim Local binary pattern and VGG-Face. After completing feature extraction using any of the feature extraction method, the extracted features were used to calculate the Euclidean distance for the face authentication job. Using the reference photo vectors, the SVM was used to conduct classification tasks. The MAD of digital photographs was the focus of this study; however, print-scan photographs were not considered.

Premised on an examination of Photo Response Non Uniformity (PRNU), Debiasi [10] suggested a morphing detection technique. It is based on a spectral study of the morphing-induced fluctuations within the PRNU. The wavelet-based denoising filter was used to extract the PRNU for each image. The frequency distortion removal (FDR) PRNU improvement is next applied to the retrieved PRNU. The Discrete Fourier Transform (DFT) was used to recover the frequency spectrum of the PRNU in each cell as part of the feature extraction process. The magnitude spectrum that results illustrates the morphing-induced changes in the PRNU signal. To quantify these impacts, a histogram of DFT Magnitudes was produced to depict the spectrum's magnitude distribution. Because picture post-processing tasks like sharpening, contrast enhancement, and blurring can have a significant impact on PRNU features, this proposed method investigated the impact of several image post-processing strategies on detection performance. The suggested detection system was resistant to image scaling and sharpening, with the exception of histogram equalization. To combat the failure to recognize altered photos, a deeper analysis and improved detection methodologies are required (histogram equalization). A summary of the related reviews are given in Table 1.

Table 1 Summary of related works

Author	Approaches used	Findings	Limitations
Makrushin [18]	Benford features	The upside of the suggested system is that it had the option to perform well even on JPEG-compacted transformed pictures	The strategy could not distinguish morphed pictures in the wake of performing print and scan operation on the images
Debiasi [10]	PRNU-based technique	The proposed method investigated the effects of various image post-processing techniques on detection efficiency, finding that the proposed detector was resistant to image sharpening and scaling	The system failed for MAD on morphed images processed with histogram equalization
Kenneth [15]	Neighborhood component analysis (NCA) and local binary pattern	In this paper it was identified that image sharpening operation on morph images could alter the morph artefacts making the altered morphed images difficult for a MAD system to detect	A drawback of this study is that the accuracy of the sharpening image sources is low in comparison with the normal images. This result shows that more accurate methods are needed for improved MAD in the presence of image sharpening operation
Raghavendra [20]	VGG19 and AlexNet	When comparing digital photographs with an equal error rate (EER) of 8.223% to print-scan photos with an EER of 12.47%, the proposed method achieves a better outcome for digital photographs	The MAD could not attain an outstanding EER for MAD in the print-scan image compared to the digital images

(continued)

Table 1 (continued)

Author	Approaches used	Findings	Limitations
Ortega-Delcampo [19]	Convolutional neural network (CNN)	The significant contribution of this work is the enhancement of picture quality and visual aspects accomplished following the de-morphing procedure, as well as the discovery of the impostor's concealed identity	The CNN model extracted a large number of features that made training or learning of these features slow
Singh [17]	CNN (pre-trained AlexNet)	The authors provide a novel database of morphing photos and trusted live capture probing images collected in a realistic border crossing scenario with ABC gates Proposed a new method for detecting morphing attacks that uses a combination of scores from a quantized normal-map phase and dispersed reconstructed image features to exploit the intrinsic border crossing situation	The problem with this paper is that the suggested algorithm ignores image post-processing tasks including sharpening, print-scan, compression, and blurring
Wandzik [21]	faceNet, Dlib, VGG-face, high-dim local binary pattern, and SVM	Instead of adding new components, the proposed approach makes use of an established feature extraction pipeline for face recognition systems. It doesn't need any fine-tuning or changes to the current recognition scheme, and it can be trained with a small dataset	This study only looked at the MAD of digital photos, not the print-scanned photos that are utilized for authentication in some nations

(continued)

Table 1 (continued)

Author	Approaches used	Findings	Limitations
Raghavendra [1]	BSIF	The system attained a good performance with APCER of 1.73% that shows its applicability to a real-world scenario	The downside of this work is robustness with respect to the dataset used. The dataset was generated using a single morphing tool (GNU Image Manipulation Program), limiting its performance

3 Methodology

This section presents the description of the methods used to carry out this research [22]. These techniques includes data collection and generation, image post-processing and pre-processing, feature extraction, dimensionality reduction, feature fusion and data classification. The proposed system is illustrated in Fig. 1.

3.1 Data Collection

Using different facial photographs from 100 persons, a new morphed facial data set comprising 200 morphed photos and 150 bona-fide images was developed, totaling 350 image. To assist diversify the database, female and male of diverse complexion were employed in the facial photos. The images for the subjects came from a variety

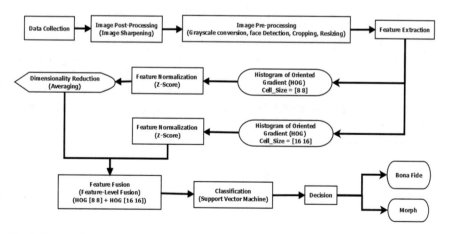

Fig. 1 Proposed system

of electronic sources, including the Yale face database [23] and the 123RF photo website [24]. The altered face images were created with the help of 2 morphing tools:

1. **Magic morph tool**: This is a free Windows-based program that acts as a useful image and graphics utility for users. It is high-speed morphing and warping program. Magic Morph allows users to animate their still images into SWF, GIF, or AVI files with morphing effects. It is easy and straightforward to use. Quick and multithread pyramid methods, expert-quality warping and morphing techniques, and real-time visualization functions are all included in the app. TIFF, BMP, JPEG, J2K, PNG, ICO, GIF, TGA, WBMP, PCX, WMF, and JBG are among the file formats supported by this method. Its compatible output files, on the other hand, are AVI, GIF and SWF Movie, JPEG and BMP Sequence.
2. **FantaMorph tool**: The FantaMorph utility is a transforming software that may be used to create photo transformations and current transform activity effects. It assists users in locating facial features such as the nose, eye, and mouth, and then combines these features from multiple real-life faces to produce a virtual face. FantaMorph comes in three different editions: Standard, Professional, and Deluxe, and it works on both Windows and Mac computers.

The images were adjusted by hand and blended in an antique shell. The morphing software creates an overview of the transition from one subject to another. The last picture that has been transformed is manually taken by showing its similarity to the participating subjects' faces during transformation. The generated transformations are now high in calibre and have little to no identifiable artefacts.

3.2   Image Sharpening

A morphed image that undergoes a sharpening procedure loses part of its artefacts, causing MAD to be problematic. Since human perception is extremely sensitive to edges and details in a picture, imaging sharpening is typically used as an image post-processing technique. As images are primarily made up of high-frequency portions, high-frequency distortion can degrade graphic quality [15]. Improved quality of the visual image is achieved through improving the high-frequency segments of the picture. Thus, sharpening of the morphed images can emphasize the edges of a picture and modify the subtleties that can also change the morph highlights that make the morphed photo challenging to detect.

3.3   Face Pre-processing

In the face pre-processing stage, four operations were carried out. These operations include Facial landmark detection, cropping, image resizing and Grayscale conversation. Each of these processes is discussed in details in the subsections below.

3.3.1   Facial Features Detection

Facial features, such as the nose, eyes, lips, brows, and jawline, are employed to restrict and signify significant regions of focus [13]. The Viola-Jones approach was used to detect face features in this study. The Viola-Jones computation employs Haar-basis filtering, a scalar object in the middle of the image, and various Haar-like structures [25]. Haar feature selection, integral photo screening, Adaboost training, and a cascading classifier are the four phases of this approach for face recognition [26]. The input image is first converted into an integral image via the Viola-Jones face detection algorithm. The integral image is a method for generating the pixel sum intensities in a square of a picture in an operational manner. A more detailed discussion of the Viola-Jones technique is found in Wang [17], Viola and Jones [18], and Jensen [20].

3.3.2   Image Cropping

Cropping is the removal of a photographic image from unnecessary external areas. The approach typically involves removing some of the peripheral regions of an image to eliminate extra trash from the image, enhance its framing, alter the aspect ratio, or highlight or separate the subject matter from its background. After detecting the facial features using the Voila Jones algorithm, the face images were cropped to magnify the primary subject (face) and further reduce the angle of view to a dimension of 150×150 pixels established on the identified features to make sure that the MAD algorithm is only applied to the face.

3.3.3   Grey-Scale Conversion

A grey-scale image in digital image processing is one in which a single sample representing only a quantity of light is the value of each pixel; that is, it holds only intensity values. The pictures in grey-scale, a kind of grey monochrome, are made entirely of shades of grey. At the lowest intensity, the contrast varies from black to white at the highest [27]. In this phase, the cropped RGB or coloured face images were converted to a grey-scale image to prepare the images for feature extraction.

3.3.4 Image Resizing

The graphic primitives that constitute a vector graphic image can be sized using geometric transformations without losing image quality when resizing it in digital image processing. Image resizing can be interpreted as image resampling or image reconstruction [28, 29]. The image size can be changed in several ways, but the Nearest-Neighbour Interpolation (NNI) algorithm was adopted in this paper. The NNI algorithm is one of the more straightforward ways of increasing image size. This deals with exchanging every pixel in the output with the nearest pixel; this ensures there will be several pixels of the same colour for up-scaling. Pixel art may benefit from this because sharp details can be preserved [30]. In this step, the cropped images were all resized to the same scale to enable the extraction of the same number of feature vectors.

3.4 Feature Extraction

Image features contain essential information such as points and edges that are vital for image analysis. Images are extracted using several techniques. In this paper, the Histogram of Oriented Gradient (HOG) extractor was utilized to extract gradient features from the bona-fide and morphed images.

3.4.1 Histogram of Oriented Gradient (HOG)

The HOG is a function extractor utilized for object discovery in image processing and computer vision. The method counts how many times a gradient orientation occurs in a particular image area [31]. HOG is invariant to photometric and geometric transformations; This makes it very well adapted for human detection [32]. The HOG features are generated as follows: After pre-processing and resizing the image, the variance in the x and y directions for each image pixel is determined. The magnitude and directions are computed using the equations in Eqs. 1 and 2, correspondingly.

$$\text{Total Gradient Magnitude} = \sqrt{(G_x)^2 + (G_y)^2} \tag{1}$$

where G_y denotes the y-direction gradient and G_x denotes the x-direction gradient.

$$\text{Direction} = \tan(\theta) = {G_y}/{G_x} \tag{2}$$

The value of the angle (θ) is presented in Eq. 3

$$\theta = a\tan(G_y/G_x) \tag{3}$$

The HOG gradient descriptor was used in this research because the image morphing process reduces the changes in the high frequency of the image and decreases the morphed images' gradient steepness, enhancing MAD.

There are no optimal values or scale in the HOG feature description to extract the best feature for classification. For example, cell size 8 × 8 is HOG with fine cells. But perhaps it is not the best scale (because the cells are too small and noise might just be observed) (On the other hand, too large cells, like cell size 16 × 16, may be too large, and there will have uniform histograms everywhere). The best way to obtain the best features is by extracting HOG at various scales and combining them. Hence the 8 × 8 pixels cell size was used to capture small-scale spatial information from the images, while 16 × 16 pixels cell size was used to capture large-scale spatial details from the pictures. The HOG with 16 × 16 pixels cell sizes consists of 648 feature vectors. The HOG with 8 × 8 pixels cell sizes consists of 4320 feature vectors.

3.5 Feature Normalization

Normalization is the process of converting values measured on multiple scales to a nominally standard scale, which is commonly done before averaging. The approach of feature normalization is used to normalize the range of independent variables or data characteristics. Normalization is the process of converting qualities to a scale that is equivalent. This improves the models' performance and training reliability [33]. The retrieved features were normalised to ensure that the HOG (8 × 8) and (16 × 16) weight feature vectors were all on the same scale and contributed equally to the classification outcome. The Z-Score approach was used to normalize the extracted features.

3.5.1 Z-Score

A prevalent technique to normalise features to zero mean and unit variance is the Z-score normalisation technique [34]. The Z-Score is an arithmetic statistic for comparing a score to the average of a collection of scores [35]. Z-Score is calculated based on the formula in Eq. 4.

$$Z_{\text{score}}(i) = \frac{x_i - \mu}{s} \tag{4}$$

where μ is the mean of the distribution, s is the standard deviation and x_i = number of object in the distribution. Equation 5 provides the formula for calculating the standard deviation:

$$s = \sqrt{\frac{1}{n-1} \sum_{i=1}^{n} (x_i - \mu)^2} \tag{5}$$

3.6 Averaging Dimensionality Reduction and Feature Fusion

Dimensionality reduction is the translation of data from a high-dimensional space into a low-dimensional space such that some significant features of the source data are preserved by the low-dimensional representation, preferably similar to its fundamental dimension [36, 37].

3.6.1 Averaging Dimensionality Reduction

In this study, dimensionality reduction was performed as an intermediate step to facilitate the feature fusion process. Dimensionality reduction was performed only on HOG 8×8 pixels cell size. The HOG 8×8 pixels cell size consisting of 4320 features was reduced to 648 features. Dimensionality was performed using averaging. Averaging was used to ensure that every initial HOG 8×8 features contributed to the new reduced features. The averaging dimensionality reduction was performed on HOG 8×8 features because the summation feature-level technique can only be applied to different features of the same dimension. Hence, the HOG 8×8 features' dimensionality must be equal to the HOG 16×16 features' dimensionality.

To average the HOG 8×8 features, the features were divided by the number of HOG 16×16 features. The result of this division was used to group the features for averaging. For example, the 4320 (the number of HOG (8×8) features) was divided by 648 (number of GLCM features), which results in approximately 7. This means that the first seven (7) features will be averaged first, followed by the next seven features. Averaging of each seven feature continues till the 4320th feature. The algorithm for dimensionality reduction and feature-level fusion technique is presented in algorithm 1.

ALGORITHM 1: Averaging Dimensionality Reduction and Feature Fusion

---

Inputs:
H8[] ={$X_1, X_2, X_3, ..., X_n$} //HOG (8 x 8 cell size) feature vectors
H16[] ={$Y_1, Y_2, Y_3, ..., Y_n$} //HOG (16 x 16 cell size) feature vectors

Output:
F[]   //Fused feature of H8[] and H16[]

1:      get number of rows for H8[],n_rows
2:      get number of columns for H8[], C1
3: get number of columns for H16[], C2
4: M1 = floor $\left(\frac{C_2}{C_1} \right)$ // get floor value for C2 divided by C1
5: N1= (C2 – M1) * C1 + M1 //to get first set of columns to average
6: for i = 1 to n_rows //Looping through rows
7:      Q1 = mean(H8(1 to N1)   // to average the selected first set of columns for H8[] features
8:      for t=N1 + 1 to C2 //Loop through the group columns and /increment by M1
9:            Q1 =Concatenate (Q1, mean(H8(t to t + M1 – 1)) //merge the average of the
first,   second, third, fourth, ...C1 selected set of columns for H[] features
10:        end
11:    end
12:    $\sum H_{16}[] Q_1$ //average Feature fusion for H16[] and reduced dimension H8[]
13:    F[] = Concatenate ($H_{16}[], Q_1$)
14:    return F[]

3.6.2   Feature Fusion

Feature Fusion is a technique for combining similar data extracted from a collection of training and testing images without losing any detail [38–40]. Summation feature-level technique was used for fusion. The summation feature-level method is expressed in Eqs. 6 and 7. The summation feature-level fusion formula to fuse H8 and H16 is presented in Eq. 6.

$$\sum H_{16}Q_1 \tag{6}$$

The fused H_8 and H_{16} were then concatenated to produce the final feature vectors using the formula in Eq. 7.

$$F = \text{Concatenate } (H_6, \ Q_1) \tag{7}$$

The new set of concatenated feature vectors (648 features) generated using the summation feature level fusion technique were used to train the SVM classifier.

The main advantage of feature-level fusion is that it detects correlated feature values produced by different feature extraction methods, resulting in a compact collection of salient features that can boost detection accuracy [41].

3.7 Image Classification

Classifications algorithms simply categorize the input to correspond to the desired output [42]. Image classification is the technique of categorizing and marking sets of pixels inside an image according to a set of laws [43]. In this paper, the SVM classifier was utilized to classify the face photos into morph or bona-fide photos.

3.7.1 Support Vector Machine (SVM)

SVM is a supervised learning framework with related learning algorithms for regression and classification analysis [44]. The SVM technique seeks to locate a hyper-plane in a D-dimensional domain (where D denotes the number of features) that distinguishes between data sets [45]. Hyperplanes are decision lines that help categorize data. Data points on either side of the plane can be allocated to various groups. The number of functions determines the hyperplane's dimension as well. The decision function of a binary SVM is described in the input space by Eq. 8:

$$\gamma = h(x) = \text{sign}\left(\sum_{j=1}^{n} u_j y_j K\left(x, x_j\right) + v\right) \tag{8}$$

where x is the to-be-categorised feature vector, j indexes the training instances, n is the number of training instances, y_j is the label (1/−1) of training example j, $K(,)$ is the kernel function, and u_j and v are fit to the data to maximise the margin. Training vectors for which $u_j \neq 0$ are called support vectors [46]. A significant advantage of SVM is that it is versatile (that is, different Kernel functions can be specified for the decision function) [47].

3.8 Performance Metrics

The following evaluation metrics were used to assess the performance of the suggested technique.

1. **Attack Presentation Classification Error Rate (APCER)**: This is the same as the false acceptance rate. It is defined as a percentage of morphing attack that is classified as bona-fide images [48]. Equation 9 contains the formula for APCER:

$$APCER = \text{False positive}/(\text{True Positive} + \text{False Positive}) \tag{9}$$

2. **Bona fide Presentation Classification Error Rate (BPCER)**: This is the same as the false rejection rate. In a particular context, the BPCER is described as the

fraction of genuine presentations wrongly categorized as presentation attacks, or the fraction of genuine photos incorrectly labelled as morphing attacks [21]. In Eq. 9, the BPCER formula is presented

$$BPCER = False\ Negative/(True\ Positive + False\ Negative) \qquad (10)$$

3. **Accuracy (ACC)**: This is a statistic for evaluating the consistency of categorization models. The percentage of correctly classified instances for an independent testing data set can be precisely described as accuracy [49]. Equation 11 shows the accuracy:

$$ACC = \frac{True\ Positive + True\ negative}{True\ Positive + True\ negative + False\ Positive + False\ negative} \qquad (11)$$

4 Results and Discussion

In this paper, experimentation were carried out based on three techniques: HOG (8 × 8) + SVM, HOG (16 × 16) + SVM and HOG (16 × 16) + HOG (8 × 8) + SVM (proposed system) algorithms with respect to the sharpened images and non-sharpened images. The following six categories of investigations have been carried out:

1. Classification images that have not been post-processed using HOG (8 × 8) + SVM technique.
2. Non-post-processed images classification using HOG (16 × 16) + SVM technique.
3. Non post-processed images classification using a fusion of HOG (8 × 8) and HOG (16 × 16) features algorithm.
4. Post-processed images classification using HOG (8 × 8) + SVM algorithm.
5. Post-processed images classification using HOG (16 × 16) + SVM algorithm.
6. Post-processed images classification using a fusion of HOG (8 × 8) and HOG (16 × 16) features algorithm.

Table 2 consists of the MAD experimentation result for post-processed images.

Table 2 shows that the suggested system had the highest accuracy for the sharpened images, with a value of 95.71%, as opposed to HOG (8 × 8) + SVM and HOG (16 × 16) + SVM, which had accuracy of 94.29 and 90.00%, correspondingly. According on the BPCER figures shown in Table 2, the suggested system has the best result, with BPCER of 3.36% at APCER = 5% and BPCER of 1.68% at APCER = 10%. Also, the suggested system performed better when compared with a previous work by Kenneth [15] on MAD after performing an image sharpening operation.

Figure 2 shows a chart comparing the performance of four different MAD techniques, namely HOG (8 × 8), HOG (16 × 16), proposed technique (HOG (8 × 8)

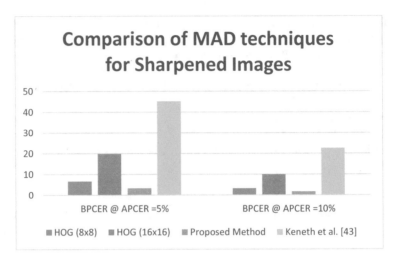

Fig. 2 Comparison of MAD techniques for Sharpened Images

+ HOG (16 × 16)) and Kenneth [15]. Figure 2 is a visualization of the performance measures in Table 2.

The proposed system's improved performance can be attributed to the concatenation of gradient features HOG (8 × 8) and HOG (16 × 16) extracted from the post-processed morphed and bona-fide images.

Table 3 shows that the suggested system had the best performance for non-sharpened images, with a rating of 97.14%, contrasted to HOG (8 × 8) + SVM and HOG (16 × 16) + SVM, both of which had a value of 94.29%. The proposed solution also has the best result, with BPCER of 1.63% at APCER = 5% and BPCER of 0.82% at APCER = 10%, according to the BPCER estimates shown in Table 3. Whereas the methods offered by Ramachandra [48] and Ramachandra [50] have BPCER of 45.76% and BPCER = 7.59% at APCER = 5% and BPCER of 13.12% and BPCER = 0.86% at APCER = 10%.

Figure 3 shows a chart comparing the performance of seven different MAD techniques, namely HOG (8 × 8), HOG (16 × 16), proposed technique (HOG (8 × 8) +

Table 2 MAD Classification Result for Sharpened Images

Sharpened images			
Algorithm	Accuracy (%)	BPCER (%) @	
		APCER = 5%	APCER = 10%
HOG (8 × 8) + SVM	94.29	6.67	3.33
HOG (16 × 16) + SVM	90.00	19.85	9.93
Kenneth [15]	85	45.19	22.59
Proposed method (HOG (8 × 8) + HOG (16 × 16) + SVM)	95.71	3.36	1.68

Table 3 MAD classification results for non-sharpened images

Non-sharpened images			
Techniques	Accuracy (%)	BPCER (%) @	
		APCER = 5%	APCER = 10%
HOG (8 × 8) + SVM	94.29	4.93	2.47
HOG (16 × 16) + SVM	94.29	4.93	2.47
Proposed method (HOG (8 × 8) + HOG (16 × 16) + SVM)	**97.14**	**1.63**	**0.82**
Steerable textures [48]	–	45.76	13.12
Laplacian pyramid + LBP [50]	–	7.59	0.86
Deep color residual noise [51]	–	3.00	1.50
Transferable deep-CNN [20]	–	14.38	7.53

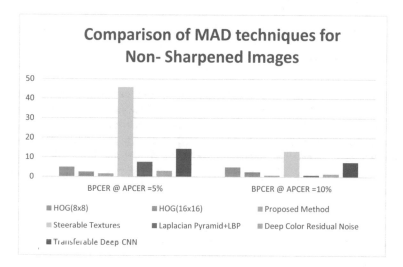

Fig. 3 Comparison of MAD techniques for non-sharpened images

HOG (16 × 16)), steerable texture, Laplacian Pyramid + LBP, Deep colour residual noise and Transferable deep-CNN techniques. Figure 3 is also a visualization of the performance measures in Table 3.

This paper was competent to accomplish MAD more reliably than earlier MAD research. This is due to the system's capacity to recognize altered images utilizing a fusion of powerful and resilient feature descriptors, after image sharpening functionality has been performed on those images.

5   Conclusion

This study conducted MAD after the sharpening operation was applied on both the morphed and bona-fide images based averaging dimensionality reduction and summation feature-level fusion of the HOG (8 × 8) and HOG (16 × 16) gradient features using the SVM classifier. These extracted features were normalized to adjust the features measured on different scales to a notionally standard scale. These normalized features were fused using the feature-level fusion method. The SVM classifier learned these fused features, which classified the features into two categories: morphed or bona-fide. The proposed method was compared to existing MAD techniques and single feature descriptor methods. From the results obtained, it can be concluded that the proposed system has a high-performance accuracy of 97.14 and 95.71% for non-sharpened images and sharpened images, respectively as compared to the existing MAD methods.

6   Future Works

This study used only two morphing applications to create the altered photos. To improve MAD's robustness, several morphing algorithms should be utilized to build a more resilient morphed datasets. The morphed datasets utilized in this study was created in-house using morphing software that was readily available. There was no currently accessible extensive database for MAD. As a result, it is suggested that in the future, a extensive publicly accessible morph database containing both real and morphed photographs be created as a baseline for MAD algorithms.

References

1. Raghavendra R, Raja KB, Busch C (2016) Detecting morphed face images. In: 2016 IEEE 8th international conference on biometrics theory, applications and systems (BTAS), Niagara Falls, NY, USA, pp 1–7. https://doi.org/10.1109/BTAS.2016.7791169
2. Olanrewaju L, Oyebiyi O, Misra S, Maskeliunas R, Damasevicius R (2020) Secure ear biometrics using circular kernel principal component analysis, Chebyshev transform hashing and Bose—Chaudhuri—Hocquenghem error-correcting codes. Signal Image Video Process 14(5):847–855. https://doi.org/10.1007/s11760-019-01609-y
3. Damer N, Saladie AM, Braun A, Kuijper A (2018) MorGAN: recognition vulnerability and attack detectability of face morphing attacks created by generative adversarial network. In: 2018 IEEE 9th international conference on biometrics theory, applications and systems (BTAS), Redondo Beach, CA, USA, pp 1–10. https://doi.org/10.1109/BTAS.2018.8698563
4. Ferrara M, Franco A, Maltoni D (2014) The magic passport. In: IEEE international joint conference on biometrics, Clearwater, FL, USA, pp 1–7. https://doi.org/10.1109/BTAS.2014.6996240
5. Mislav G, Kresimir D, Sonja G, Bozidar K (2021) Surveillance cameras face database. In: SCface—Surveillance cameras face database. https://www.scface.org/. Accessed 02 Feb 2021

6. Bharadwaj S, Dhamecha TI, Vatsa M, Singh R (2013) Computationally efficient face spoofing detection with motion magnification. In: 2013 IEEE conference on computer vision and pattern recognition workshops, OR, USA, pp 105–110. https://doi.org/10.1109/CVPRW.2013.23

7. Tolosana R, Gomez-Barrero M, Busch C, Ortega-Garcia J (2020) Biometric presentation attack detection: beyond the visible spectrum. IEEE Trans Inf Forensics Secur 15:1261–1275. https://doi.org/10.1109/TIFS.2019.2934867

8. Chingovska I, Mohammadi A, Anjos A, Marcel S (2019) Evaluation methodologies for biometric presentation attack detection. In: Marcel S, Nixon MS, Fierrez J, Evans N (eds) Handbook of biometric anti-spoofing. Springer International Publishing, Cham, pp 457–480. https://doi.org/10.1007/978-3-319-92627-8_20

9. Damer N et al (2019) Detecting face morphing attacks by analyzing the directed distances of facial landmarks shifts. In: Brox T, Bruhn A, Fritz M (eds) Pattern recognition, vol 11269. Springer International Publishing, Cham, pp 518–534. https://doi.org/10.1007/978-3-030-12939-2_36

10. Debiasi L, Scherhag U, Rathgeb C, Uhl A, Busch C (2018) PRNU-based detection of morphed face images. In: 2018 international workshop on biometrics and forensics (IWBF), Sassari, pp 1–7. https://doi.org/10.1109/IWBF.2018.8401555

11. Kramer RSS, Mireku MO, Flack TR, Ritchie KL (2019) Face morphing attacks: investigating detection with humans and computers. Cogn Res Princ Implic 4(1):28. https://doi.org/10.1186/s41235-019-0181-4

12. Scherhag U, Rathgeb C, Merkle J, Breithaupt R, Busch C (2019) Face recognition systems under morphing attacks: a survey. IEEE Access 7:23012–23026. https://doi.org/10.1109/ACCESS.2019.2899367

13. Seibold C, Samek W, Hilsmann A, Eisert P (2017) Detection of face morphing attacks by deep learning. In: Kraetzer C, Shi Y-Q, Dittmann J, Kim HJ (eds) Digital forensics and watermarking, vol 10431. Springer International Publishing, Cham, pp 107–120. https://doi.org/10.1007/978-3-319-64185-0_9

14. Venkatesh S, Ramachandra R, Raja K, Spreeuwers L, Veldhuis R, Busch C (2019) Detecting morphed face attacks using residual noise from deep multi-scale context aggregation network, p 10

15. Kenneth OM, Sulaimon AB, Opeyemi AA, Mohammed AD (2021) Face morphing attack detection in the presence of post-processed image sources using neighborhood component analysis and decision tree classifier. In: Misra S, Muhammad-Bello B (eds) Information and communication technology and applications. ICTA 2020, vol 1350, pp 340–354. https://doi.org/10.1007/978-3-030-69143-1_27

16. Weng Y, Wang L, Li X, Chai M, Zhou K (2013) Hair interpolation for portrait morphing. Comput Graph Forum 32(7):79–84. https://doi.org/10.1111/cgf.12214

17. Singh JM, Ramachandra R, Raja KB, Busch C (2019) Robust morph-detection at automated border control gate using deep decomposed 3D shape and diffuse reflectance. http://arxiv.org/abs/1912.01372. Accessed 01 Sep 2020

18. Makrushin A, Neubert T, Dittmann J (2017) Automatic generation and detection of visually faultless facial morphs. In: Proceedings of the 12th international joint conference on computer vision, imaging and computer graphics theory and applications, Porto, Portugal, pp 39–50. https://doi.org/10.5220/0006131100390050

19. Ortega-Delcampo D, Conde C, Palacios-Alonso D, Cabello E (2020) Border control morphing attack detection with a convolutional neural network de-morphing approach. IEEE Access 1–1. https://doi.org/10.1109/ACCESS.2020.2994112

20. Raghavendra R, Raja KB, Venkatesh S, Busch C (2017) Transferable deep-CNN features for detecting digital and print-scanned morphed face images. in 2017 IEEE conference on computer vision and pattern recognition workshops (CVPRW), Honolulu, HI, USA, pp 1822–1830. https://doi.org/10.1109/CVPRW.2017.228

21. Wandzik L, Kaeding G, Garcia RV (2018) Morphing detection using a general-purpose face recognition system. In: 2018 26th European signal processing conference (EUSIPCO), Rome, pp 1012–1016. https://doi.org/10.23919/EUSIPCO.2018.8553375

22. Misra S (2021) A step by step guide for choosing project topics and writing research papers in ICT related disciplines. In: Misra S, Muhammad-Bello B (eds) Information and communication technology and applications, vol 1350. Springer International Publishing, Cham, pp 727–744. https://doi.org/10.1007/978-3-030-69143-1_55
23. Yale face database. In: Yale face database. http://vision.ucsd.edu/content/yale-face-database. Accessed 11 Nov 2020
24. 23RF (2020) Black man face stock photos and images. In: Black man face stock photos and images. https://www.123rf.com/stock-photo/black_man_face.html?sti=lo3vts77wcrg1jyzpbl. Accessed 07 Dec 2020
25. Wang Y-Q (2014) An analysis of the viola-jones face detection algorithm. Image Process Line 4:128–148. https://doi.org/10.5201/ipol.2014.104
26. Viola P, Jones M (2001) Rapid object detection using a boosted cascade of simple features. In: Proceedings of the 2001 IEEE computer society conference on computer vision and pattern recognition. CVPR 2001, Kauai, HI, USA, vol 1, p I-511–I-518. https://doi.org/10.1109/CVPR. 2001.990517
27. Saravanan C (2010) Color image to grayscale image conversion. In: 2010 second international conference on computer engineering and applications, Bali Island, Indonesia, pp 196–199. https://doi.org/10.1109/ICCEA.2010.192
28. Dong W-M, Bao G-B, Zhang X-P, Paul J-C (2012) Fast multi-operator image resizing and evaluation. J Comput Sci Technol 27(1):121–134. https://doi.org/10.1007/s11390-012-1211-6
29. Malini MS, Patil M (2018) Interpolation techniques in image resampling. Int J Eng Technol 7(34):567. https://doi.org/10.14419/ijet.v7i3.34.19383
30. Parsania Mr PS, Virparia Dr PV (2016) A comparative analysis of image interpolation algorithms. IJARCCE 5(1):29–34. https://doi.org/10.17148/IJARCCE.2016.5107
31. Suard F, Rakotomamonjy A, Bensrhair A, Broggi A (2006) Pedestrian detection using infrared images and histograms of oriented gradients. In: 2006 IEEE intelligent vehicles symposium, Meguro-Ku, Japan, pp 206–212. https://doi.org/10.1109/IVS.2006.1689629
32. Dalal N, Triggs B (2005) Histograms of oriented gradients for human detection. In: 2005 IEEE computer society conference on computer vision and pattern recognition (CVPR'05), San Diego, CA, USA, vol 1, pp 886–893. https://doi.org/10.1109/CVPR.2005.177
33. Kumar BS, Verma K, Thoke AS (2015) Investigations on impact of feature normalization techniques on classifier's performance in breast tumor classification. Int J Comput Appl 116(19):11–15. https://doi.org/10.5120/20443-2793
34. Shalabi LA, Shaaban Z, Kasasbeh B (2006) Data mining: a preprocessing engine. J Comput Sci 2(9):735–739. https://doi.org/10.3844/jcssp.2006.735.739
35. Kolbaşi A, Ünsal PA (2015) A comparison of the outlier detecting methods: an application on turkish foreign trade data. J Math Stat Sci 5:213–234
36. Sembiring RW, Zain JM, Embong A (2011) Dimension reduction of health data clustering, p 10
37. Yang W, Wang K, Zuo W (2012) Neighborhood component feature selection for high-dimensional data. J Comput 7(1):161–168. https://doi.org/10.4304/jcp.7.1.161-168
38. Gawande U, Zaveri M, Kapur A (2013) A novel algorithm for feature level fusion using SVM classifier for multibiometrics-based person identification. Appl Comput Intell Soft Comput 2013:1–11. https://doi.org/10.1155/2013/515918
39. Sudha D, Ramakrishna M (2017) Comparative study of features fusion techniques. In: 2017 international conference on recent advances in electronics and communication technology (ICRAECT), Bangalore, India, pp 235–239. https://doi.org/10.1109/ICRAECT.2017.39
40. Vaidya AG, Dhawale AC, Misra S (2016) Comparative analysis of multimodal biometrics. Int J Pharm Technol 8(4):22969–22981
41. Bhardwaj SK (2014) An algorithm for feature level fusion in multimodal biometric system. Int J Adv Res Comput Eng Technol 3(10):5
42. Olaleye T, Arogundade O, Adenusi C, Misra S, Bello A (2021) Evaluation of image filtering parameters for plant biometrics improvement using machine learning. In: Patel KK, Garg D, Patel A, Lingras P (eds) Soft computing and its engineering applications, vol 1374. Springer, Singapore, pp 301–315. https://doi.org/10.1007/978-981-16-0708-0_25

43. Shinozuka M, Mansouri B (2009) Synthetic aperture radar and remote sensing technologies for structural health monitoring of civil infrastructure systems. In: Structural health monitoring of civil infrastructure systems, Elsevier, pp 113–151. https://doi.org/10.1533/9781845696825. 1.114
44. Meyer D, Leisch F, Hornik K (2003) The support vector machine under test. Neurocomputing 55(1–2):169–186. https://doi.org/10.1016/S0925-2312(03)00431-4
45. Chih-Wei H, Chih-Jen L (2002) A comparison of methods for multiclass support vector machines. IEEE Trans Neural Netw 13(2):415–425. https://doi.org/10.1109/72.991427
46. Sopharak A et al (2010) Machine learning approach to automatic exudate detection in retinal images from diabetic patients. J Mod Opt 57(2):124–135. https://doi.org/10.1080/095003409 03118517
47. Crammer K, Singer Y (2001) On the algorithmic implementation of multiclass Kernel-based vector machines, p 28
48. Ramachandra R, Venkatesh S, Raja K, Busch C (2020) Detecting face morphing attacks with collaborative representation of steerable features. In: Chaudhuri BB, Nakagawa M, Khanna P, Kumar S (eds) Proceedings of 3rd international conference on computer vision and image processing, vol 1022. Springer, Singapore, pp 255–265. https://doi.org/10.1007/978-981-32-9088-4_22
49. Ferrara M, Franco A, Maltoni D (2016) On the effects of image alterations on face recognition accuracy. In: Bourlai T (ed) Face recognition across the imaging spectrum. Springer International Publishing, Cham, pp 195–222. https://doi.org/10.1007/978-3-319-28501-6_9
50. Ramachandra R, Venkatesh S, Raja K, Busch C (2019) Towards making Morphing attack detection robust using hybrid scale-space colour texture features. In: 2019 IEEE 5th international conference on identity, security, and behavior analysis (ISBA), Hyderabad, India, pp 1–8. https://doi.org/10.1109/ISBA.2019.8778488
51. Venkatesh S, Ramachandra R, Raja K, Spreeuwers L, Veldhuis R, Busch C (2019) Morphed face detection based on deep color residual noise. In: 2019 ninth international conference on image processing theory, tools and applications (IPTA), Istanbul, Turkey, pp 1–6. https://doi.org/10.1109/IPTA.2019.8936088

A Systematic Literature Review on Forensics in Cloud, IoT, AI & Blockchain

N. S. Gowri Ganesh⊙, **N. G. Mukunth Venkatesh, and D. Venkata Vara Prasad**

Abstract In the growing diversified software applications, cybersecurity plays a vital role in preserving and avoiding the loss of data in terms of money, knowledge, and assets of businesses and individuals. The Internet of Things and cloud computing are nowadays the integral part of most software applications that assist in acquiring and storing data seamlessly. It provides the convenience of accessibility for the end-user like home automation, storage of huge streams of data, giving elasticity for increasing or decreasing the volume of data. When it comes to decentralized behavior, applications need to be transformed into blockchain technology. Blockchain technology offers value-added features to applications in terms of enhanced security and easier traceability. The blockchain's unchangeable and incorruptible nature protects it from tampering and hacking. Forensics requires the collection, preservation, and analysis of digital evidence. Artificial Intelligence is predominant in many areas and momentum is gaining to utilize it in the field of forensics. This chapter reviews the application of forensics using Artificial Intelligence in the field of Cloud computing, IoT, and Blockchain Technology. To fulfill the study's goal, a systematic literature review (SLR) was done. By manually searching six (6) well-known databases, documents were extracted. Based on the study topic, thirty three (33) primary studies were eventually considered. The study also discovered that (1) highlights several well-known challenges and open-Issues in IoT forensics research, as it is dependent on other technologies and is crucial when considering an end-to-end IoT application as an integrated environment with cloud and other technologies. (2) There has been less research dedicated to the use of AI in the field of forensics. (3) Contributions on forensic analysis of attacks in blockchain-based systems is not found.

N. S. G. Ganesh (✉)
MallaReddy College of Engineering and Technology, Hyderabad, India
e-mail: gowriganesh@mrcet.ac.in

N. G. M. Venkatesh
Panimalar Engineering College, Chennai, India

D. V. V. Prasad
SSN College of Engineering, Chennai, India
e-mail: dvvprasad@ssn.edu.in

© The Author(s), under exclusive license to Springer Nature Switzerland AG 2022
S. Misra and C. Arumugam (eds.), *Illumination of Artificial Intelligence in Cybersecurity and Forensics*, Lecture Notes on Data Engineering and Communications Technologies 109, https://doi.org/10.1007/978-3-030-93453-8_9

Keywords Forensics · Cloud computing · Blockchain technology · Provenance ·
IOT · AI · Machine learning

1   Introduction

Internet Technology and cyber applications is growing amazingly great which sur-
prises us in terms of how it solves many problems with its applications impacting
individual life and various industries. The internet applications like ecommerce,
social media, internet entertainment tools like Netflix started to use various cyber
technology such as cloud computing, block chain technology. These applications
stores, process data in the services offered by the cloud computing and blockchain
technology which is actually available in the cyberspace with the help of end devices
termed as Internet of Things such as sensors. As there is increase in the processing
and storing of the data in the cyberspace there is large possibilities of these data being
stolen, replaced, misused which can be commonly termed as cybercrime. Cybercrime
[28] is the attack against the computer hardware and software. Though the technology is great for the intended user, the cybercrime is a great threat and it requires to be
prevented for not occurring and also need to be tracked if one happened. Nowadays
software applications are built with the confluence of various technologies like cloud
computing, IOT and Blockchain Technology for the assisting the end users in terms
of easy to use, desired computing and networking components, large storage space
without investing in the storage devices, dynamic computing with end devices and
to gain distributed access. We have been witnessing in the day to day news about
theft, murder and many undesired happenings categorized by the law of the state or
country is defined as crime. The crime or offence can affect an individual or a group
of people. The severity of the crime depends on how many people affected and what
time it takes to recover to the original state of life. The person who does the crime is
termed as criminal. There are various reasons for a person to do a crime. The criminal
uses some tools and the opportunity to commit crime. Victim is the person or group
of persons affected by the crime. Cybercrime may be defined as "Any unlawful act
where computer or communication device or computer network is used to commit
or facilitate the commission of crime". According to the 2019 Internet Crime report
released by Internet Crime Complaint Centre (IC3) of the Federal Bureau of Investigation, United States, India stands third [24] as the victims of crimes among many
countries in the world. In the case of cybercrime, the tool used by the criminal to
perform crime is the resources of computer or software or internet or whatsoever
that is confined to the terminologies of cyberspace. The criminal performing the
cybercrime would take the advantages of his knowledge about the cyberspace tools.
In some cases the criminal may not be the computer expert but does crime such as
drug trades, child abuse with the aid of the information available in the social media
like facebook and instagram. Such crimes are better described as cyber aided crimes.
When a complaint regarding the cybercrime is registered the concerned authorities
will look for the digital traces so that it can be used as the evidence for the happened

crime. As per the definition of Interpol Digital forensics refers to the scientific process of identification, preservation, collection and presentation of digital evidence so that it is admissible in the court of law. Many preventative approaches have been adapted to aid in the prevention of cybercrime. Spam detection systems [1] assisted mobile users to avoid receiving unwanted sms spam, which is creating annoyance and unwanted cybercrime activities. The Interpol also defines Forensic science as the systematic and coherent study of traces to address questions of authentication, identification, classification, reconstruction, and evaluation for a legal context. Forensic tools have been developed for a variety of environments, including mobile devices, memory, and SIM cards [5], to cope with the collecting of digital evidence. Earlier the cyberspace was restricted to the computer and software, as the information and communication technologies grow, the cybercrime occurs in the advanced technologies such as application implemented with cloud computing, block chain technology and Internet of Things. In the context of distributed systems, the concept of cloud computing emerged for the sake of ubiquitous access to the computing, storage, development and software availability. Further to the introduction of sensor, mobile devices, IoT due to the limitation in its computing and storage capacity, these smart devices are attached to the cloud. Blockchain technology has introduced the concept of decentralized system which focused on data integrity in various applications. The chapter deals in the pursuit towards the goal for the better study of forensics using the artificial intelligence and in the fields of cloud computing, blockchain technology, IoT (hereinafter referred as CIIotB) that explores the interplay between these interrelated areas. The primary contributions based on the primary studies are to:

1. Present the strategies of forensics in cloud, IoT and Blockchain Technology and the application of AI in these areas
2. Present Findings and Discussion, for Artificial Intelligence Assisted forensics.
3. Present Challenges and Open Issues in the application of AI in the forensics of the distributed system concerned to Cloud, IoT and Blockchain Technology (CIIoTB).

The remainder of the chapter is structured as: background in Sect. 2, prior research in Sect. 3, Problem definition in Sect. 4, Systematic Literature review process in Sect. 5, Findings in Sect. 6 and Results and Discussion in Sect. 7 followed by challenges and open issues.

2 Background

There is a wide variety of cybercrime incidents happening throughout the world. As the world is moving towards automation of activities most of the application domain in government, businesses keep their valuable records in the form of digital documents and databases. As the use of online activities increases, these servers are the targets for the cybercrime causing huge damage or disruption. Cyber forensics comes to the rescue for finding these breaches and attacks. An organization working with the best practices of cyber forensics in fact can reduce the risk of attack and improve security. The most severely affected country by cybercrime in 2018 is United States [15], with respect to damage in financial sector. The experts in the industry estimate that the U.S. government is attributed to the costs of over 13.7 billion U.S. dollars due to cyberattacks. The cybercrime has no border and India is also one of the victim for these type of attacks. To have a taste of the work of cyber forensics experts in India, we could check with the news [57] that Rs.6 crore fraud unearthed, license of duty free shop suspended wherein which the forensics experts of C-DAC has established the fraud by finding the desired evidences from the computers. Organizations [71] take assistance of the computer forensics when they face problems such as the theft of Intellectual property, disputes in employment, cheating fraud enabled by phishing mails, forgeries, regulatory compliance. Computer forensics [28] has mainly 3 process 1. Collect the digital evidence 2. Analyze the data for the crime happened 3. Reporting is the final outcome of the above steps. This is clearly illustrated in the following figure depicting the process involved in forensics. Cloud computing offers delivery models Infrastructure as a Service (IaaS), Software as a Service(SaaS) and Platform as a Service(PaaS) with the benefits of less cost, faster computing, secured access to the data with the better performance. The distributed and virtualization technologies led to the evolution of the cloud computing. The cloud services are generally offered to the cloud customers by the cloud service providers (CSP). The main characteristics of cloud computing as per NIST definition is On-demand self-service, Ubiquitous network access, Location independent resource pooling, rapid elasticity, and pay per use. Cloud computing forensic science [47] is the application of scientific principles, technological practices and derived and proven methods to reconstruct past cloud computing events through identification, collection, preservation, examination, interpretation and reporting of digital evidence Cloud forensics [55] deals with the analysis of the cybercrime in the cloud computing. It deals with the three dimensions namely, technical, organizational and legal. The software/hardware tools to perform the forensic process. The various stakeholders/artifacts involving in the cloud operation such as cloud service providers, customers, Service level Agreement (SLA) are dealt in organizational aspect. Legal comprises of the involved documents that do not violate the legality/regulations of the various organizations/regions. Blockchain Technology allows the decentralization of the distributed software system in which components of the shared system agree on the shared system states without the requirement of trust or any centralized system. The design approaches of the blockchain technology are used in various applications. The stakeholders of application developers started to realize that the technology can be used in banking and finance such as international payments, in capital markets such as fast clearing and settlement, improvement in operations, performing audit trail, supply chain management, healthcare, media and energy. Blockchain technology [37] comes into rescue when the digital evidence is to be stored while safeguarding confidentiality, on online file storage with encryption that interacts with the private implementation of Hyperledger Fabric. Chain of Custody [38] is the one that maintains the log the details of how the evidence is gathered, analyzed and stored for producing the information about when, where and who came into the contact with the subject matter of evidence. It maintains the history of handling digital evidence in the chronological order. It is vital to preserve the digital evidence without any com- promise to the integrity so that it can be produced before the court of law. The preservation of the Digital evidence is automated with the help of the blockchain technology hyperledger composer. It prevents the digital assets to be corrupted by the untrusted participants. Many real world applications use IOT devices to capture data of the environment. The edge devices distribute the load of collecting the data and are scalable. The IOT devices nowadays contribute to receive a huge amount of data for processing and are stored in centralized server. The security and space problems occurred due to this is resolved with the decentralized blockchain database. The IoT [63] and blockchain along with AI are used in the applications such as healthcare, military, government, smart home and agriculture. These applications take decision automatically as it is incorporated with the Artificial intelligence to posses the decision making capabilities. Thus the combination of AI, blockchain and IOT is emerging as a better ecosystem for preserving the digital evidence and for other process in the field of forensics. This chapter also will discuss about the challenges to apply the process of forensics.

3 Prior Research

Cloud computing, blockchain technology and IoT are the frequently used technologies and are interrelated in many applications such as smart city, smart home. The systematic review of the application of forensics using Artificial intelligence in the combination of these fields appeared to be none or very limited to the best of our knowledge. Smart Homes are connected to many smart devices which are classified as the Internet of Things (IoT). These devices are connected to cloud as the IoT's are characterized by low computational capability and storage capacity. Cloud offers various benefits in terms of computational and storage capacity in the form of Infras tructure as a Service (IaaS) and Storage as a Service. Applications using IoT can be hosted on the cloud to provide the uninterrupted service to its end user/applications. This is supported by the features availability and resilience of the cloud. These applications can be built with the blockchain technology to enhance data integrity in decentralized connectivity. The relationship between the Internet of Things and other developing technologies such as big data analytics, cloud, and fog computing is examined in this [21] article. The research [40] identified cloud forensics taxonomy, investigation tools and summary of digital artifacts. The survey [65] Identified and explained major challenges that arise during the complicated process of IoT-based investigations Frameworks that aim to extract data in a privacy-preserving manner or secure evidence integrity using decentralized blockchain-based technologies are given special consideration. The work [26] contributed the anatomy of several proposed combination platforms, apps, and integrations, as well as summarizing significant areas to enhance cloud and IoT integration in future works. The authors also emphasize the importance of improved horizontal integration among IoT ser-

Table 1 Summary of prior research

S. no	Survey articles	No. of primary studies	Year span of the references
1	Al-Fuqaha [21]	195	2007–2014
2	Jahantigh [26]	38	2011–2019
3	Manral et al. [40]	80	2009–2019
4	Stoyanova [65]	201	2011–2019
5	Agustín Salas-Fernández et al. [59]	41	1975–2020
6	Taylor [66]	42	2015–2018

vices. The paper [8] establishes AI Forensics as a new discipline under AI Safety. Explainable domain AI provides a number of methods that could be useful in deciphering the decision-making process in malevolent models. Because computational intelligence [43] is based on human intellect, it is intended to do tasks at or above human ability. It is based on a number of fundamental concepts, including evolutionary algorithms, neural networks, fuzzy systems, and multi-agent systems. There are numerous examples of multi-agent system [70] uses in forensic investigations. Artificial intelligence approaches are used to detect threats and optimize attacks by employing specific measures in the detection and attack techniques. The reduction of characteristics in the training stage accounts for a major amount of the optimization in threat identification. The systematic review in this article Metaheuristics [59] are important in decreasing these characteristics. The systematic study [66] also offers light on future research, teaching, and practice approaches in the blockchain and cyber security space, such as blockchain security in IoT, blockchain security for AI data, and sidechain security. All of the preceding studies address questions about the broader usage of cloud computing, IoT and blockchain technology, but none of them particularly look at its application of AI in forensics in the distributed systems environment (CIIoTB). Overview of the prior research is summarized in the following Table 1.

4 Problem Definition

This section will define the problem that the study is attempting to solve, as well as the study's aim and its limitations. The term Forensics has been use around the time with the inception of usage of word criminal investigation and its related activities. While digital forensics techniques are employed in a variety of situations other than criminal investigations, the ideas and procedures are largely the same. While the method of investigation may differ, the sources of evidence do not. Computer-generated data is the source of digital forensic examinations. This has traditionally been limited to magnetic and optical storage medium, although snapshots of memory from running systems are increasingly being investigated. In the recent situations, the cloud computing acts as the storage medium where in which the data to be investigated are stored. The security incidents happen in the cloud and IoT environments governed by blockchain technology need to be monitored. In this regard these environments need to be aware about the threats that can happen. Forensic aware environment in the application involving cloud, IoT and blockchain is the need of the hour to handle the forensic analysis in the pre-incident and post-incident scenarios. The ecosystem of IoT forensics actually involves IoT devices along with the cloud. Any incident happened at the devices level in the IoT environments, the data that need to be considered for investigation is at all the levels of the IoT layers which includes the related device, connected cloud storage and computing locations including client systems.

4.1 Aim

Recent study [18] shows that there are various applications that are implemented with the Blockchain integrated IoT using cloud computing. The survey in this paper [69] discusses in entirety about the advantages of the integration of Blockchain Technology with cloud and Internet of Things. Cloud of Things offers elasticity and scalability characteristics to improve the efficiency of blockchain operations, while blockchain delivers unique ways to address difficulties in decentralization, data privacy, and network security. As a result, blockchain and Cloud of Things connection model has been widely considered as a promising facilitator for a variety of application scenarios. There are varieties of applications refer Fig. 1 that are benefited by such integration of technologies like smart home, smart city, and smart university [44].

The growth of various types of Blockchain integrated cloud and IoT has led to a new type of attack on vital infra-structures that might be regarded new weapons in the new linked applications. Data in a networked environment grows at an exponential rate, necessitating the use of robust analysis tools capable of real-time event analysis and decision assistance. In addition to logging capabilities, forensic capabilities such as entire session packet captures from malicious network connections aimed at turning packet data into documents, web pages, voice over IP, and other recognized files are required. Latest high-performance systems with powerful hardware as well as the various advanced paradigms of Artificial Intelligence, Machine learning, and Deep learning, in conjunction with data analyt-ics, make it possible to create data models fed by cause-and-effect analysis. Security Information and Event Man-agement (SIEM) [22] can make smarter event detection and decision making by using these technologies. Setting this context, this chapter deals about the three research questions that covers about the strategies of forensics currently applied in cloud, IoT and blockchain environment, forensics that is dealt in the integrated applications involving these technologies and about the application of artificial intelligence for the assistance of forensics.

Fig. 1 Blockchain, AI integrated cloud and IoT applications

4.2 Limitations

Considering the wide range of technologies involved, the scope of the review is based on the research questions specified. The strategies of the forensics are dealt with the articles based on the stages of digital forensics. The inter-related, inter-dependent of the cloud, IoT and blockchain technologies (ClIotB) are discussed to emphasize on the advantages of such kind applications and then the forensics in these environment. Then the application of the Artificial Intelligence (AI) is reviewed in the field of digital forensics in general computing and also in ClIotB environment. The term AI indicates its pertinent technologies Machine Learning (ML), Deep Learning (DL), Knowledge representation and Ontologies.

5 Systematic Literature Review Process

Systematic Literature Process consists of the 3 phases: (1) review planning, (2) review execution, and (3) review documentation phases.

5.1 Review Planning

Research Goals The goal of this study is to review existing studies and their findings, as well as to outline research activities in forensics employing AI in cloud computing,

Table 2 Research Goals

Research questions	Discussion
RQ1: What are the forensic strategies adapted in cloud computing, blockchain and IoT?	Cloud computing, IoT, and blockchain forensics have diversified away from traditional digital forensics. A review of the most recent forensic applications in these areas will aid in understanding the full scope of its impact in digital investigation
RQ2: How Forensics is practiced in the inter-related, interdependent Cloud computing, IoT and Blockchain (CIIoTB) environment?	Integration of promising technologies such as cloud computing, blockchain, and the Internet of Things has shown to be quite beneficial in many applications like smart home, smart factory. A study of the interrelationship between these technologies from end to end in the context of forensics is required to assist the investigator in moving forward with the investigation smoothly
RQ3: How Artificial Intelligence improves the forensic methodologies in digital forensics and CIIoTB environment ?	Artificial intelligence (AI) can be used in digital forensics. This will provide an overview of the approaches used to deploy AI in digital forensics. The studies can elaborate for incidents involving cloud, IoT infrastructure, and their services in conjunction with blockchain in both criminal and civil litigation

blockchain, and IoT. Three research questions in Table 2 are devised to help us focus our efforts.

Up till early 2021, we found 49 primary studies on the applications of digital forensics in cloud computing, blockchain, AI and IoT. This list of studies can be used by other academics to expand their research in this area. These studies can serve as useful benchmarks for comparing research of a similar nature. We do a thorough evaluation of the data in these studies and provide it in order to represent the research, thoughts, and concerns in the disciplines of forensics employing AI in cloud computing, blockchain, and IoT.

Research Methodology In order to attain the purpose of answering the study questions, we conducted the SLR in accordance with the criteria established by Kitchenham [31] and Charters. To allow for a full examination of the SLR, we endeavored to progress through the review's planning, conducting, and reporting steps in iterations.

Defining Resource Databases IEEE Xplore Digital Library, ScienceDirect, SpringerLink, Wiley Library, ACM Digital Library, Microsoft Academic Search, MDPI and Google Scholar were among the platforms examined. Other journals indexed by Scopus and web of science collected through the Google search engine are also available through search platforms. The keywords were chosen to facilitate the discovery of study findings that would help answer the research questions.

Selection of Primary Studies

Research Terms Passing terms into a publication's or search engine's search function brought up primary studies. The keywords were chosen to encourage the publication of research findings that would help answer the research questions. Kitchenham [31] recommends experimenting with different search phrases generated from the research topic and compiling a list of synonyms, abbreviations, and variant spellings. For improved results, Kitchenham [31] proposes utilizing advanced search strings built with Boolean AND's and OR's. The search terms were as fol-lows:

("IoT" OR Internet of Things) AND cloud forensics ("IoT" OR Internet of Things) AND blockchain forensics ("cloud" OR "cloud computing" OR "IoT" OR "Internet of Things" OR "blockchain" OR "block-chain" OR "distributed ledger") AND "forensics" ("cloud" OR "cloud computing" OR "IoT" OR "Internet of Things" OR "blockchain" OR "block-chain" OR "distributed ledger") AND "investigation" ("cloud forensics" OR "blockchain forensics" OR "IoT forensics") AND "intelligence" ("cloud forensics" OR "blockchain forensics" OR "IoT forensics") AND "AI" ("cloud forensics" OR "blockchain forensics" OR "IoT forensics") AND "machine learning" ("cloud forensics" OR "blockchain forensics" OR "IoT forensics") AND "deep learning"

The searches were made using the title, keywords, or abstract, depending on the search platforms. We conducted searches and analyzed all studies that had been published up to this point. Duplicate articles are deleted based on the results of these searches, and the first batch of over 112 publications is acquired. In the second phase, these publications are examined and classified as needed. The publications are excluded from the third and final phase of the literature evaluation if they do not meet the inclusion criteria listed below.

Selection criteria After the search phrases have been utilized in the databases, the selection criteria will determine which content will be considered for study. The inclusion/exclusion criteria are used only after all of the full texts have been retrieved in order to filter out all of the irrelevant texts. Inclusion criteria The following inclusion criteria were chosen for this study:

– Must be peer-reviewed
– Must be published in journals or conferences
– Must be published until 2021
– Relevant to the Research Goals
– Written in English.

Criteria for exclusion The following characteristics were chosen as exclusion criteria:

– Does not meet the criterion for inclusion
– Found a duplicate article in a different search
– Does not meet the focus points based on the research questions devised.

Article Evaluation Further to the inclusion and exclusion criteria, the articles are evaluated based on the following focus points to match the defined research questions.

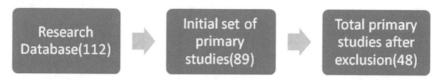

Fig. 2 Sequence for selection of primary study

Table 3 Number of primary studies considered per year

Year	<2015	2016	2017	2018	2019	2020	2021
Contributions	1	2	4	14	14	8	5

- **Criteria 1(C1)**: *Forensics*: The paper must focus on using forensics to solve a specific problem using cloud, IoT, or blockchain technology. The outcome of this criterion will enable to match with RQ1 questions.
- **Criteria 2(C2)**: *Context*: The study goals and outcomes must be placed in a sufficient context of forensics. This makes it possible to correctly interpret the research.
- **Criteria 3(C3)**: *Technology Integration*: Assessing the application of forensics in the cloud, IoT, or blockchain environment, or their mixtures, aids in answering RQ2 questions.
- **Criteria 4(C4)**: *Artificial Intelligence*: The study must provide sufficient information to provide an accurate picture of how artificial intelligence, machine learning, or deep learning has been applied for forensics in the field of cloud, IoT and blockchain technology which assists providing answers to RQ3.

Analysis Method Considering the inclusion, exclusion criteria and focus points the analysis is done separately to deal with the broad subject coverage, specific subjects dealt in a manuscript. The analysis also projected the key points reported, Challenges and the open issues in the subject discussed.

SLR analysis revealed 112 studies that needed to be considered further. After a rigorous evaluation of the 112 research that were chosen, only 49 (refer Fig. 2) publications remained that could address the SLR's study question. As a result, those 49 papers have been chosen as primary studies.

The Table 3 describes number of primary studies published each year regarding the forensics in cloud, IoT, and blockchain technology.

6 Findings

Each primary research paper was read, with relevant qualitative and quantitative data gathered and presented in the discussion section. The primary studies considered are tabulated in Table 4 with the details about the criterian, date of publication and name of the research database.

Table 4 Summary of primary studies

Sl. no	Primary study	C1	C2	C3	C4	Year	Research question?	Research database
P1	Ho [23]	✓	✓			2018	RQ-1	ScienceDirect
P2	Khan [29]	✓	✓			2016	RQ-1	ACM
P3	Pichan [50]	✓	✓			2018	RQ-1	ScienceDirect
P4	Zawoad [74]	✓				2013	RQ-1	ScienceDirect
P5	Ahsan [2]	✓	✓			2021	RQ-1	IEEE
P6	Liu [36]	✓				2019	RQ-1	IEEE
P7	Qi [52]	✓	✓			2017	RQ-1	IEEE
P8	Fu [20]	✓	✓		✓	2018	RQ-3	IEEE
P9	Ajay Kumara and Jaidhar [3]	✓	✓		✓	2017	RQ-3	ScienceDirect
P10	Pourvahab and Ekbatanifard [51]	✓	✓		✓	2019	RQ-2	IEEE
P11	Rane [53]	✓	✓		✓	2021	RQ-2	Google Scholar
P12	Zou [76]	✓	✓			2019	RQ-1	IEEE
P13	Irfan et al. [25]	✓				2016	RQ-1	Wiley
P14	Al-Masri [41]	✓	✓		✓	2018	RQ-3	IEEE
P15	Pichan [49]	✓	✓	✓		2020	RQ-2	IEEE
P16	Janjua [27]	✓	✓	✓	✓	2020	RQ-3	Google Scholar
P17	Patil and Ainapure [48]	✓	✓			2019	RQ-1	IEEE
P18	Zhang [75]	✓	✓	✓		2017	RQ-2	IEEE
P19	Cheng [11]	✓	✓			2017	RQ-1	IEEE
P20	Teing [68]					2019	RQ-1	IEEE
P21	Teing [67]	✓	✓			2018	RQ-1	Wiley
P22	Srinivasan and Ferrese [64]	✓	✓			2019	RQ-1	Wiley
P23	Ricci [56]	✓	✓	✓		2019	RQ-2	IEEE
P24	Rane and Dixit [54]	✓	✓	✓		2019	RQ-2	Springer
P25	Aljahdali [6]	✓				2021	RQ-1	Google Scholar

(continued)

Table 4 (continued)

Sl. no	Primary study	C1	C2	C3	C4	Year	Research question?	Research database
P26	Babun [7]	✓	✓		✓	2018	RQ-3	Google Scholar
P27	Shrivastava [61]	✓	✓		✓	2019	RQ-3	Springer
P28	Lutta [39]	✓	✓	✓		2020	RQ-3	Google Scholar
P29	Costantin [14]	✓	✓			2020	RQ-1	Google Scholar
P30	Chi [13]	✓	✓			2018	RQ-1	IEEE
P31	Meffert [42]	✓	✓			2018	RQ-1	ACM
P32	Nieto [46]	✓	✓			2018	RQ-1	Google Scholar
P33	Le [33]	✓	✓			2019	RQ-1	ScienceDirect
P34	Nieto [45]		✓	✓		2020	RQ-1	IEEE
P35	Billard [9]					2018	RQ-1	ACM
P36	Li [34]	✓	✓			2021	RQ-1	ScienceDirect
P37	Ryu [58]	✓	✓	✓		2019	RQ-2	Springer
P38	Kirrane and Di Ciccio [30]	✓	✓		✓	2020	RQ-3	IEEE
P39	Le [33]	✓	✓	✓		2018	RQ-2	IEEE
P40	Li [35]	✓	✓	✓		2019	RQ-2	IEEE
P41	Dasaklis [16]	✓	✓	✓		2019	RQ-2	Google Scholar
P42	Duy [19]	✓	✓			2019	RQ-2	IEEE
P43	Datta [17]	✓	✓		✓	2018	RQ-3	Springer
P44	Sikos [62]	✓	✓	✓	✓	2021	RQ-3	Wiley
P45	Schneider and Breitinger [60]	✓	✓	✓	✓	2020	RQ-3	Google Scholar
P46	Xu [72]	✓		✓		2018	RQ-3	IEEE
P47	Chhabra [12]	✓	✓		✓	2020	RQ-3	Springer
P48	Bonomi [10]	✓	✓			2018	RQ-1	Google Scholar
P49	Kumar et al. [32]	✓	✓		✓	2017	RQ-3	IEEE

7 Results and Discussion

Forensic technologies used for computer has found a tremendous growth based on the demand that is presently available with the latest software and hardware applications that is built around various distributed systems such as Cloud, IoT and Blockchain. Digital forensics generically follows the process of collection of data, Examination and analysis of the relevant records collected and reporting to the desired authority as per law.

7.1 Results

This section deals with the results of the research questions

RQ1: What are the forensic strategies adapted in cloud computing, blockchain and IoT? Forensics strategies are adapted in various environments cloud computing, blockchain and IoT depending on the requirements. The same is tabulated in the Tables 5, 6 and 7.

RQ2: How Forensics is practiced in the inter-related, interdependent Cloud computing, IoT and Blockchain (CIIoTB) environment? The following Table 6 gives a list of Forensics strategies that are practiced in the inter-related, interdependent Cloud computing, IoT and Blockchain (CIIoTB) environment.

RQ3: How Artificial Intelligence improves the forensic methodologies in digital forensics and CIIoTB environment? Artificial Intelligence (AI) and Machine Learning (ML) have been around for a long time, but now we have the processing capacity to create strong artificial neural networks (ANN) in a reasonable amount of time with the help of robust hardware and software. In this section we tried to find the researches considering the application of AI and its allied techniques for the forensics in the field of cloud, IoT and Blockchain technology. AI and machine learning offer numerous benefits and have a promising future. However, the same technology can be used to plan, automate, and carry out a variety of violent crimes that can be harmful to individuals. As AI becomes more freely available to a larger number of individuals, the technology's potential for harmful usage grows considerably.

7.2 Discussion

This section deals with the discussion about the Research questions raised. The study resulted into following Progress arrow towards AI assisted forensics for CIIoTB environment (Fig. 3).

Table 5 Summary of research question 1

Stages	Strategies	No. of papers	Articles	Forensics
Evidence collection	Time stamp analysis	1	P1	Cloud
	Privacy preserving of cloud data	1	P12	Cloud
	Framework using SIEM	1	P13	Cloud
	Assessing information quality produced by IoT devices	1	P29	IoT
	Integrity and authenticity of digital evidence using blockchain	2	P33, P34	Blockchain
	Blockchain based evidence management	1	P48	Blockchain
	Weighted evidence for confidence rating using blockchain	1	P35	Blockchain
Examination and analysis	Cloud log forensics	5	P2, P3, P4, P5, P17	Cloud
	Live forensics for volatile data	2	P6, P7	Cloud
	Cloud IP traceback	1	P19	Cloud
	Cloud forensics investigation with big data	2	P20, P22	Cloud

The progress arrow on AI forensics depicts the enablers AI, cloud, IoT and Blockchain. The necessary parameters that are found in the findings were added for the reader to understand the concepts in quick glance.

RQ1:What are the forensic strategies adapted in cloud computing, blockchain and IoT? The primary goal of the forensic strategies adapted in cloud computing, IoT and Blockchain environment is devised in such a way that any trace of the crime detected can be easily detected. The complexity of these distributed systems is the consequence in the intricacies of the forensic process. The digital forensic process in includes the collection of evidence, Examination and Analysis and Presentation. The Electronic evidence can be expressed in a variety of ways, including speech, text, graphics, photographs, and so on. It also possesses concealment, electronic, accuracy, dispersion, vulnerability, and mass features. Knowledge of the suspected

Table 6 Summary of research question 2

Stages	Strategies	No. of papers	Combination of technology	Articles
Evidence collection	Evidence collection in IaaS cloud using SDN and blockchain technology	1	Cloud and blockchain	P10
	Blockchain secure logging as a service	3	Cloud and blockchain	P11, P18, P24
	Framework for IoT data in the cloud for timeline analysis	1	IoT and cloud	P30
	Process provenance for privacy preserva-tion using blockchain in cloud	1	Cloud and blockchain	P18
Examination and analysis	Logging model in IoT-cloud	1	IoT and cloud	P15
	Blockchain based distributed cloud	1	Cloud and blockchain	P23
	IoT forensic chain for analyzing, traceability of IoT data using blockchain	1	IoT, cloud and blockchain	P40

network's system structure and software and hardware configuration information, including network topology, server, workstation, gateway, switch, router, and other hardware information, as well as network operating system and relevant application software con-figuration information. Electronic evidence can be dynamically sent in network equipment deployed across multiple geographical areas.

Cloud Forensics Cloud forensics addresses cloud-based incidents and their services in both criminal and civil proceedings. In most cases, the forensic process is a post-incident investigation. However, given the prevalence of cloud services and the ever-changing technological and cyber threat landscape, having forensically friendly cloud in place for continuous evidence gathering, aggregation, and storage is critical. The field of post-incident forensics has received a lot of attention. Cloud forensics processes will differ depending on the cloud computing service and deployment model. We have extremely limited control over process or network monitoring in SaaS and PaaS. In contrast, in IaaS, we can acquire more control and implement forensic-friendly logging mechanisms.

Time Stamp Analysis In cloud environments, the study [23] of significance of times-tamps as a type of file metadata discloses traces of digital evidence. File-based meta-data is examined, and timestamps are compared and validated across various cloud

Table 7 Summary of research question 3

Stages	Strategies	No. of papers	Articles	Forensics
Evidence collection	Semantic search on encrypted dataset	1	P8	Cloud
	Virtual machine analysis	1	P9	Cloud
	Log preservation in fog using agent based machine learning	1	P16	cloud
	Comparison of machine learning techniques for validated eata	1	P49	cloud
	Evidence collection in data intensive IoT	1	P14	IoT and Cloud
Examination and analysis	Digital forensic framework for smart environments	1	P26	Cloud
	Big data analytics with IoT with machine learning	1	P47	IoT and cloud
	Ontology for forensic investigation	1	P44	Cloud
	Modeling for predicting cyber hack breaches	1	P46	Cloud
	Investigating of AI systems that are hostile	1	P44	Cloud
	Evidence collection in data intensive IoT	1	P14	IoT

access behavioral patterns. To examine timestamps and distinguish between trends based on different types of cloud access operations, researchers used direct observation and cross-sectional analysis. This research contributes to the cloud forensics investigation of data breach occurrences, which collects, identifies, and analyzes the crime clues, features, and evidence of the incidents.

Cloud Log Forensics Cloud Logs: "Logging" is the process of recording events in a file when using an operating system, process, system, network, virtual machine, or application and the file is known as a "log file." The log file records the steps that were taken in a specific order during an execution. "Cloud logs" refers to all logs created in a cloud computing environment.Cloud log files are used to investigate various cyberattacks by documenting various system and network events. A feasible alternative is to search the cloud log files for malicious behavior and evaluate them using log analysis tools. The process of studying cloud log data through cloud computing

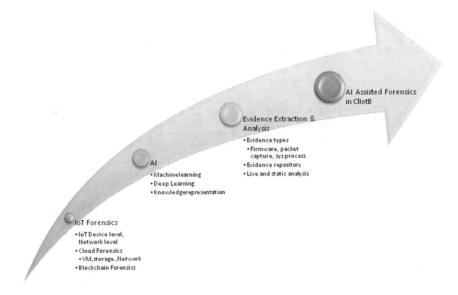

Fig. 3 AI forensics for CIIoTB

or through third-party analysis services is known as cloud log forensics. Cloud log forensics (CLF) [29] helps investigators by discovering attackers' malicious conduct through in-depth cloud log analysis. Cloud log accessibility qualities include concerns such as cloud log access, cloud log file selection, cloud log data integrity, and cloud log trustworthiness. As a result, cloud log forensic investigators rely on cloud service providers (CSPs) to gain access to various cloud logs. A dishonest investigator can tamper with the log before handing it over to the authorities. Furthermore, the investigator can compromise the user's privacy in coordination with unethical CSP or CSP employee(s). This can take various forms, including the insertion of incorrect entries, the removal of critical entries, the modification of existing data, and the rearrangement of log entries to deceive investigators and hide malevolent activity. Many users are hosted on cloud servers. This opens the door for a hostile cloud user to deny that the log files under review are the result of another user's activities. A log file leak can expose information, resulting in a privacy infringement. To show the correctness of logs, cloud logs should display the accurate history of a system's event with the occurring time. Tamper-resistant logs are required. An incorrect log entry cannot be introduced as a valid one by anyone other than the true logger. Both the user whose activity is documented in the log and the investigating entity should be able to verify it. A secure cloud log that has the properties of correctness, tamper proof, chain of custody, and forward secrecy should be kept in such a way that it may be used in a criminal court of law. The CFLOG [50] application operates on the virtual stack of computing environment at the Hypervisor level in the CFLOG framework. It creates one or more log files for each cloud system user, allowing for the separation of distinct user activity. A set of Log File Per User can be supplied as a log file. In a multitenancy situation, this makes evidence segregation easier. SecLaaS [74] encrypts the log with the investigating agency's public key and creates a log chain to ensure the cloud user's confidentiality and integrity. In order to limit the risk associated with data volatility, it collects data from one or more log sources, parses the data, and then stores the parsed data in persistent storage. Logs are encrypted using the particular user's public key in Cloud Log Assuring Soundness and Secrecy (CLASS) [2], so only the user may decipher the material. We produce proof of past log (PPL) using Rabin's fingerprint and Bloom filter to prevent unauthorized log change. This method considerably lowers verification time. Log files provide information such as user log-in and log-out attempts, network events, and an explanation of when and why logs are generated. Snort and Hadoop are used to analyze logs[48] in Eucalyptus. Cluster logs to detect capability or information exchange issues with clusters, fault logs to know specifics of known remedies and details error code, and debug logs generated from the level of debugging-based logging are all part of the Eucalyptus or Snort IDS in a cloud environment. This record will be exported to the Syslog server and used for map-reduce in accordance with the forensics investigation's requirements. Flume can be used to solve log location and format concerns when analyzing datasets. Cipher-text policy attribute-based encryption(CP-ABE) techniques coupled with forward and backward security in multi authority cloud storage systems allows [73]. In a cross-tenant access context, a CSP's cloud resource mediation service (CRMS) [4] can provide users with safe access control services. Cloud forensics requires the secure archiving and investigation of various records. Untrustworthy cloud stakeholders and malevolent actors from the outside can work together to change logs after the event and remain undetectable. forensic-aware blockchain-assisted secure logging-as-a-service [54] for cloud environments to securely store and process logs while addressing multi-stakeholder collusion and maintaining integrity and confidentiality.

Live Forensics for Volatile Data Forensic Data Acquisition: To speed up the forensic investigation process in the cloud, large amounts of volatile data must be recorded, transmitted, and assessed quickly LiveForen [36], is that the integrity of a system that enables a secure forensic data collection and transmission process in the cloud has been validated. To convey forensic material as a data stream, a unique fragile watermark is inserted into the data stream without changing the data itself. This is implemented in the cloud infrastructure itself. According to the findings, Live Forensics can achieve low overhead for attestation, data integrity verification, and scalability in an IaaS cloud environment. In cloud computing, live forensic tools run either in the target Operating System (OS) or as an external hypervisor, which is untrustworthy since it may be misled by the hacked OS. ForenVisor [52], focused to dependable live foren-sics. It uses a lightweight architecture to decrease the size of the Trusted Computing Base (TCB) that takes evidence directly from hardware and uses the File-safe module to protect the evidence and other sensitive files. It is used to collect process data, raw memory, and I/O data like keystrokes and network traffic. Many cloud forensics tools look for suspicious activities in data while ignoring specifics about privacy breaches and behavioral traits. In this paper [76], we use a multi-granularity privacy leakage forensics tool to investigate privacy breaches induced by malware in the cloud. Forensic inquiry is carried out using continuous RAM mirroring tech-nology and dynamic taint analysis. Security information and event management solutions are often set up to collect logs from cloud infrastructure equipment and store them in a central location. The database's logs are then normalized so that they may be interpreted. Correlation directives or a correlation engine are used to correlate the normalized logs, and events are generated as a result. The virtual infrastructure was subjected to a number of attacks, all of which were identified by SIEM solution framework [25]. Correlation necessitates extra calculations. Some shortcomings in static analysis can be addressed by live analysis. For the reconstruction of the crime scene, volatile data such as memory and I/O data, including keyboard input and network packets, is crucial. For live analy-sis, more volatile information such as the process list, kernel objects, and network traffic is provided. Live analysis raises trust issues because the analysis tool could be hacked by malware, and the visible volatile data could be forged by a rootkit. The nature of cloud-enabled large data storage solutions makes it difficult to identify residual evidence. Syncany is a popular Java-based open source and cross-platform cloud storage solution that offers data storage on a variety of backing media. This seeks to cut down on the amount of time and resources needed for a real investigation. The data remnants recovered by using the Syncany private cloud storage service as a backbone for massive data storage are discussed [68]. These files may contain the sync, file management, authentication, and encryption metadata required for identifying synced files and cloud hosting instances, as well as connecting cloud transaction receipts. CloudMe is a software-as-a-service cloud model. Users of CloudMe can share content with each other and with the public via email, text messaging, Facebook, and Google sharing. Client-side residual artefacts that give core-evidences serve as a starting point for CloudMe research in a big data environment [67]. The cache database, web caches, log and configuration files should all be examined by forensic practitioners researching the CloudMe cloud application. Memory dumps may have the ability to provide alternate techniques for retrieving application cache, logs, configuration files, and other information of forensic importance, according to the physical memory captures. Migrating the forensic analysis process to the cloud, which currently offers metered services, will result in better resource use while lowering expenses. The benefit of FaaS [64] will be that it will provide accreditation and certification organizations with easy access to validation and certification tools and processes. Network forensics, security auditing, network problem diagnosis, and performance testing are just a few of the uses for IP traceback. It's utilized in cyber-investigation processes to figure out where packets come from and how they get there. The goal is to prevent unauthorized users from requesting cloud information traceback for nefarious purposes. For authenticating traceback service requests, a temporal token-based authentication system called FACT [11] verifies that the entity seeking traceback service is an actual recipient of the packets to be traced.

The study must provide sufficient details to provide an accurate picture of how artificial intelligence, machine learning, or deep learning has been applied to a given problem.

IoT Forensics One of the most difficult issues in IoT-forensics is the analysis and correlation of diverse digital data to enable efficient interpretation of complex scenarios. This is the method for extracting unique objects representing persons or devices from case files, creating the context of the digital inquiry, and gradually improving knowledge by employing more case data such as network captures. External searches using open source intelligence sources are used as necessary, and the JSON Users and Devices Analysis tool [45] is used to construct the context from JSON files, complete it, and display the entire context using dynamic graphs. This article [6] examines two IoT forensic investigation models, each with its own set of stages and procedures for forensic investigators. They are IoT digital forensic investigation model and application-specific digital forensic investigation model. is a form of IoT application that is very specialized. The use of IoT apps allows for greater flexibility and consistency in the research of specific types of applications. The IoT digital forensic investigation model can perform analysis and inquiry in a relatively obvious path. IoT Forensics is investigated in terms of data quality, with a theoretical model that incorporates data quality, technology resources, and human factors. The collected information must be accessible and intelligible as digital data, as well as acquired and stored, to be acceptable for forensics purposes. A priori data discarding is not suggested in forensics due to the possibility of removing essential details. IoT Forensics is investigated [14] in terms of data quality, with a theoretical model that incorporates data quality, technology resources, and human factors. The collected information must be accessible and intelligible as digital data, as well as acquired and stored, to be acceptable for forensics purposes. A priori data discarding is not suggested in forensics due to the possibility of removing essential details. The goal of forensic state acquisition from the Internet of Things [42] is to address the acquisition of the state of IoT devices so that a clear picture of events can be produced. It comprises of three states of a centralized Forensic State Acquisition system to achieve this goal. Controllers were used to connect to IoT devices, the cloud, and other controllers. The digital witness technique uses a mechanism that allows citizens to submit their data while maintaining some level of anonymity In IoT applications, a digital witness is an innovative way to acquire digital evidence. However, there are some severe privacy concerns with this method. To include privacy aspects, a privacy-aware IoT-Forensics model [46] was defined in. A set of standard privacy principles is regarded a vital aspect of the digital witness solu-tion's lifespan. Citizens are likely to cooperate more in digital investigations conducted in dynamic IoT environments as a result of their faith in the system's handling of sensitive information kept on their own devices.

Blockchain Forensics The accuracy of automated forensic instruments has a significant impact on the reliability of inquiry outcomes. There is a need for a robust automated system for preserving evidence artifacts and providing auditing capabilities to ensure the validity of forensic tools in order to maintain evidence integrity. The forensic-chain [37] approach is based on Hyperledger Composer, a permissioned Blockchain that runs in a controlled environment overseen by the consortium or single entity that deploys it. This eliminates the need for a trusted third party to verify evidence transfer and reach an agreement. Evidence collection, preservation, and validation can all be improved. Despite the fact that various entities control the evidence, the Chain of Custody (CoC) must ensure that it is not tampered with during the inquiry. Dematerialize the Chain of Custody process with blockchain-based Chain of Cus-tody [10], ensuring auditable integrity of collected evidence and owner traceability. It is built on a permissioned and private blockchain. It prevents unauthorised and untrustworthy persons from handling digital evidence. Each piece of digital evidence is given a confidence level, which aids jurors and magistrates in their work. An im-mutable e-evidence blockchain [9] is used to collect e-evidence. The Digital Evidence Inventory provides the ability to construct a digital evidence chain. ability to assign a level of confidence to each piece of evidence Provide a Global Digital Timeline to specialists. The DEI, which also provides traceability, is accessible to all parties in a study. The "chain of evidence" is enforced by the blockchain: any change to the evidence can be tracked back. From evidence collection to access, a blockchain-based [34] lawful evidence management scheme to supervise the complete evidence flow and all court records. To ensure the privacy of witnesses, brief randomizable signatures are employed to anonymously validate their identities. To record evidentiary transactions, a 'consortium blockchain' based on a local ethereum network has been created. The ability to study and evaluate the security of the entire network system is provided by SDN's digital forensics mechanisms. It should be able to recognize, gather, and analyze log files as well as specific information about net-work devices and traffic. Hackers, on the other hand, can change, even delete log files after breaking into a machine or device to erase proof of their presence and actions in the system. SDNLog-Foren [19] is a network forensics, a blockchain-based technique for log management in SDN has been developed. Log collectors and analyzers, as well as Blockchain-based storage, are all part of the SDN log collection agents.

7.3   How Forensics is Practiced in the Inter-Related, Interdependent Cloud Computing, IoT and Blockchain (CIIoTB) Environment?

The demand for interconnected technologies varies depending on the application's requirements. Some applications require Cloud and IoT, while others require IoT and blockchain, or all three technologies combined. The following are some of the numerous studies that have been done in the field of forensics in such interconnected technologies.

Evidence collection in blockchain and cloud The evidence is collected and stored in a blockchain that is distributed among several peers in this forensic architecture [51]. All data is encrypted and saved in the cloud server, depending on the sensitivity level. A block is produced in the SDN controller for each piece of cloud data, and the data's history is logged as metadata. By installing Fuzzy based Smart Contracts, the system allows users to track their data. Finally, the construction of a Logical Graph of Evidence using blockchain data enables evidence analysis.

Application of logs for trust Powered by the blockchain Secure Logging-as-a-Service [53] is a technique for safeguarding logs, ensuring their trustworthiness and retrieval in the event of tampering, allowing for a calm forensic examination. For storing blockchains A decentralized off-chain data storage platform called Interplanetary File System is used. It combines distributed processing with an efficient lookup approach, resulting in readily accessible data for forensics. The Ethereum blockchain is being utilized. The Internet of Things generates events and log data. IoT forensics examines events and data in relation to them. What went wrong? When did the incident take place? How did the incident take place? What happened and/or who did it?. As a solution, an integrated IoT-Cloud computing environment [49] is built with an event recording model and architecture to facilitate digital forensics in IoT. It specifies when these parameters should be included in the logs. The fact that all log data is stored on the Cloud allows for data aggregation and manipulation. To the Cloud repository, IoT log data is embedded with essential information. Cloud computing and IoT are two comparatively challenging technologies, with Cloud computing resolving the majority of IoT issues.

Data for provenance, acquisition and analysis It is presented a process provenance in cloud [75] that uses blockchain and cryptography group signature technologies to give proof of existence and privacy preservation for process records. The provenance system will allow for the auditing of process records. Because each node of the forensic data gathering process generates a submission record and a blockchain receipt, all cloud forensics stakeholders may validate the process record. For the acquisition of data from blockchain-based distributed storage systems like STORJ [56], forensically sound procedures and tools are required. It's a peer-to-peer cloud storage network with end-to-end encryption that lets users transfer and share data without relying on a third-party service like Google Drive. A data gathering and forensic analysis framework [13] for IoT devices connected to mobile devices and the cloud This aims to provide a consolidated evidence format for IoT investigations and a picture of how events unfolded in a cloud-based environment. To show the connected facts in digital admissibility, a timeline analysis was performed. On the basis of blockchain technology, a framework [58] for the IoT environment has been developed. All IoT device communications are saved as transactions in the blockchain, making the existing chain of custody process easier. Participants in the forensic investigation, such as device users, manufacturers, investigators, and service providers, can certify the investigation process publicly using a public distributed ledger. IoT forensics framework [33] built on blockchain to improve the integrity, authenticity, and non-repudiation features of acquired evidence. It is designed using a cryptographic-based strategy to address the issue of identity privacy. To address the issue of identity privacy, the evidence submitter's identity is hidden from the public using a modified Merkle signature system. The Internet of Things (IoT) forensic chain [35] provides forensic inquiry with high levels of authenticity, immutability, traceability, robustness, and distributed confidence between evidentiary entities and examiners. It can provide traceability assurance and track the origins of evidence items. The details of the evidence are preserved in block chains. By making the audit train transparent, it can boost the trust of both evidence items and examiners. After evaluating the links between evidence items, provenance, traceability, and auditability of each evidence item, the fundamental idea is to retrieve artifacts from IoT devices and write them to a blockchain-based framework. Blockchain technology has lately been offered as a potential alternative for building reliable digital forensics techniques. As indicated in the available literature [16], there is a link between blockchain technology, cloud technologies and current digital forensics methodologies.

7.4 How Artificial Intelligence Improves the Forensic Methodologies in Digital Forensics and in ClIoTB Environment?

Machine learning has the potential to revolutionize the way forensic scientists assess pattern evidence. Combating fraud necessitates a diverse set of algorithms, data sources, and methodologies, as well as a set of processes that can serve as a successful methodology for machine learning forensics investigators.For attackers, AI systems present new opportunities: it may be easier to control AI systems to commit harmful acts than it is to write native software. They could, for example, train an existing AI system to identify a victim instead of inventing a face-detection algorithm. This contributes to the idea that a system is "malicious by design."

AI in cloud, IoT and blockchain systems: Following are some research work related to AI in cloud, IoT and blockchain systems. As cloud computing becomes more widely adopted, an increasing number of users are outsourcing their datasets to the cloud. Datasets are frequently encrypted before being outsourced to protect privacy. However, because it is difficult to search for specific keywords in encrypted datasets, the frequent practice of encryption makes optimal data exploitation difficult. ECSED [20] is a new semantic search technique based on the encrypted datasets' concept hierarchy and semantic relationships between ideas.

Virtual Machine introspection using ML: Virtual Machine Introspection has developed as a fine-grained, out-of-VM security solution for detecting malware by introspecting and recreating the live guest Operating System's volatile memory state. At the hypervisor, the reconstructed semantic data collected through introspection are available in a mix of benign and harmful states. Existing out-of-VM security solutions require substantial human analysis to discern between these two states. By using introspection, Memory Forensics Analysis, and machine learning techniques at the hypervisor an advanced VMM-based, guest-assisted Automated Internal-and-

External introspection system [3] is created. The machine learning model is usually trained by the cloud service provider who owns it. Because the cloud is unable to divide the encrypted dataset according to the best attributes chosen, a new approach [32] for decision tree training without dataset splitting is proposed. The cloud service provider learns nothing about the user's input and classification result, while the trained model is kept hidden from the user, who can only learn the classification result. It assures that both the user and the cloud service provider's privacy is protected, while simultaneously lowering the user's compute and communication expenses.

Application of Intelligence for logs: Determining what type of data should be collected from IoT devices and how forensic investigators might use traces from such devices is getting increasingly difficult. FoBI [41] is a fog-based IoT forensic platform that seeks to address this problem using user-agent. It makes use of the fog computing paradigm, which allows intelligence to be pushed to the network's edge via a gateway. This is appropriate for data-intensive IoT systems. The fog node or gate-way can be programmed with intelligence to filter the data that needs to be transmitted. The data connection between an IoT device and a fog node can then be used to recover forensic evidence. Logs from IoT-based smart environments are collected and stored in cloud-connected storage. Cloud-assisted models provide log secrecy and confidentiality, but they are vulnerable to multi-stakeholder collusion. By addressing the multi-stakeholder issue in a fog enabled cloud privacy preservation automation of log probing [24] via non-malicious command and control botnets in the container environment, the security and privacy-aware distributed edge node log preservation may be achieved. By carefully collecting data from the local immutable blockchain, Holochain provides an agent-centric and relativistic environment for creating the underlying validity of data and ensuring data integrity for distributed applications. A digital forensic framework for a smart environment is referred in this work [7], such as smart homes and smart offices, in the smart app programming platforms. Modifier and Analyzer are the two primary components. Modifier analyzes the source code of smart apps at compile time, detecting forensically relevant information and automatically inserting trace logs. The logs are then stored in a database at runtime. The Analyzer then uses data processing and machine learning techniques to extract useful and usable forensic information from the devices' behavior in the case of a forensic inquiry. A honeypot [61] can detect new types of IP addresses or attacks. An improved honeypot aids in the detection of these attack fingerprints. Analysts can categorize these assaults and enhance or update the firewall rules by analyzing the system logs. All communicated sessions are saved in log files by Cowrie honeypot. These log files reveal the attackers' fingerprints as well as the commands they used. Data from external attackers is classified using a machine learning technique. SVM has been determined as the best classifier of all classifiers.

Machine learning models for cloud, iot and blockchain By providing a Cloud Malicious Actor Identifier paradigm, it primarily focuses on this reasonable requirement of cloud forensic investigators. Using a well-known machine learning technique called Gradient Boosting, this model detects the malicious actors associated with a specific crime scene and ranks them according to their likelihood of being malicious. The primary goal of this methodology [17] is to reduce the time spent probing each and every IP address during an inquiry. To investigate the opposing features of IoT forensics, a quadrant model is given. The model compares the effectiveness of forensic investigative processes to the admissibility of evidence integrity, while also considering user privacy and service providers' adherence to laws and regulations. A semi-automated forensic procedure [39] based on machine learning could eliminate the human aspect from profiling and surveillance processes, resolving data security concerns. This gives the data owner confidence in their privacy and confidentiality, ensuring that only authorized users have access to the data. Authors [12] have concentrated on the examining phase for big data in a distributed environment for processing rising data from IoT-based environment systems. The architecture of forensics analysis is separated into four key parts. Data collector and information generator, analytics and extraction module, machine learning model design, and model analysis on several efficiency matrices are all included. The models utilized are Naive Bayes and Random Forest. The execution of arbitrary computations is supported by smart contract platforms like Ethereum and Hyperledger Fabric. The usefulness of these platforms for defining and enforcing data and service usage limitations, as well as providing compliance guarantees, has yet to be determined. In such an environment, symbolic artificial intelligence techniques [30] such as semantic technology-based policy languages and business process compliance tools and methodologies can be utilized together to give assurances. In the realm of expanding numbers of computing devices and IoT networks, as well as the data storage and processing requirements of digital forensics, automation is widely desired. Due to a lack of semantics defined for digital forensic investigation concepts, it has unstructured forensic data obtained from many sources. Purpose-built ontologies enable integrity checks via automated reasoning and facilitate anomaly detection for the chain of custody in digital forensic investigations by formally characterizing digital forensic concepts and attributes. In the integration layer of a framework [62], semantically enriched with Linked Data, digital evidence annotated with ideas from this ontology is used. For automated reasoning, OWL, RDF and SWRL are used. The number of smart devices on the market is continually increasing. AI systems are making decisions that affect our daily lives in an increasingly independent manner. Their acts may cause accidents, harm, or, more broadly, breach regulations, whether intentionally or unintentionally, and as a result, they may be held responsible for a variety of occurrences. It investigates AI systems that are intentionally hostile. The research [60] looks into how a CNN-based camera-based object recognition system could be used in a drone that is suspected of attacking a person. Forensic work requires fresh methodologies, according to research. The use of common explainability approaches like LIME or SHAP is limited. Analyzing cyber incident data sets is an important way to have a better understanding of how the threat environment is evolving. A statistical study of a span of year's data collection of breach incidents is carried out. Because of autocorrelations, both hacking breach incidence inter-arrival periods and breach sizes should be described using stochastic processes rather than distributions.

Both qualitative and quantitative trend analysis on the data set are under-taken to gain deeper insights [72] into the evolution of hacking breach episodes. The research [32] satisfies neighborhood outlier query processing, model reconstruction for specific data anomalies, random data verification, and root outlier detection in a network segment, among other things. It provides a method for determining and testing evidence that meets a set of criteria.

8 Challenges and Open Issues

8.1 Cloud Forensics

Despite the advantageous feature of having ubiquitous computing and storage facilities, Challenges from various researches can be consolidated and listed out at each stage of cloud forensics: (1) Physical resource location (Identification) (2) Evidence Integrity (Preservation) (3) Jurisdiction (Collection) (4) Timestamp disparity (Analysis) (5) Compliance (Reporting). Also in addition following are encountered:

– Data that is constantly changing
– Data integrity and chain of custody
– Data provenance
– Inadequate forensic tools
– Loss of data due to machine restart
– Inadequate knowledge of cloud complexity and other technological knowledge.

Open Issues still need to be addressed can be enumerated as:

– Practicality of conducting live and remote analysis on physical systems are less feasible or impossible.
– Logging capabilities and potential evidence data for the purpose of forensics need to be provided by cloud service provider to the investigators. Dependency on the relevant data completely lies on CSP.
– Trust in CSP is an important issue as the artifacts are vulnerable with malicious administrator and current or former employee.
– Evidence isolation and privacy concerns conflict in cloud.
– Dependability on Cloud service provider
– Data analysis from multiple sources due to its distributed technology.
– Access to Registry entries, temporary files, and memory is impossible because of virtualization technology.
– Incident of cybercrime is not bound to occur at a particular time and place due to the existence of cloud in geographically distributed characteristics - cannot quickly progress in the purview of government law of a country.
– Timestamp mappings of different file systems are complex.

8.2 IoT Forensics

IoT devices are continuously manufactured in various sizes, power and range based on the requirements by the in-dustry and appliances market. Attacks on these devices are obvious and tracking for the same requires the IoT forensics. Challenges in this field is listed out as below:

- Evidence identification is hugely difficult in the IoT context, because examiners may not even know where the investigated data is physically housed in some circumstances.
- In IoT-centric instances, the source of evidence could be diverse, ranging from an autonomous vehicle that caused a deadly accident to a smart toaster that came on in the middle of the night and started a fire in the home.
- There is no advice or established approach for forensically sound evidence collection from an IoT device.
- Evidence data must be obtained from various remote servers in IoT Forensics, making the goal of maintain-ing correct Chain of Custody substantially more difficult.
- IoT nodes are always on, and they generate a massive amount of data, making end-to-end analysis challenging.
- Geolocation data from a particular IoT device indicates that it was present at the crime scene at the time of the incident. Examiners must ensure that the device's location and time settings are correct. It's also im-portant to figure out if the gadget was used by another individual at the same time.

The absence of security in cyberspace is reflected in the profusion of issues in IoT forensics.

- Forensic tool limitations: Researchers and forensics experts must identify techniques and solutions that allow for accurate evidence gathering and storage.
- Legal authorities, cloud service providers, and device manufacturers must all work together to solve IoT se-curity issues.
- Need for standardization and certification.

8.3 Blockchain Forensics

The recovery of files and metadata that can be valuable in a prosecution is one of the numerous issues with blockchain-based distributed storage forensics. There's no certainty that such information can be recovered from a suspect's local storage. Identifying the location of file shards is a challenge that investigators may face.

8.4   AI Forensics

Forensics using AI has led to many questions which are: How can other AI approaches like reinforcement learning, alternative scenarios, and other deep learning network architectures like LSTMs be used to identify suspects? How may explainability methods be used to help with grey-box model analysis? How can data mining techniques be used to discover erroneous decisions? How may operational data or other sorts of proof be used? How can you tell if your system has been reset? While ontology-based digital forensic investigation systems have substantial benefits in terms of querying, data aggregation, and data fusion, they can only reach their full potential with structured data. However, given the vast and ever-expanding spectrum of computer devices, data capture frequently yields forensically sound data without capturing the semantics. As a result, one of the primary issues is figuring out how to efficiently generate structured data from hybrid data. To support IoT Fog/Edge systems are now established for the ease of supports. But attacks to these devices are expected and forensics to these devices with AI techniques is yet to be explored. This might include using machine learning techniques to improve bot operation and using log analysis at the fog level to discover rouge edge devices.

9   Conclusion

This Systematic Literature Review (SLR) informs scholars and industry practitioners about modern AI, cloud, IoT, and Blockchain based forensics methodologies. While there are articles that we had considered for discussion on forensics on AI , cloud and IoT, we were unable to locate any contributions on attacks and forensics studies of assaults in blockchain-based systems. Few of the papers were found on attacks related to bitcoin currency which is not our focus and not considered for our study in this review. The review [66] demonstrates that how blockchain can be used for security purposes. Similarly, the SLR reveals that fewer studies have been dedicated to the application of AI in the field of forensics. SLR also illustrates the importance of a number of well-known difficulties in IoT forensics research, as it is dependent on other technologies and is critical when considering an end-to-end IoT application as an integrated environment with cloud and other technologies.

References

1. Abayomi-Alli O et al (2019) A review of soft techniques for SMS spam classification: methods, approaches and applications. Eng Appl Artif Intell 86:197–212. ISSN: 0952-1976. https://doi.org/10.1016/j.engappai.2019.08.024. https://www.sciencedirect.com/science/article/pii/S0952197619302155. Accessed 21 Sep 2021

2. Ahsan MAM (2021) CLASS: cloud log assuring soundness and secrecy scheme for cloud forensics. IEEE Trans. Sustain. Comput 6(2):184–196. https://doi.org/10.1109/TSUSC.2018. 2833502
3. Ajay Kumara MA, Jaidhar CD (2017) Leveraging virtual machine introspection with memory forensics to detect and characterize unknown malware using machine learning techniques at hypervisor. Dig Investig 23:99–123. https://doi.org/10.1016/j.diin.2017.10.004
4. Alam Q (2017) A cross tenant access control (CTAC) model for cloud computing: formal specification and verification. IEEE Trans Inf Forensics Secur 12(6):1259–1268. https://doi. org/10.1109/TIFS.2016.2646639
5. Alhassan JK et al (2018) Comparative evaluation of mobile forensic tools. In: Rocha A, Guarda T (eds) Proceedings of the international conference on information technology & systems (ICITS 2018). Advances in intelligent systems and computing. Springer International Publishing, Cham, pp 105–114. ISBN:978-3-319-73450-7. https://doi.org/10.1007/978-3-319-73450-7_11
6. Aljahdali A (2021) IoT forensic models analysis. Rev Rom Inform Si Autom 31(2):21–34. https://doi.org/10.33436/v31i2y202102
7. Babun L (2018) IoTDots: a digital forensics framework for smart environments. arXiv:180900745
8. Baggili I, Behzadan V (2019) Founding the domain of AI forensics. arXiv:191206497
9. Billard D (2018) Weighted forensics evidence using blockchain. In: Proceedings of the 2018 international conference on computing and data engineering. Association for Computing Machinery, New York, NY, USA, pp 57–61. https://doi.org/10.1145/3219788.3219792
10. Bonomi S (2019) B-CoC: a blockchain-based chain of custody for evidences management in digital forensics. https://doi.org/10.4230/OASIcs.Tokenomics.2019.12
11. Cheng L (2017) FACT: a framework for authentication in cloud-based IP traceback. IEEE Trans Inf Forensics Secur 12(3):604–616. https://doi.org/10.1109/TIFS.2016.2624741
12. Chhabra GS (2020) Cyber forensics framework for big data analytics in IoT environment using machine learning. In: Multimed Tools Appl 79(23):15881–15900. https://doi.org/10. 1007/s11042-018-6338-1
13. Chi H (2018) A framework for IoT data acquisition and forensics analysis. In: 2018 IEEE international conference on big data, pp 5142–5146. https://doi.org/10.1109/BigData.2018. 8622019
14. Costantini F (2020) Assessing information quality in iot forensics: theoretical framework and model implementation. arXiv:201214663
15. Cycles T, Text provides general information S. assumes no liability for the information given being complete or correct D. https://www.statista.com/topics/3387/us-government-and-cyber-crime/
16. Dasaklis T (2020) SoK: blockchain solutions for forensics
17. Datta S (2018) An automated malicious host recognition model in cloud forensics. In: Perez GM (ed) Networking communication and data knowledge engineering. Springer, Singapore, pp 61–71. https://doi.org/10.1007/978-981-10-4600-1_6
18. Duan R, Guo L (2021) Application of blockchain for internet of things: a bibliometric analysis. Math Probl Eng e5547530. https://doi.org/10.1155/2021/5547530
19. Duy PT (2019) SDNLog-foren: ensuring the integrity and tamper resistance of log files for SDN forensics using blockchain. In: 2019 6th NAFOSTED conference on information and computer science (NICS.2019, pp. 416-421. https://doi.org/10.1109/NICS48868.2019.9023852
20. Fu Z (2018) Semantic-aware searching over encrypted data for cloud computing. In: IEEE Trans Inf Forensics Secur 13(9):2359–2371. https://doi.org/10.1109/TIFS.2018.2819121
21. Al-Fuqaha A (2015) Internet of things: a survey on enabling technologies, protocols, and applications. In: IEEE Commun Surv Tutor 17(4):2347–2376. https://doi.org/10.1109/COMST. 2015.2444095
22. González-Granadillo G (2021) Security information and event management (SIEM): analysis, trends, and usage in critical infrastructures. In: Sensors 21(14):4759. https://doi.org/10.3390/ s21144759

23. Ho SM (2018) Following the breadcrumbs: timestamp pattern identification for cloud forensics. Dig Investig 24:79–94. https://doi.org/10.1016/j.diin.2017.12.001
24. India stands third among top 20 cyber crime victims, says FBI report. https://www.newindianexpress.com/nation/2020/feb/23/indiastands-third-among-top-20-cyber-crime-victims-says-fbireport-2107309.html. Accessed 16 Aug 2020
25. Irfan M, Abbas H, Sun Y (2016) A framework for cloud forensics evidence collection and analysis using security information and event management. Wiley. ISSN:9:3790-3807. https://doi.org/10.1002/sec.1538. https://onlinelibrarywiley.com/doi/10.1002/sec.1538. Accessed 08 Aug 2021
26. Jahantigh MN (2020) Integration of internet of things and cloud computing: a systematic survey. IET Commun 14(2):165–176. https://doi.org/10.1049/iet-com.2019.0537
27. Janjua K (2020) Proactive forensics in IoT: privacy-aware log-preservation architecture in fog-enabled-cloud using holochain and containerization technologies. In: Electronics 9(7):1172. https://doi.org/10.3390/electronics9071172
28. Kävrestad J (2018) Fundamentals of digital forensics: theory, methods, and real-life applications. Springer International Publishing. https://doi.org/10.1007/978-3-319-96319-8
29. Khan S (2016) Cloud log forensics: foundations, state of the art, and future directions. ACM Comput Surv 7:1–7. https://doi.org/10.1145/2906149
30. Kirrane S, Di Ciccio C (2020) BlockConfess: towards an architecture for blockchain constraints and forensics. In: 2020 IEEE international conference on blockchain, pp 539-544. https://doi.org/10.1109/Blockchain50366.2020.00078
31. Kitchenham B (2004) Procedures for performing systematic reviews. Technical report TR/SE-0401. Department of Computer Science, Keele University, UK
32. Kumar N, Keserwani PK, Samaddar SG (2017) A comparative study of machine learning methods for generation of digital forensic validated data. In: 2017 ninth international conference on advanced computing (ICoAC), pp 15–20. https://doi.org/10.1109/ICoAC.2017.8441495
33. Le D-P (2018) BIFF: a blockchain-based IoT forensics framework with identity privacy. In: TENCON 2018—2018 IEEE region 10 conference, pp 2372–2377. https://doi.org/10.1109/TENCON.2018.8650434
34. Li M (2021) LEChain: a blockchain-based lawful evidence management scheme for digital forensics. Future Gener Comput Syst 115:406–420. https://doi.org/10.1016/j.future.2020.09.038
35. Li S (2019) Blockchain-based digital forensics investigation framework in the internet of things and social systems. IEEE Trans Comput Soc Syst 6(6):1433–1441. https://doi.org/10.1109/TCSS.2019.2927431
36. Liu A (2019) LiveForen. ensuring live forensic integrity in the cloud. IEEE Trans Inf Forensics Secur 14(10):2749–2764. https://doi.org/10.1109/TIFS.2019.2898841
37. Lone AH, Mir RN (2019) Forensic-chain: blockchain based digital forensics chain of custody with PoC in hyperledger composer. Dig Investig 28:44–55. https://doi.org/10.1016/j.diin.2019.01.002
38. Lusetti M (2020) A blockchain based solution for the custody of digital files in forensic medicine. Forensic Sci Int Digit Investig 35. https://doi.org/10.1016/j.fsidi.2020.301017
39. Lutta P (2020) The forensic swing of things: the current legal and technical challenges of IoT forensics. Int J Comput Inf Eng 14(5):159–165
40. Manral B et al (2019) A systematic survey on cloud forensics challenges, solutions, and future directions. ACM Comput Surv 124:1–124. https://doi.org/10.1145/3361216
41. Al-Masri E (2018) A fog-based digital forensics investigation framework for IoT systems. In: 2018 IEEE international conference on smart cloud, pp 196-201. https://doi.org/10.1109/SmartCloud.2018.00040
42. Meffert C (2017) Forensic state acquisition from internet of things (FSAIoT): a general framework and practical approach for IoT forensics through IoT device state acquisition. In: Proceedings of the 12th international conference on availability, reliability and security. Association for Computing Machinery, New York, NY,USA, pp 1–11. https://doi.org/10.1145/3098954.3104053

43. Muda AK (2014) Computational intelligence in digital forensics: forensic investigation and applications. Springer International Publishing. https://doi.org/10.1007/978-3-319-05885-6
44. Nguyen DC (2020) Integration of blockchain and cloud of things: architecture, applications and challenges. IEEE Commun Surv Tutor 22(4):2521–2549. https://doi.org/10.1109/COMST.2020.3020092
45. Nieto A (2020) Becoming JUDAS: correlating users and devices during a digital investigation. IEEE Trans Inf Forensics Secur 15:3325–3334. https://doi.org/10.1109/TIFS.2020.2988602
46. Nieto A (2018) IoT-forensics meets privacy: towards cooperative digital in-vestigations. Sensors 18(2):492. https://doi.org/10.3390/s18020492
47. NIST CCFSW (2014) NIST cloud computing forensic science challenges. National Institute of Standards and Technology
48. Patil MS, Ainapure B (2019) Analysis of dataset in private cloud for cloud forensics using eucalyptus and hadoop. In: 2019 international conference on smart systems and inventive technology (ICSSIT), pp 767–772. https://doi.org/10.1109/ICSSIT46314.2019.8987923
49. Pichan A (2020) A logging model for enabling digital forensics in IoT, in an interconnected IoT, cloud ecosystems. In: 2020 fourth world con-ference on smart trends in systems, security and sustainability (WorldS4), pp 478–483. https://doi.org/10.1109/WorldS450073.2020.9210366
50. Pichan A (2018) Towards a practical cloud forensics logging framework. J Inf Secur Appl 42:18–28. https://doi.org/10.1016/j.jisa.2018.07.008
51. Pourvahab M, Ekbatanifard G (2019) Digital forensics architecture for evidence collection and provenance preservation in IaaS cloud environment using SDN and blockchain technology. IEEE Access 7:153349–153364. https://doi.org/10.1109/ACCESS.2019.2946978
52. Qi Z (2017) ForenVisor: a tool for acquiring and preserving reliable data in cloud live forensics. IEEE Trans Cloud Comput 5(3):443–456. https://doi.org/10.1109/TCC.2016.2535295
53. Rane S (2019) Blockchain driven secure and efficient logging for cloud forensics. Int J Comput Digit Syst
54. Rane S, Dixit A (2019) BlockSLaaS: blockchain assisted secure logging as a service for cloud forensics. In: Nandi S (ed) Security and privacy. Springer, Singapore, pp 77–88. https://doi.org/10.1007/978-981-13-7561-3_6
55. Reddy N (2019) Cloud forensics. In: Reddy N (ed) Practical cyber forensics: an incident-based approach to forensic investigations. Apress, Berkeley, CA, pp 241–275. 978-1-4842-4460-9. doi: https://doi.org/10.1007/978-1-4842-4460-9_8.Accessed 15 Aug 2020
56. Ricci J (2019) Blockchain-based distributed cloud storage digital forensics: where's the Beef? IEEE Secur Priv 17(1):34–42. https://doi.org/10.1109/MSEC.2018.2875877
57. Rs.6 crore fraud unearthed. https://www.cyberforensics.in/Downloads/dutyfree.jpg
58. Ryu JH (2019) A blockchain-based decentralized efficient investigation framework for IoT digital forensics. Int J Supercomput 75(8):4372–4387. https://doi.org/10.1007/s11227-019-02779-9
59. Salas-Fernández A et al (2021) Metaheuristic techniques in attack and defense strategies for cybersecurity: a systematic review. In: Misra S, Tyagi AK (eds) Artificial intelligence for cyber security: methods, issues and possible horizons or opportunities. Studies in computational intelligence, vol 972. Springer International Publishing, Cham, pp 449–467. ISBN:978-3-030-72235-7, 978-3-030-72236-4. https://doi.org/10.1007/978-3-030-72236-4_18. Accessed 30 Sep 2021
60. Schneider J, Breitinger F (2020) AI Forensics: did the artificial intelligence system Do It? Why?. arXiv:200513635
61. Shrivastava RK (2019) Attack detection and forensics using honeypot in IoT environment. In: Fahrnberger G (ed) Distributed computing and internet technology. Springer International Publishing, Cham, pp. 402–409. https://doi.org/10.1007/978-3-030-05366-6_33
62. Sikos LF (2021) AI in digital forensics: ontology engineering for cybercrime investigations. WIRES Forensic Sci e1394. https://doi.org/10.1002/wfs2.1394
63. Singh SK (2020) Blockiotintelligence: a blockchain-enabled intelligent IoT architecture with artificial intelligence. Future Gener Comput Syst 110:721–743. https://doi.org/10.1016/j.future.2019.09.002

64. Srinivasan A, Ferrese A (2019) Forensics-as-a-service (FaaS) in the state- of-the-art cloud. In: Security, privacy, and digital forensics in the cloud. Wiley, pp 321–337. https://doi.org/10.1002/9781119053385.ch16
65. Stoyanova M (2020) A survey on the internet of things (IoT) forensics: challenges, approaches, and open issues. IEEE Commun Surv Tutor 22(2):1191–1221. https://doi.org/10.1109/COMST.2019.2962586
66. Taylor PJ (2020) A systematic literature review of blockchain cyber security. In: Dig Commun Netw 6(2):147–156. https://doi.org/10.1016/j.dcan.2019.01.005
67. Teing Y-Y (2018) CloudMe forensics: a case of big data forensic investigation. Concurr Comput Pract Exp 30(5). https://doi.org/10.1002/cpe.4277
68. Teing Y-Y (2019) Greening cloud-enabled big data storage forensics: syncany as a case study. IEEE Trans Sustain Comput 4(2):204–216. https://doi.org/10.1109/TSUSC.2017.2687103
69. Villegas-Ch W (2020) Integration of IoT and blockchain to in the processes of a University Campus. Sustainability 12(12):4970. https://doi.org/10.3390/su12124970
70. Wang D (2010) Application of adaptive particle swarm optimization in computer forensics. In: 2010 WASE international conference on information engineering, pp 147–149. https://doi.org/10.1109/ICIE.2010.131
71. What Is Computer Forensics?. Type: (GUIDE). https://www.forensiccontrol.com/what-is-computer-forensics
72. Xu M (2018) Modeling and predicting cyber hacking breaches. IEEE Trans Inf Forensics Secur 13(11):2856–2871. https://doi.org/10.1109/TIFS.2018.2834227
73. Yang K (2013) DAC-MACS: effective data access control for multiauthority cloud storage systems. In: IEEE Trans Inf Forensics Secur 8(11):1790–1801. https://doi.org/10.1109/TIFS.2013.2279531
74. Zawoad S (2013) SecLaaS: secure logging-as-a-service for cloud forensics. In: Proceedings of the 8th ACM SIGSAC symposium on information, computer and communications security. Association for Computing Machinery, New York, NY, USA, pp 219–230. https://doi.org/10.1145/2484313.2484342
75. Zhang Y (2017) A blockchain-based process provenance for cloud forensics. In: 2017 3rd IEEE international conference on computer and communications (ICCC), pp 2470–2473. https://doi.org/10.1109/CompComm.2017.8322979
76. Zou D (2019) A multigranularity forensics and analysis method on privacy leakage in cloud environment. IEEE Internet Things J 6(2):1484–1494. https://doi.org/10.1109/JIOT.2018.2838569

Predictive Forensic Based—Characterization of Hidden Elements in Criminal Networks Using Baum-Welch Optimization Technique

Mathew Emeka Nwanga, Kennedy Chinedu Okafor, Ifeyinwa Eucharia Achumba, and Gloria A. Chukwudebe

Abstract Applying Artificial intelligence tool in dissecting criminal networks is imperative for forensic criminal investigation and prediction. The first step is to identify the hidden links consisting of criminal states (CS), Active Internal Communications (AIC), time frame of attack (TFoA), and the mapped states (MS). Unfortunately, existing strategies lack the computational capacity for predictive intelligence and do not have immediate cover in existing security architectures. Motivated by this, the contributions of this paper are fourfold. First, the prediction of AIC. Second, we determined HLs new information about entities within the network which include: affiliation of the suspected individual with a network; crime characteristics; criminal trends, and/or plans in the global network. Third, Hidden Markov Model (HMM) is introduced to harness key features of CNs while predicting the probable state and timeframe of occurrence of criminal attacks. It is equally applied in determining the most probable sequence of attack vectors/payloads. The parameters are determined for parametric modeling. Fourth, using Baum-Welch Technique (BWT), the obtained parameters are optimized. The result shows that the Foot Soldiers (FS) are most vulnerable with 90% involvement in criminal attacks; the Commander carried out most strategic (high profile) attacks estimated at 2.2%. The private citizens and properties had the highest attack targets (50.6%); whereas the police and military base had 12.3% and 6.7% respectively. The results show that Boko Haram carried out the greatest level of attacks at 79.2% while Fulani extremists are responsible for 20.8% of all acts of terrorism in Nigeria from 2010 to date.

Keywords Artificial intelligence · Attack payload · Baum-Welch technique · Criminal network model · Computational intelligence · Predictive analytics

M. E. Nwanga · I. E. Achumba · G. A. Chukwudebe
Department of Electrical/Electronic Engineering, Federal University of Technology, Owerri, Nigeria
e-mail: gachukwudebe@futo.edu.ng

K. C. Okafor (✉)
Department of Mechatronics Engineering, Federal University of Technology, Owerri, Nigeria
e-mail: kennedy.okafor@futo.edu.ng

© The Author(s), under exclusive license to Springer Nature Switzerland AG 2022   231
S. Misra and C. Arumugam (eds.), *Illumination of Artificial Intelligence in Cybersecurity and Forensics*, Lecture Notes on Data Engineering and Communications Technologies 109, https://doi.org/10.1007/978-3-030-93453-8_10

1 Introduction

Artificial Intelligence is the modern way of combating crimes in most advance societies lately. It is increasingly be used for criminal forensic investigation. The adoption of AI in today's criminal network forensic and terrorist prediction has paved way for computational counterterrorism (CC). The use of predictive analytics (PA) and machine learning (ML) to detect activities and behaviors of terrorist elements is the sure way to go in providing solutions for national security and defense intelligence.

Unfortunately, the understanding and classification of crime vary greatly in time lag and space across many nations in terms of their socio-cultural and economic differences. These variations make it uneasy to have a uniform definition of crime across the various regions globally.

Application of AI in criminal networks and forensics has been found very useful within crime events discussions [1, 2]. One aspect of crime that every society uniformly frowns at irrespective of geographical, socio-cultural, and economic differences is the act of terrorism [3, 4]. Terrorism refers to violence against people and properties. It involves the facilitation of action in which the safety of the public and properties is at risk [5]. The report from the global terrorism database (GTD) by the University of Maryland, College Park, posits that more than 61,000 incidents of non-state actors have occurred globally. The available records in the GTD show that not less than 140,000 deaths have been recorded since 1970 to date. This catalyzed the momentum in the global fight against terrorism with counterterrorism measures ranging from crime analysis in other to reveal criminal trends, to sophisticated military operations.

The terms "terrorism" and "terrorist" are said to have originated during the French revolution of the late eighteenth century. The Beirut barracks bombings of 1983 gave the interesting popularity of terrorism during the reign of Ronald Reagan as the U.S. President. The popularity also grew after the attacks on New York City and Washington, D.C. in September 2001, and on Bali in October 2002 [6]. Terrorism is indeed a great threat to society that has long existed for centuries now [7]. But these Security challenges became more prominent global disturbing issues in the wake of 9/11 terrorist attacks [8]. The lethal level of attacks by terrorists poses serious challenges to national security, especially in such countries as Iraq, Afghanistan, Nigeria, Pakistan, and Syria among others. Nigeria is ranked third out of the 162 nations of the world and first in Africa concerning terrorist attacks [9, 10].

Terrorism exists in two dimensions, namely the group and lone actor form. The former is characterized by terrorist networks with two different interplay actors: covert and overt actors. Terrorist groups often exist in networks and complex forms to facilitate effective and well-coordinated attacks [11]. The nodes within the network represent the criminal elements and edges represent the communication links among the criminal elements for effective interaction and clandestine operations [12]. The word "network" has long been used in intelligence and law enforcement domains to refer to criminal or terrorist organizations. This is because offenses such as terrorist attacks, narcotics trafficking, and armed robbery depend, to a large extent, on the collective efforts of multiple and interrelated individuals. In analyzing the above crimes, the characteristics and behavior of individual offenders are not only examined but also a deep analysis is paid to the organizational structure and operation of the groups within the overall network [4], observed that terrorism studies and crime analysis generally focus on terrorist events rather than on individual attackers due to the nature of available data.

According to [3], the characteristics that are exhibited by terrorist elements such as vast and hidden geographical distribution can be modeled as a network. Ultimately, over time terrorism has gradually emerged from the characteristics of the past, characterized by geographical distribution, concealment, clandestine, and network. Indeed, the existence of terrorist elements confers it with the networking flexibility of networking with their associated networks to create greater societal mayhem.

Categorically, crime analysis can be defined as the set of systematic and analytical processes that provide timely and pertinent information about crime patterns and crime trends. Crime analysis helps to understand the occurrence of a crime and is a significant practice in law enforcement [7]. In [13], it aids the collection and analysis of data relating to a criminal occurrence, culprit, and victim, and develops information for use for crime prevention and criminal network detection activities. The work includes the use of crime-data to study criminal problems such as the characteristics of the crime, crime scenes, offenders, and victims. Crime patterns are analyzed in terms of their socio-demographic, temporal, and spatial qualities, and may be represented visually using graphs, tables, and maps. The results of crime analysis form the basis for tactical advice to police and other law enforcement agents on criminal investigations, deployment of resources, planning, evaluation, and crime prevention [14].

Analyzing criminal networks have major challenges faced by law enforcement agencies, especially concerning crime forecasting. Information relating to criminal networks is not readily available to the public due to the covert nature of the networks. Hence, the activities and dynamics of criminal networks cannot be easily determined [15]. Hidden links in the network make it difficult to detect direct relationships between criminal actors in real-world situations. The immediate consequence can be found in the lack of desired information and datasets needed for criminal analysis. Revealing critical links in criminal networks helps to capture the key elements of the network such as long-wanted criminal commanders [16]. The only possible option is to build on existing linkages in available network data. There is incognito in criminal activities, hence there could be many hidden links from entity to entity, which makes the publicly available criminal networks incomplete. New information relating to crime characteristics and trends in the network is desirably introduced as the hidden links are been revealed [17]. Therefore, sophisticated techniques are required to reveal these hidden links in criminal networks with high precision and forecast the network's inherent trends. To improve the efficiency and accuracy of the anti-terrorism method, an effective risk assessment and prediction method involving machine learning/AI are sometimes needed [18].

As part of the research solution, this work leveraged a predictive computational scheme called Hidden Markov Model (HMM) driven by Baum-Welch Algorithm as an AI optimization technique. This is used in constructing terrorist behavioral models that form the basis of any criminal forensic investigation. The approach is based on its relative mathematical ease of tractability and immense versatility. This makes it suitable for the analysis of sequential data. Clearly, HMM is a powerful mathematical tool used by engineers and computer scientists for prediction and recognition. The underlying assumption in HMM as applied in this context is that the future attack of any criminal element is independent of the past attacks given the present actions. This implies that once the present state of the terrorist group attack is known for any predictive investigation, then there will be no need for any past information to predict future action.

Generally, HMMs as a machine learning tool find application in smart grids [19], criminal networks [20, 21], reservoir prediction [22], pandemic spreading and prediction [23], filter design in the control system [24, 25], and accident prediction in vehicular Ad-hoc network [26]. This present study determines hidden links in criminal networks to develop a model of criminal behavior with HMM. The use case study scenario is on the Boko-Haram and Fulani extremist networks terrorist attacks in Nigeria within the period of the year 2010–2016.

This paper contributes to knowledge by providing a terrorist computational model which helps determine the most probable state and timeframe for the occurrence of terrorist attacks. It also provides the most probable sequence of active internal communications (AICs) that led to such attacks. The Baum-Welch optimization is applied in this research to improve the intelligence and predictive accuracy of the activities of criminal elements. This solution is adaptable to any country around the globe.

The entire paper is organized as follows. Section 2 focused on existing research on criminal networks, crime prediction, criminal network forensic, and HMM & Baum-Welch optimization approaches to criminal network analysis were discussed. Section 3 discussed the methodology. The System model is presented in Sect. 4. Section 5 presented the Forensic Predictive Algorithmic Implementation. Section 6 discussed the results and findings. Section 7 concludes the work with future directions.

2 Literature Review

Several works of literature on network characterization, criminal network forensic, Markov models, criminal networks' features, and models exist. A review of some recent studies is presented in this Section.

The growing trend of terrorist networks calls for the characterization of its dynamic systems as a non-linear time series phenomenon. The authors [4], optimized the global terrorism database (GTD) using mathematical methods with hierarchical analysis, fuzzy clustering analysis, and regression analysis of terrorist attacks. The authors were able to screen the main influencing factors of the harmfulness of terrorist incidents and used the analytic hierarchy process to carry out weight analysis. Finally, the terrorist activities were properly analyzed and computed accordingly and the authors classified the terrorist organization into 13 categories which became vital for further studies.

The authors [27], proposed a computational characterization-based model on magnetic resonance imaging. A proof of principle statistical comparison was performed on hemodynamics characteristics. The flow velocity was 0.028, blood pressure 0.016, and vascular extraction rate 0.040. The mathematical operators of the machine learning algorithm were generated; the ROC 1.0, sensitivity 1.0, and specificity 1.0. The authors concluded that the characterization image-based model was a new way to map flow and pressure fields. The work in [28], detected time delay attacks on cyberattacks by applying characterization with a deep learning-based method. The solution achieved 92% accuracy in the power plant control system when compared with 81% by random forests and 72% by k-nearest neighbors (KNNs). The work [29] used a computational characterization approach to covariance estimation using a ridge-type operator for a precision matrix estimation on a large-scale set of NMR ensembles. Structural and sequential characteristics were used to build novel elastic network model variation, with parameter estimation using particle swarm optimization (PSO). The work [30], proposed the use of motifs to represent the structure of the network as applied from statistical physics. Network characterization was applied here and the authors did conduct numerical experiments for synthetic and real-world data sets and evaluated the entropy from the motifs functions. The work [31], further applied characterization to develop multistage complex contagion models that took place over multilayer or multiplex networks. A node was randomly chosen to derive the initiate propagation that reached a fraction of the whole propagation with positive associated results.

The authors [15], studied the characterization of dynamical systems based on time series in a complex network. The authors focused on the nonlinear time series analysis and the theory of complex networks on phase space-based recurrence networks, visibility graphs, and Markov chain-based transition. The authors [32], proposed the characterization of micro traces images with deep neural networks. The methods were used to characterize the existence of criminal elements. The developed system reached a classification accuracy of 75% among five evaluated entomological families. The work [33], carried out the comparative performance of a criminal network analysis hidden link prediction model developed using deep reinforcement learning techniques against classical machine learning models such as gradient boosting machine random forest and support vector machine. The work [34], further proposed key structural attributes of a network-oriented dataset for the development of link prediction models. The deep reinforcement learning classification technique was developed to predict links/edges even on relatively small-scale datasets and adequate predictive accuracy was achieved.

Similarly, the authors [18], carried out research and stated that anti-terrorism is on the new dimension of stage5 with multiple challenges. According to the authors, the new challenge was on how to extract useful and valuable information from massive data efficiently. The work [35], further noted the difficulty of obtaining the activities of terrorists or suspect by law enforcement agencies. The authors ensured that a rule-based scoring system on physical and official identity attributes was developed with a graph-based analysis on social identity attributes for organized crime. Augmentation schemes that support identity hypotheses were also proposed by [14]. Research work [13], proposed a new forensic analysis for identifying influential criminals members and their communication channels within the network communication. The authors stated that their method can be used to build a network from crime incident reports. A prediction performance evaluation was done with another system and superior performance was confirmed.

The work [36], proposed the use of clustering and classification machine learning algorithms on evolving database files to achieve intelligence in the digital investigation process. They further built a comprehensive framework for intelligent forensic investigations.

Authors in [37], highlighted terrorism events that occurred within the last four decades in Egypt. This was done with a statistical technique that used the Global Terrorism Database (GTD). The developed algorithms generated deeper meaningful insights with composite rules mining the database. The work [38], proposed social network analysis, wavelet transform, and the pattern recognition approach that investigated the dynamics of a terrorist network. The research work eventually predicted the attack behavior of terrorist groups. The authors concluded that certain dynamic characteristics of terrorist groups are vital in predicting terrorist attacks. In [39], the authors used hidden Markov models (HMMs) for DNA methylation by describing the occurrence of spatial methylation patterns over time and proposed several models with different neighborhood dependencies. The authors performed a numerical analysis of the HMMs applied to comprehensive hairpin and non-hairpin bisulfite sequencing measurements and accurately predicted wild-type data. In [40], the authors investigated the performance of a hidden Markov models-based detection rule with a Bayesian network. The developed theory of HMMs was based on Markov chain representation for the likelihood ratio and r-quick convergence for Markov random walks. The authors in [41], in the same process used HMM to perform passivity analysis criteria and established the Markov jump singularly perturbed systems with partial unknown probabilities.

In [25, 42], methods to design an asynchronous controller with a hidden Markov model were developed and its application in computational neuroscience. The method applied HMM to drive the closed-loop system with stochastic stability. In the work, the random controller gain fluctuation and system modes hiding for the controller were the expected performance. The authors in [43], built a precise equivalence between the inference equations of HMMs with time-invariant hidden variables. The authors established that the rate of a neural network can implement the posterior influence of HMMs with accurate inference results.

The work [44], proposed the use of HMMs for the secret key establishment. The author used HMMs and based on the same underlying Markov chain to derive a computational-efficient asymptotic converse bound for the secret key capacity of the correlated—HMM scenario. The authors [21], further proposed the use of a hidden Markov model for a sibling hidden Markov model (SHMM). The authors studied the secret key capacity for various types of SHMM and used the joint probability of the observations as the norm of a Markov random matrix and used its convergence to a Lyapunov exponent. The work [45], described a hidden Markov model for the evolution of an advanced persistent threat (APT). The model aimed to validate whether the evolution of the partially reconstructed attack campaigns was indeed consistent with the evolution of an APT. The model was validated with data obtained from experts and had a good measure of representation. The authors in [46], also proposed an effectively nonparametric approach to fitting hidden Markov models to time series of counts. The method involved the use of state-dependent distributions to estimate a completely data-driven way of fitting hidden Markov Models to time series of counts without the need to specify a parametric family of distributions. The methodology was implemented in the accompanying R package count, with two real-data applications available on CRAN. The authors [23], in the same dimension, proposed the use of the hidden Markov Chain for the evolution of COVID-19 in Morocco. The authors provided both the recorded and forecast data matrices of the cumulative number of the confirmed, recovered, and active cases. The authors in [47], built a hidden Markov model and Naïve Bayes for attack prediction. The authors used the KDDCUP'99 model with a network intrusion dataset. A comparative study of the two models based on the accuracy of prediction was done.

HMM gave higher accuracy when compared to Naïve Bayes. The authors in [48], further proposed Hidden Markov Model with Baum-Welch Algorithm to classify the future generated sensor data. The study was compared with the HMM-based on Viterbi algorithm and Baum-Welch outperformed the Viterbi algorithm with a prediction accuracy of 88.86%.

A research gap in existing literature was fixed by developing a terrorist (criminal) computational model using Hidden Markov Model (HMM). This work focused on the development of a predictive mathematical characterization of criminal network elements. The leveraged computational model would facilitate the prediction and detection of criminal attacks by terrorist elements in Africa/Nigeria and any other region with terrorism challenges.

3   Methodology

In this Section, this paper characterized and modeled the hidden features of criminal networks in Nigeria from 2010 to 2016 for forensic predictive analytics. It categorized the hidden features that aid in the proper identification of the criminal states viz: Commander (Cd), Gatekeeper (Gk), and Foot-Soldiers (Fs). It also modeled the criminal Active Internal Communications (AICs) that often lead to criminal attack(s). The system architecture, represented in Fig. 1, is made up of the following components: Data Extraction (DE), Data Analytics (DA), Crime Computational Model (CMM) building, and Optimization and Analysis (OA).

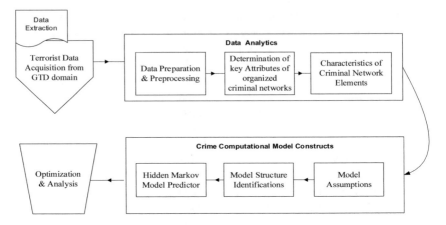

Fig. 1 Predictive computational system architecture

3.1 Data Collection Technique

The critical data used in this research was obtained from the global terrorism database [49]. The Nigerian criminal activities data for terrorist groups were collected from the global terrorism database metadata for the period of 2010–2016. The crime data contains the following information: names of suspected criminal groups that carried out the attack, the attack targets, the target sub-type, description of the attack, and the geographic location of the attack among others. The timeframe of an attack and terrorist states that carried out the attack were also sieved out of the dataset.

3.2 Characterization of Hidden States

One major challenge in criminal networks study is identifying the hidden links: the criminal states (CS) and the AIC within the mapped state. This is particularly the case where the states seem to be hidden in all ramifications. (i.e., there is no known clue to identifying the criminal state and its features). Figure 1 shows the proposed predictive computational architecture for computing criminal network behavior. The success, survival and effective coordination of the terrorist (criminal) network depend on the intelligence and core competence of two extreme elements (the commander and the foot-soldiers). Attack(s) of Cd, Gk and Fs are both coherent and action dependent. The strength of the hidden elements was considered as a coherent trace and action dependence. The terrorist trace is coherent if the preconditions of every action taken by each hidden element are a logical consequence from the initial state and the post-conditions of all the preceding actions [50]. Attacks in the criminal network always precede its AIC. The criminal states were properly identified and characterized with numerical coding of 0, 1, and 2 accordingly.

Consequent upon the assumption of the states above, three unique attack criteria were employed in identifying the hidden states:

i.   Number of attack personnel
ii. Crime profile
iii. Nature of crime.

The coding technique is predicated on the fact that every criminal attack is preceded by active internal communication (AIC) in the form of instruction from superiors to subordinates. It is assumed that every criminal communication is initiated mainly by the commander to/through the Gatekeeper to the foot-soldiers.

Accordingly, if an attack was assumed to be done by a commander, Cd is coded as 1, while Gk and Fs is coded as 0. This is known as active internal communication. If a crime is assumed to be carried out by a Gatekeeper, the coding is $Cd = 1$; $Gk = 2$ and Fs as 0. Similarly, if an attack is carried out by a foot-soldier, the coding is $Cd = 1$; $Gk = 2$ and $Fs = 1$. This is because Cd and Fs is the extreme criminal state actor. A similar coding method is employed in linking the criminal states (CS) to the timeframe of attacks. This is also predicated on the fact that when a crime happens within a given time frame, the criminal active internal communication that led to the crime would have occurred in the previous timeframe. Therefore, if a crime is assumed to have happened in a particular timeframe, the timeframe is coded as $V_i = 2$ and the preceding timeframe as $V_{i-1} = 1$ and 0 for the timeframe after it has happened.

This implies that a criminal AIC begins in the timeframe before the attack and gets to the peak in the timeframe of attack, and in a practical sense, just before the attack in the same timeframe. The outcome of this coding technique is used to compute the emission probabilities matrix.

4 System Model

4.1 Hidden Markov Model for Criminal States Characterization

In this section, let's consider the first-order Markov's chain be defined as in Eq. 1.

$$\theta = (S, \ M, \ \pi) \tag{1}$$

where S, M and π are the states, transition matrix, and initial probabilities respectively. The HMM assumption is that the states (S) are hidden, but at every point in time, the states emit some observable symbols, v, with certain probabilities. Thus, v is visible. The probabilities of these emissions depend only upon the underlying states. These additional features (emission symbol, $\sum$, and emission probability) add up to give the HMM. Hence, to mathematically characterize the criminal networks using Hidden Markov's Model (HMM), the following definitions are given in Eqs. 1 to 10.

4.2 Terrorist (Network) States

Let Eq. 2 be the states of the criminal network

$$S = s_1, s_2, s_3. \tag{2}$$

where S_1 represents the commander (Cd), S_2 represents the Gatekeeper (Gk) and S_3 represents the Foot-soldiers (Fs). Hence, a three-State (3-S) Markov Model is deemed appropriate for the work being presented.

4.3 Transition Probabilités Matrix

Let Eq. 3 be the transition probabilities, that is, the conditional probabilities that a criminal attack was carried out by *state j* given that an AIC was made by *state i*.

$$p = (p_{ij}). \tag{3}$$

It is the probability that the stochastic event (terrorist action) changes current states s_i to the next state s_j. Statistically, the sum of the probabilities of transitioning from any given state to another next state is 1.

Explicitly,

$$\forall s_i \in S, \sum_{s_j \in S} p_{ij} = 1. \tag{4}$$

where s_i and S represent the terrorist states variables and the criminal computational space, respectively.

Furthermore,

$$p_{ij} = p\left(s_i/s_j\right). \tag{5}$$

Then the transition probability matrix is given as:

$$M = p_{ij} = \left[p\left({}^{s_i}/_{s_j} \right) \right] = \begin{pmatrix} p\left({}^{s_1}/_{s_1} \right) & p\left({}^{s_1}/_{s_2} \right) & p\left({}^{s_1}/_{s_3} \right) \\ p\left({}^{s_2}/_{s_1} \right) & p\left({}^{s_2}/_{s_2} \right) & p\left({}^{s_2}/_{s_3} \right) \\ p\left({}^{s_3}/_{s_1} \right) & p\left({}^{s_3}/_{s_2} \right) & p\left({}^{s_3}/_{s_3} \right) \end{pmatrix}. \tag{6}$$

where $p(S_1/S_1)$ is the conditional probability of attack by a commander given that AIC was from a commander. This is assumed to be a non-zero value ($\neq 0$) because the commanding group consists of more than one individual, i.e. the commander, deputy (ies), and/or personal assistants (PAs). Also in the criminal network, actors' relationships/ties with themselves and with other actors are deeply considered in the analysis. This differentiates criminal networks from social networks. The commanding group can initiate AICs within the group that might lead to criminal attacks. Also in Markov's Model assumption, an event can choose to remain in the same state after n-steps. In this case, we term it an active internal communication within the states and in Bayesian notation, it is the probability of carrying out an attack(s) by the Cd, given that Cd made the communication within the network. Essentially, $p(S_3/S_1)$ is the conditional probability of attack by Cd given that communication came from an Fs. This is equal to zero (0) as Fs are assumed not to communicate directly with the commanders since they do not have a direct link with the commanders. It is assumed that there is no feedback/backward communication in this context. All outgoing communication is from a superior to a subordinate.

4.4 Initial Probabilities Matrix

Let Eq. (7) be a vector of initial probabilities, that is, the probabilities of criminal AIC by each state in the network.

$$\pi = (\pi_i) = \{ p(s_1), p(s_2), p(s_3) \} \tag{7}$$

It is the probability that the stochastic event begins in a state s_i.
Also,

$$\sum_{s_i \in S} \pi_i = 1. \tag{8}$$

4.5 Emission Symbols

Let Eq. (9) denote the vector of the timeframe of AICs from the criminal states representing the emission symbols,

$$\sum v = v_i = \{v_1, v_2, v_3\}. \tag{9}$$

where: v_1 = Day, that is, the proportion of criminal active internal communication (AICs) made in the day.

  v_2 = Evening, that is, the proportion of criminal AICs made in the evening.

  v_3 = Night, is the proportion of criminal AICs made at night.

4.6 Emission (Observation) Probabilities Matrix

The probability of emission (observation) is the conditional probability that internal communication was done in a particular timeframe given that it came from a particular state. This is given as:

$$\delta = p\left(v_i/s_i\right) = \begin{pmatrix} p\left(v_1/s_1\right) & p\left(v_1/s_2\right) & p\left(v_1/s_3\right) \\ p\left(v_2/s_1\right) & p\left(v_2/s_2\right) & p\left(v_2/s_3\right) \\ p\left(v_3/s_1\right) & p\left(v_3/s_2\right) & p\left(v_3/s_3\right) \end{pmatrix}. \tag{10}$$

$$\lambda = (M, \delta, \pi). \tag{11}$$

Equation 11 is the general equation or compact notation of the Hidden Markov Model and contains the transition, observation, and initial probabilities matrix of this research. Considering the features of the Hidden Markov Model and the characteristics of terrorist elements and according to [51], the probability of a state path S joint to an observation sequence $\sum$ in the context of this study is defined and stated in Eq. 12.

$$p(S, \sum) = p\left[\left(\prod_{i=1}^{n} p\left(s_i/s_j\right)\right)\left(\prod_{i=1}^{n} p\left(v_i/s_i\right)\right)\right] \tag{12}$$

where $\prod_i^n p\left(s_i/s_j\right)$ and $\prod_i^n p\left(v_i/s_i\right)$ are the transition and observation probabilities, respectively of the terrorist network. The equation represents the computational model of the behavior of terrorist elements by the joint probability of Eqs. 6 and 10. This is because the objective of HMM is to determine the most probable sequence of states S_i that initiates a criminal AIC that leads to terrorist attacks in a given timeframe. Thus, Eq. 12 is the most important equation of this research and represents the model of criminal (terrorist) behavior using HMM. The work [52] has been explored to structure this paper while leveraging Soft Computing [53] in looking at security in communication networks [54, 55], and malware system interfaces [56] as well as novel Spine-leaf model [57].

5 Forensic Predictive Algorithmic Implementation

5.1 HMM for Criminal AICs and Attacks

In this Section, the forensic computational procedure for the HMM parameters (transition, initial, and emission probabilities) is designed using Algorithm I. This is implemented with R-Statistics software due to the cumbersome nature of HMM.

Algorithm I. HMM, Parametric Computation for Terrorist Network Characterization.

```
1:   Inputs:          Predictive Attack-CallSchedule
                      History of Attack States ( )
     Output:          Attack prediction profile ( )
     int i←0;
     While true do
                hmm = initHMM(c("Cd","Gk","Fs"),c(" v₁ "," v₂ "," v₃ "),
                if hmm = null then
                        transProbs=matrix(c(.934,.3416,0,.066,.3416,.5,0,.3168,.
                        5),3),
                           emissionProbs=matrix(c(.222,.333,.105,.556,.246,.320,.2
                        22,.421,.575),3))
                while hmm not null do
                        # Sequence of observation
                        cd = sample(c(rep(" v₂ ",4),rep(" v₂ ",10),rep(" v₃ ",4)))
                        gk = sample(c(rep(" v₁ ",19),rep(" v₂ ",14),rep(" v₃ ",24)))
                        fs = sample(c(rep(" v₁ ",77),rep(" v₂ ",234),rep(" v₃
                ",421)))
                        observation = c(cd,gk,fs)
                        /* Ensure none of the parameters is negative */
                        if cd < 0 or gk < 0 or fs < 0 then
                                # Initialise HMM
                                hmm = initHMM(c("Cd","Gk","Fs"),c(" v₁
                        "," v₂ "," v₃ "))
                        else      /* Given valid parameter values then simulate
                from the HMM */
                                simHMM(hmm, 100)
     end while
                /* Compute the value of Baum-Welch */
                bw = baumWelch(hmm,observation,100)
                print(bw$hmm) /* Display the result */
                break /* Exit the loop */
     End while
     Return
```

6   Results and Findings

The descriptive statistics of the hidden features of criminal networks are identified and presented in subsequent tables from the data. Table 1, presents the timeframe of terrorist attacks for the studied groups in Nigeria between 2010 and 2016 [49]. On the whole, 13.8% of the attacks within the study period were carried out in the day, 9.7% were done in the evening while 76.6% of the attacks were at night.

This shows that terrorist groups have a greater attack on citizens and properties. The target attraction of terrorist elements on the destruction of citizens and properties is considered the major action of every group in both national and continental terrorism. The data analytics applied in the analysis of terrorist activities from the global terrorism database (GTD) within the Boko-Haram and Farmers Herdsmen network for this research period offers this insight of hierarchical understanding. The predictive analytics from Fig. 2 provides a guide on the characterization of the hidden elements within these groups of criminal networks from the commander to the gatekeeper and foot-soldiers. Further details and actions are represented in Fig. 2.

Table 1 Timeframe of terrorist attacks

		Frequency	Percent	Cumulative percent
Valid	Day	37	13.8	13.8
	Evening	26	9.7	23.4
	Night	206	76.6	100.0
	Total	269	100.0	

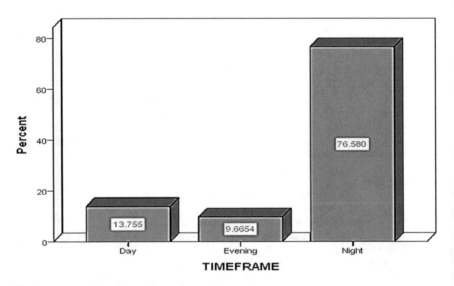

Fig. 2 Percentage timeframe of terrorist attacks

Table 2 Descriptive statistics of target-type

		Frequency	Percent
Valid	Business	9	3.3
	Educational institution	26	9.7
	Government (General)	18	6.7
	Military	18	6.7
	Police	33	12.3
	Private citizens & property	136	50.6
	Religious figures/institutions	22	8.2
	Others	7	2.6
	Total	269	100.0

Table 2, presents the descriptive statistics for the terrorist's target types from the study carried out with the data. On the whole, 3.3% of the targets are business outfits, 9.7% educational institutions, 6.7% general government institutions. 6.7% of the attacks were targeted on the military, 12.3% on the police, 50.6% on private citizens and properties, and 8.2% on religious figures/institutions while the other target types constitute 2.6% of the total attacks.

Table 3, presents the cross-tabulation of the hidden states and time frame of attack. Out of the whole attacks by the terrorist groups, none of the attacks were done in the day by the commander, 1.5% of the attacks were done in the evening by the commander while 0.7% were done at night. For the foot-soldiers, 10.8% of the total attacks were done in the day, 7.1% in the evening while 72.9% were carried out at night. 3.0% of the attacks were carried out by the gatekeepers in the day, 1.1% in the evening, and 7.1% at night. A critical look at the table shows that majority of the attacks were done at night. The behavior and state-level attack perpetrations of the terrorists are represented graphically in Fig. 3 using the values in Table 3. Foot-soldiers perpetuate the highest level of attack and as multiple cells, dwell at the

Table 3 Timeframe * criminal state

			State			Total
			Cd	Fs	Gk	
Timeframe	Day	Count	0	29	8	37
	Evening	% of total	0.0%	10.8%	3.0%	13.8%
		Count	4	19	3	26
		% of total	1.5%	7.1%	1.1%	9.7%
	Night	Count	2	196	8	206
		% of total	0.7%	72.9%	3.0%	76.6%
Total		Count	6	244	19	269
		% of total	2.2%	90.7%	7.1%	100.0%

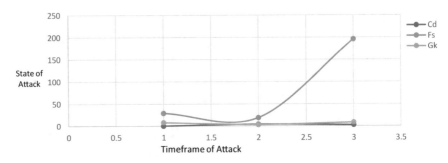

Fig. 3 Timeframe of attack versus the criminal states

bottom of the criminal network pyramid (CNP). Points 1, 2, and 3 at the timeframe axis of the graph represent a day, evening, and night respectively.

Generally, in criminal networks, the commander communicates with the gate-keeper without feedback for purposes of secrecy and efficiency (SAE) of actions and to avoid been uncovered. Figure 4 represents the interaction of Cd and Gk based on centrality. The values of these two elements concerning their timeframe of attacks in Table 3 are used to plot this graph. The nature and profile of the attack target deter-mine the mission strategy and the grade of criminal elements to be used for the attack. The strength of every criminal network lies in the intelligence of the commander and the capabilities of the foot-soldiers and not necessarily the gatekeeper.

Table 4 presents the cross-tabulation of the timeframe of attack and terrorist group. Boko-Haram on the whole carried out a total of xxx attacks as follows: 12.3% of their attacks in the daytime, 9.3% in the evening, and 57.6% at night. It indicates that the Boko-Haram group has a more lethal level of attack in Nigeria than other terrorist groups as shown in Fig. 5.

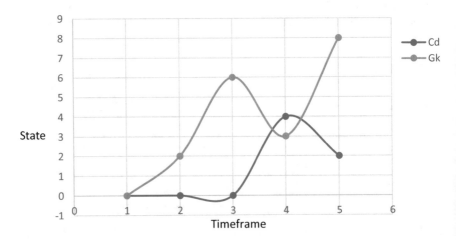

Fig. 4 Interaction of commander versus gatekeeper

Table 4 Timeframe * group-name

			Group name		Total
			Boko-Haram	Fulani-extremists	
Time frame	Day	Count	33	4	37
		% of total	12.3%	1.5%	13.8%
	Evening	Count	25	1	26
		% of total	9.3%	0.4%	9.7%
	Night	Count	155	51	206
		% of total	57.6%	19.0%	76.6%
Total		Count	213	56	269
		% of total	79.2%	20.8%	100.0%

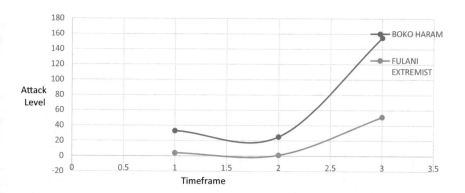

Fig. 5 Terrorist type and timeframe of attack

Using Algorithm I, the results of HMM for Criminal AICs and attacks are shown in Table 4.

The transition probability matrix is computed from Eq. 6 as presented in Eq. 13.

$$M = p_{ij} = \begin{pmatrix} 0.934 & 0.066 & 0 \\ 0.341 & 0.3416 & 0.3168 \\ 0 & 0.5 & 0.5 \end{pmatrix}. \qquad (13)$$

The initial probability is computed from Eq. 7 as presented in Eq. 14.

$$\pi = (0.1863, 0.4981, 0.3157). \qquad (14)$$

The Emission probabilities Matrix is computed from Eq. 10 as presented in Eq. 15.

$$\delta = p\left(v_i/s_i\right) = \begin{pmatrix} 0.222 \ 0.556 \ 0.222 \\ 0.333 \ 0.246 \ 0.421 \\ 0.105 \ 0.320 \ 0.575 \end{pmatrix}. \tag{15}$$

$$\lambda_1 = (M, \ \delta, \ \pi). \tag{16}$$

The results of the HMM as implemented in the R-Statistics Console (GUI, 64 bit) package are presented in Table 5. Equation 16 is the test Hidden Markov Model (HMM_T) which is optimized with Baum Welch Algorithm (BWA) to generate Eq. 20. The BWA is also called a forward–backward algorithm. In HMM, the problem of learning the parameters (that is the transition and emission probabilities) does exist. Thus, HMM has to be trained. BWA is an optimization algorithm that enables the training of both parameters. It works in form of expectation-maximization (EM) and as an iterative algorithm, it does the computation of an initial estimate for the probabilities, and then uses the estimates to compute a better estimate, thereby iteratively improving the probabilities that it learns. Hence, the maximum likelihood of the parameters is achieved. In this context, 100 iterations were recorded.

Table 5 presents the predicted sequence of criminal AICs and the corresponding timeframes that most probably led to criminal attacks. The commander was predicted to be involved in 36 AICs, gatekeeper 28, and the Foot-soldiers 36. On the most probable timeframe of communication, v_1(day) was found to be the timeframe with the highest AICs, closely followed by v_2(evening) and then v_3(night). The frequency of the predicted communications done by the criminal states and their corresponding timeframe is specifically presented in Table 6. The results indicate that the commander and the foot soldiers have the same amount of AICs thereby confirming the initial coding that the strength of the network lies in the intelligence and competencies of both elements.

It is very vital to note that attacks within the criminal network precede AIC. Attack and AIC do not take place at the same timeframe. Hence, the results will serve as a guide to the security agencies in their daily planning and deployment of resources for combating terrorism within the nation. It will also help them understand the dynamics of the terrorists.

The result of the optimized HMM parameters computed using the Baum-Welch HMM algorithm is presented specifically from Eqs. 17 to 20.

$$\hat{\pi} = \left(0.333 \ 0.333 \ 0.333\right). \tag{17}$$

Also, the optimized transition matrix is presented as:

$$\hat{M} = \begin{pmatrix} 0.6666667 \ 0.1666667 \ 0.1666667 \\ 0.1666667 \ 0.6666667 \ 0.1666667 \\ 0.1666667 \ 0.1666667 \ 0.6666667 \end{pmatrix}. \tag{18}$$

Table 5 Predicted sequence of terrorist active internal communications (AICs)

Sequence of communication	Predicted CS and corresponding timeframe of AIC	Sequence of communication	Predicted CS and corresponding timeframe of AIC
1	Gk, v_3	15	Cd, v_1
2	Gk, v_1	16	Cd, v_2
3	Gk, v_1	17	Cd, v_1
4	Gk, v_1	18	Cd, v_2
5	Gk, v_1	19	Cd, v_1
6	Gk, v_2	20	Cd, v_3
7	Gk, v_1	21	Cd, v_3
8	Cd, v_1	22	Cd, v_2
9	Gk, v_1	23	Fs, v_3
10	Fs, v_1	24	Cd, v_2
11	Fs, v_3	25	Fs, v_2
12	Fs, v_2	26	Fs, v_2
13	Fs, v_2	27	Fs, v_3
14	Fs, v_1	28	Fs, v_1
29	Fs, v_3	54	Fs, v_3
30	Fs, v_1	55	Fs, v_3
31	Fs, v_1	56	Gk, v_1
32	Fs, v_2	57	Fs, v_1
33	Cd, v_3	58	Fs, v_1
34	Fs, v_1	59	Fs, v_2
35	Fs, v_1	60	Fs, v_1
36	Gk, v_1	61	Fs, v_2
37	Gk, v_3	62	Cd, v_1
38	Gk, v_2	63	Gk, v_2
39	Gk, v_2	64	Gk, v_3
40	Gk, v_3	65	Gk, v_2
41	Gk, v_3	66	Gk, v_2
42	Fs, v_3	67	Cd, v_1
43	Cd, v_1	68	Cd, v_1
44	Cd, v_1	69	Cd, v_2
45	Cd, v_2	70	Cd, v_3
46	Cd, v_2	71	Gk, v_2
47	Cd, v_3	72	Gk, v_2
48	Cd, v_1	73	Cd, v_2
49	Cd, v_2	74	Cd, v_2

(continued)

Table 5 (continued)

Sequence of communication	Predicted CS and corresponding timeframe of AIC	Sequence of communication	Predicted CS and corresponding timeframe of AIC
50	Fs, v_1	75	Cd, v_2
51	Fs, v_2	76	Gk, v_1
52	Cd, v_1	77	Gk, v_1
53	Cd, v_3	78	Fs, v_2
79	Cd, v_3	90	Gk, v_3
80	Fs, v_1	91	Cd, v_2
81	Fs, v_1	92	Cd, v_2
82	Cd, v_3	93	Gk, v_3
83	Cd, v_1	94	Fs, v_2
84	Cd, v_1	95	Fs, v_3
85	Gk, v_3	96	Gk, v_3
86	Gk, v_3	97	Fs, v_2
87	Fs, v_2	98	Fs, v_3
88	Cd, v_1	99	Fs, v_2
89	Cd, v_2	100	Fs, v_3

Table 6 Frequency of predicted active internal communications

State	Timeframe of AICs			
–	v_1	v_2	v_3	Total
Cd	14	14	8	36
Gk	10	8	10	28
Fs	13	13	10	36
Total	37	35	28	100

The optimized transition matrix presents the right diagonal elements (internal communication) as the most probable events in Markov's sequence.

While the optimized emission probability matrix is given as;

$$\hat{\delta} = \begin{pmatrix} 0.1239157 & 0.3197026 & 0.5563817 \\ 0.1239157 & 0.3197026 & 0.5563817 \\ 0.1239157 & 0.3197026 & 0.5563817 \end{pmatrix}. \tag{19}$$

$$\lambda_1^* = \left(\hat{M}, \ \hat{\delta}, \ \hat{\pi} \right). \tag{20}$$

Equation 20 is the result of the Baum-Welch algorithm which optimizes the HMM_T parameters (Transition, emission, and initial probabilities) to attend the optimal value (s). Thus, $\hat{M}$, $\hat{\delta}$ and $\hat{\pi}$ are the optimized transition probability matrix, optimized emission probability matrix, and the optimized initial probability matrix respectively. The result presented in Eq. 19 shows that the optimized emission matrix values have unique characteristics with all v_3(night) internal communications equal and is the most probable timeframe of communications that leads to criminal attacks. This is followed by v_2(evening) with the second-highest values of the AICs and v_1(day) having the lowest values.

7 Conclusion and Future Directions

This study identified and modeled the states and active internal communication features of criminal networks. Key features such as the criminal states (CS) and AIC were determined as hidden entities. Numerical coding was used to bridge some missing links and computational parameters needed to effectively model features of a criminal network. The work expanded key idea behind HHM and showed how observed events can have no one-to-one associations. These are linked to states via probability distributions. The scheme provides a double process (stochastic) which has the Markov chain as the key stochastic process. With this scheme, state transitions and various stochastic event processes offer statistical linkage within the entire security ecosystem. Results show that the timeframe of attacks has a significant association with the terrorist states. Also, the HMM model results highlighted the most probable state and timeframe for the occurrence of criminal attacks. It further shows the most probable sequence of such attacks. The HMM parameters obtained were optimized using the Baum-Welch technique. This is to ensure efficient and reliable results. The incidence of Boko-Haram and Fulani extremists as terrorists (criminal elements) poses the deadliest threat to national security with Boko-Haram attack level at 79.2% and Fulani extremist at 20.8%. As observed, the highest attack target is on the private citizens and properties with 50.6%. This is followed by police and military base with 12.3% and 6.7% respectively.

The work applied HMM based Baum-Welch optimization algorithm as an AI tool to understudy and model the behaviors of terrorist elements for criminal forensic purposes within the Nigeria security spectrum. The result and findings obtained in the characterization of the hidden features of the groups will be applied to the security agencies in intelligence gathering for fighting the menace of terrorism in Nigeria. Future work will focus on big data use-case algorithm for complex terrorist network prediction using HMM and Python networked robotics in AI cloud platform. Investigation and comparisons with other complex optimization schemes will be validated also.

References

1. Chen J, Chen Y, Chen L, Zhao M, Xuan Q (2021) Multiscale evolutionary perturbation attack on community detection.IEEE Trans Comput Soc Syst
2. Liu M, Qian P (2021) Automatic segmentation and enhancement of latent fingerprints using deep nested unets. IEEE Trans Inf Forensics Secur
3. Li G, Hu J, Song Y, Yang Y, Li HJ (2019) Analysis of the terrorist organization alliance network based on complex network theory. IEEE Access
4. Zuo B, Zhu W, Li F, Zhuo J (2020) Modeling and quantitative analysis of terrorist attack task list. In: Proceedings of the 2020 IEEE 4th information technology networking, electronic and automation control conference, ITNEC 2020
5. Cocarascu O, Toni F (2017) Identifying attack and support argumentative relations using deep learning. In: EMNLP 2017—Conference on empirical methods in natural language processing, proceedings
6. (2019) Study of the 9–11 Hijackers Network. IEEE Access
7. Hora A, Bari A, Rawat S (2020) Machine learning approaches to uncover terrorism network in India. In: 2020 international conference for emerging technology, INCET 2020
8. Crawford B, Keen F (2020) The Hanau terrorist attack: how race hate and conspiracy theories are fueling global far-right violence. CTC Sentin
9. West J (2018) Global Terrorism Index (2018) Available Online: https://www.hsdl.org/c/global-terrorism-index-2018/
10. I. for E. and peace IEP (2019) Global peace index. Available Online: https://rel iefweb.int/sites/reliefweb.int/files/resources/GPI-2019-web003.pdf#:~:text=GLOBAL%20PEACE%20INDEX%2020192019%20The%20Institute%20for%20Economics,   and %20tangible%20measure%20of%20human%20wellbeing%20and%20progress.
11. Luan M, Sun D, Li Z, Xu F (2018) Analyzing core structure and role transition features of terrorist organizations based on meta-network take al qaida as an example. In: 2018 IEEE 4th international conference on computer and communications, ICCC 2018
12. Lim M, Abdullah A, Jhanjhi N, Supramaniam M (2020) Indexed metrics for link prediction in graph analytics. Int J Adv Comput Sci Appl
13. Taha K, Yoo PD (2019) Shortlisting the influential members of criminal organizations and identifying their important communication channels. IEEE Trans Inf Forensics Secur
14. Mason J, Esterline A (2020) Security and a framework for identity. In: 2020 IEEE symposium series on computational intelligence, SSCI 2020
15. Zou Y, Donner RV, Marwan N, Donges JF, Kurths J (2019) Complex network approaches to nonlinear time series analysis. Phys Rep
16. Taha K, Yoo PD (2017) Using the spanning tree of a criminal network for identifying its leaders. IEEE Trans Inf Forensics Secur 12(2):445–453
17. Lim M, Abdullah A, Jhanjhi N (2019) Performance optimization of criminal network hidden link prediction model with deep reinforcement learning. J King Saud Univ Comput Inf Sci
18. Lin Z, Dou Y, Li J (2020) Analysis model of terrorist attacks based on big data. In: Proceedings of the 32nd chinese control and decision conference, CCDC 2020
19. Kumar D, Bhowmik PS (2019) Hidden markov model based islanding prediction in smart grids. IEEE Syst J
20. Granstrom K, Willett P, Bar-Shalom Y (2016) Asymmetric threat modeling using HMMs: Bernoulli filtering and detectability analysis. IEEE Trans Signal Process
21. Khalili Shoja MR, Amariucai GT, Wang Z, Wei S, Deng J (2019) On the secret key capacity of sibling hidden Markov models. IEEE Trans Inf Forensics Secur
22. Fjeldstad T, More H (2020) Bayesian inversion of convolved hidden markov models with applications in reservoir prediction. IEEE Trans Geosci Remote Sens
23. Marfak A et al (2020) The hidden Markov chain modelling of the COVID-19 spreading using Moroccan dataset. Data Br
24. Dong S, Wu ZG, Su H, Shi P, Karimi HR (2019) Asynchronous control of continuous-time nonlinear Markov jump systems subject to strict dissipativity. IEEE Trans Automat Contr

25. Ren C, He S, Luan X, Liu F, Karimi HR (2021) Finite-time L2-gain asynchronous control for continuous-time positive hidden Markov jump systems via T-S fuzzy model approach. IEEE Trans Cybern
26. Aung N, Zhang W, Dhelim S, Ai Y (2018) Accident prediction system based on hidden Markov model for vehicular Ad-Hoc network in urban environments. Inf
27. Wu C et al (2020) Patient-specific characterization of breast cancer hemodynamics using image-guided computational fluid dynamics.IEEE Trans Med Imaging
28. Ganesh P et al (2021) Learning-based simultaneous detection and characterization of time delay attack in cyber-physical systems. IEEE Trans Smart Grid
29. Zhang H, Shan G, Yang B (2020) Optimized elastic network models with direct character-ization of inter-residue cooperativity for protein dynamics. IEEE/ACM Trans Comput Biol Bioinformat
30. Zhang Z, Chen D, Bai L, Wang J, Hancock ER (2020) Graph motif entropy for understanding time-evolving networks. IEEE Trans Neural Netw Learn Syst
31. Zhuang Y, Yağan O (2020) Multistage complex contagions in random multiplex networks. IEEE Trans Control Netw Syst
32. Da Cruz ACA, Schwab FA, Maia OMA, De Faria RA, Borba GB, Pilla V (2019) Charac-terization of entomological micro traces images with deep neural networks. In: 2019 IEEE symposium series on computational intelligence, SSCI 2019
33. Lim M, Abdullah A, Jhanjhi NZ, Supramaniam M (2019) Hidden link prediction in criminal networks using the deep reinforcement learning technique. Computers
34. Lim M, Abdullah A, Jhanjhi NZ, Khurram Khan M (2020) Situation-aware deep reinforcement learning link prediction model for evolving criminal networks. IEEE Access
35. Phillips M, Amirhosseini MH, Kazemian HB (2020) A rule and graph-based approach for targeted identity resolution on policing data. In: 2020 IEEE symposium series on computational intelli, SSCI 2020
36. AdamI Y, Varol C (2020) Intelligence in digital forensics process. In: 8th Int'l sympos on digital forensics and security, ISDFS 2020
37. Khalifa NEM, Taha MHN, Taha SHN, Hassanien AE (2020) Statistical insights and association mining for terrorist attacks in Egypt. Adv Intell Syst Comput 921:291–300
38. Li Z, Sun D, Li B, Li Z, Li A (2018) Terrorist group behavior prediction by wavelet transform-based pattern recognition, vol 2018
39. Luck A, Giehr P, Nordstrom K, Walter J, Wolf V (2019) Hidden Markov modelling reveals neighborhood dependence of Dnmt3A and 3b activity. IEEE/ACM Trans Comput Biol Bioinformat
40. Der Fuh C, Tartakovsky AG (2019) Asymptotic Bayesian theory of quickest change detection for hidden markov models. IEEE Trans Inf Theory
41. Li F, Xu S, Shen H, Ma Q (2020) Passivity-based control for hidden Markov jump systems with singular perturbations and partially unknown probabilities. IEEE Trans Autom Contr
42. Li F, Xu S, Zhang B (2020) Resilient asynchronous H_control for discrete-time Markov jump singularly perturbed systems based on hidden Markov model. IEEE Trans Syst Man Cybern Syst
43. Yu Z et al (2020) Emergent inference of Hidden Markov models in spiking neural networks through winner-take-all. IEEE Trans Cybern
44. Shoja MRK, Amariucai GT, Wang Z, Wei S, Deng J (2017) Asymptotic converse bound for secret key capacity in hidden Markov model. In: IEEE international symposium on information theory
45. Brogi G, Di Bernardino E (2019) Hidden Markov models for advanced persistent threats. Int J Secur Netw 14(4):181–190
46. Adam T, Langrock R, Weiß CH (2019) Penalized estimation of flexible hidden Markov models for time series of counts. Metron
47. Ingale S, Paraye M, Ambawade D (2020) Enhancing multi-step attack prediction using hidden Markov model and Naive Bayes. In: International conference on electronics and sustainable communication systems, ICESC 2020

48. Anandhalekshmi AV, Rao VS, Kanagachidambaresan GR (2020) HMM based on Baum-Welch algorithm for predicting critical data packets in IoT network. In: 11th international conference on computing, communication, and networking technologies, ICCCNT 2020
49. NC, for the study of T, R to T START (2016) Global terrorism database. Available online: https://www.economicsandpeace.org/wp-content/uploads/2016/11/Global-Terrorism-Index-2016.2.pdf
50. Argamon S, Howard N (2009) Computational methods for counterterrorism
51. Turek D, de Valpine P, Paciorek CJ (2016) Efficient Markov chain Monte Carlo sampling for hierarchical hidden Markov models. Environ Ecol Stat
52. Misra S (2021) A step by step guide for choosing project topics and writing research papers in ICT related disciplines. In: Misra S, Muhammad-Bello B (eds) Information and communication technology and applications. ICTA 2020. Commun Comput Inf Sci 1350. Springer, Cham. https://doi.org/10.1007/978-3-030-69143-1_55
53. Nwankwo KE et al (2020) A Panacea to soft computing approach for Sinkhole attack classification in a wireless sensor networks environment. In: International conference on futuristic trends in networks and computing technologies. Springer, Singapore, pp 78–87
54. Bameyi OJ et al (2020) End-to-end security in communication networks: a review. In: International conference on innovations in bio-inspired computing and applications. Springer, Cham, pp 492–505
55. Azeez NA et al (2020) Identifying phishing attacks in communication networks using URL consistency features. Int J Electron Secur Dig Forensics 12(2): 200–213
56. Subairu SO et al An experimental approach to unravel effects of malware on system network interface. In: Advances in data sciences, security and applications. Springer, Singapore, pp 225–235
57. Okafor KC, Achumba IE, Chukwudebe GA, Ononiwu GC (2017) Leveraging fog computing for scalable IoT datacenter using spine-leaf network topology. J Electr Comput Eng 2017:1–11, Article ID 2363240 (Egypt)

An Integrated IDS Using ICA-Based Feature Selection and SVM Classification Method

Roseline Oluwaseun Ogundokun⃝, **Sanjay Misra**⃝, **Amos O. Bajeh,
Ufuoma Odomero Okoro, and Ravin Ahuja**

Abstract The continuous development of computer networks has created serious worries about vulnerability and security. Network administrators have embraced Intrusion Detection Systems (IDS) to offer vital network security. Commercial IDS in the market are incapable of detecting fresh threats and instead produce false alarms for the typical user activity. Artificial Intelligence (AI) may be used to address these difficulties and enhance accuracy. ICA-based feature selection (FS) ranks features based on the attribute-class label correlation. The authors suggested an ICA-based feature selection algorithm combined with a support vector machine (SVM) classifier for detecting anomalies in network connections. The KDDCUP 99 datasets, which is a benchmark dataset for intrusion detection with current threats, were used in the experiments. In contrast to several state-of-the-art approaches, the suggested model outperforms them in terms of accuracy, sensitivity, detection rate (DR) false alarm, and specificity. IDS may be used to secure wireless payment systems. It is possible to establish secure integrated network management that is error-free, therefore boosting performance.

Keywords ICA · Artificial intelligence · Intrusion detection system · KDDCUP 99 · Feature selection

R. O. Ogundokun (✉)
Center of ICT/ICE Research, Covenant University, Ota, Nigeria
e-mail: ogundokun.roseline@lmu.edu.ng

S. Misra
Department of Computer Science and Communication, Ostfild University College, Halden, Norway

A. O. Bajeh
Department of Computer Science, University of Ilorin, Ilorin, Nigeria
e-mail: bajeh.amos@lmu.edu.ng

U. O. Okoro
Department of Geography, University of Ibadan, Ibadan, Nigeria

R. Ahuja
Vishwakarma Skill University, Gurgaon, India

© The Author(s), under exclusive license to Springer Nature Switzerland AG 2022
S. Misra and C. Arumugam (eds.), *Illumination of Artificial Intelligence in Cybersecurity and Forensics*, Lecture Notes on Data Engineering and Communications Technologies 109, https://doi.org/10.1007/978-3-030-93453-8_11

1 Introduction

Due to the extensive expansion and dissemination of network connectivity, network security and protection against cyber-attacks are becoming increasingly important [1]. In the field of network security, an IDS acts as a protector, identifying and blocking attackers from gaining entry to the network. When an impostor attempts to access a network, the IDS's principal function is to prevent the intrusion, and this act must be taken before there is any harm or access to cognizant information. Aside from that, the major aim of an IDS is to protect the network system's availability, integrity, and confidentiality. Intelligent IDS is now thought to be a productive resolution for network security and defense opposing extrinsic invaders [2]. The question of what way to improve the proficiency of intelligent IDS, which has resulted in a fundamental point of network security, has recently been of serious concern [3]. Furthermore, several concerns have been explored in the literature [4], for instance, a poor DR in comparison to novel assaults, early integration, inoperativeness, and the fact that audit data is severely overloaded.

Misappropriation and anomaly IDSs are the two most common forms of IDSs [5]. Instructions are detected based on system fault limitations and designated as attack signatures for an IDS misappropriation type. Nonetheless, it is unable to detect new or unknown threats. Anomaly IDSs, on the other hand, rely on common patterns and use them to recognize any activity that differs considerably from the norm. Misappropriation IDS identifies interruptions with earlier recognized patterns by comparing existing interruption patterns into consideration for examination; otherwise, anomaly IDS identifies patterns built on the examination of data obtained from normal practice [6]. The entire anomalous patterns must be treated as potential assaults, despite not being identified as such; hence, the likelihood of receiving false-positive findings in anomaly-based detection may be higher than intended.

Several researchers have viewed Intrusion Detection (ID) to be a difficult task in respect to classification and FS [7]. To increase the performance of IDSs, numerous FS approaches have been made known and implemented into them. In recent decades, a wide range of filter and wrapper techniques have been used to build intelligent IDSs for FS that may improve network security while also detecting contemporary assaults [8]. Identify the inherent relationships among the involved features and substantial class in the filter methods, and then eliminate unnecessary features from the inputted features. Several evolutionary computations (EC) approaches have been implemented for ID as wrapper techniques in recent decades, including swarm intelligence procedures such as genetic algorithm (GA), differential evolution (DE), particle swarm optimization (PSO), and grasshopper optimization algorithm (GOA) [9].

Many optimization issues have been solved using a variety of optimization techniques [10]. The relevance of this is that we can resolve the utmost factual problems by applying correct mathematical models and algorithms. Researcher [11] presented a competent algorithm named GOA, which is one of the newest bio-inspired algorithms that has piqued the interest of many academics in many optimization disciplines. Ibrahim et al. [12] recently presented an innovative wrapper technique built on GOA to improve SVM components and choose optimum feature subclasses. It might similarly be used to tackle actual challenges. Mirjalili et al. [13] proposed a multi-unbiassed strategy for tackling multi-objective challenges that were inspired by grasshopper navigation in nature. When tackling various realistic and worldwide unobstructed and restricted optimization problems, the GOA approach can provide superior results and experience.

To address the limitations of individual approaches, many filter and wrapper approaches have been projected in current research investigations to choose the optimum feature subsets, referred to as hybrid FS techniques [14]. The author of [8] researched an innovative hybrid technique for FS that used a filter as a linear correlation coefficient and a wrapper as a cuttlefish procedure. The fitness function in their work was a decision tree classifier. In addition, for determining the kind of network attack, many hybrid classification methods, for instance, ICA-SVM and PSO-DT have been suggested [15]. Anomalies aren't always simple to spot, and the majority of anomaly detection algorithms miss the important patterns. Several researchers have investigated various ML approaches and classifiers for training and assessing data as normal or anomalous for innumerable IDSs in this setting [16].

To minimize the high computing costs of engines, this study presented a novel technique that used ICA as a feature selection method to improve detection accuracy for accurate attack identification. A two-stage model was designed to decrease the implementation period and advance the classification performance of the SVM approach. The postulated method was similarly employed to optimize the alteration constraints of the SVM. There were two phases to the proposed method (ICA-SVM). Using the ICA technique, first eliminate the superfluous features from the original data. The reduced data is thereafter directed to the second step, where SVM classification was employed to determine if network traffic is normal or abnormal. The following are the key contributions of this study:

- We utilized a framework for spotting anomalies (attack), utilizing a feature selection technique called ICA to avoid the costly computational expenses of picking the most suited agents as well as their control parameter values.
- To address the limitations of individual methods, we presented a novel method for detecting security violations in IDSs that combines ICA FS and ML classifiers in this work.
- In the suggested technique, an SVM was employed as a suitability function to identify important characteristics that can aid in properly classifying assaults. The suggested approach is utilized in this context to advance the resulting quality by changing the SVM regulatory parameter values.
- The goal was to achieve satisfactory results for the IDS datasets, KDD Cup 99, in terms of classifier performance in noticing invasions built on the optimal number of features.

The remaining part of the article is systematized as thus: The related work on FS in intrusion detection is obtainable in Sect. 2. The current ICA FS method, as well as the SVM classifier, are discussed in Sect. 3. Section 4 shows how the suggested approach for selecting the best characteristics from the dataset and distinguishing between attacks is implemented. The dataset's experimental assessments and discussion were also presented in Sect. 4. The article was concluded with future works suggested in Sect. 5.

2 Literature Review

The IDS-related researches were discussed in this section. A machine learning method was created to address one of the most serious issues that computer networks confront. Network administrators can improve network security by employing machine learning. Any discrepancy pattern in the network data can be found by the machine. Several studies have been conducted in this sector to advance network security.

Saharkhizan et al. [17] suggested a hybrid approach to identify intrusions like User to Root (U2R) and Remote to Local (R2L). Back-propagation was used to lower the dataset's dimension and group the data. The proposed technique is an unsupervised learning technique. In the realm of pattern identification and IDS, the K-Nearest Neighbor method (KNN) [17] has widely been employed. To identify intrusion, automatic classification techniques such as SVM, and neural network (NN) [18]. Bayesian method [19–21] and k-means were employed [22, 23]. The increased dimensionality of network data has provided developers and researchers with a level of complexity that will help them improve intrusion detection. In addition, the researchers used a variety of feature selection techniques to deal with dimensionality.

Shams et al. [24] used an SVM classifier for the incorporation of the IDS into MANET. The goal of their research was to identify a Denial of Service (DoS) attack. With the SVM method, the DR was further suitable. Zargar et al. [25] investigated new methods for detecting DDoS flooding assaults in their study. They've categorized the available countermeasures in terms of how they avert, discover, and answer to DDoS inundating assaults. Horng et al. [26] also presented the SVM method, which connects a hierarchical clustering method with an IDS method. The hybrid model was put to the test using the well-known KDD Cup dataset. According to the accuracy metrics, the analysis findings showed higher performance in the discovery of DoS and Probe assaults. To improve the detection system's performance. Pandeeswari et al. [27] presented a hybrid method that integrates the Fuzzy C-Means Clustering and ANN (FCM-ANN) algorithms. This study compares the Nave Bayes (NB) classifier with the Classic ANN technique, both of which are machine learning methods. It has been discovered that the hybrid model outperforms. Shah et al. [28] utilized a hybrid SVM and Fuzzy logic method to identify the invaded system. It has been discovered that the suggested model outperforms that of the others. Zhang et al. [29] integrate gaussian NB with the PCA method. The PCA technique was employed to decrease the dimensionality of dataset features. The suggested system was found to obtain a higher DR. Peng et al. [30] presented a DR technique to discover intrusion in a network. The preprocessing stage was primarily concerned with digitizing and then normalizing the strings in the provided dataset. There was a comparison between the suggested model and current models such as the KNN and NB algorithms and it was discovered that the suggested method was superior.

The J48 classifier was used by Aldyani et al. [31] to categorize the IDS. To decrease dimensionality, the rough set theory and IG techniques were used. Soft clustering was developed to manage vague objects from ID datasets. Kamarudin et al. [32] developed a unique technique called integrated Invasion anomaly discovery for enhancing ID. The major goal of their research was to utilize a hybrid FS method to improve the classification algorithm by selecting the most important characteristics. The Logit Boost and Random Forest (RF) was used as the ML classifiers. Salunkhe and colleagues [33] employed the use of an ensemble classifier to evaluate and identify intrusion in a network. The ensemble classifier was a hybrid approach that incorporates both data-level and feature-level methods. It has been determined that the hybrid model can improve incursion. Information entropy was utilized by several studies to improve the IDS by choosing the utmost important characteristics from the imbalance network, such as [34, 35]. Alsaadi et al. advocated using computational intelligence models to enhance IDS, for instance, ACO and PSO techniques were employed as preprocessing methods for choosing the utmost important characteristics from the dataset. These characteristics are input into three ML classifiers, which are KNN, SVM, and NB, and they are used to improve IDS. Ganapathy et al. [36] examine a variety of intelligent techniques for enhancing network ID. Jaisankar et al. [37] proposed two dissimilar intelligent FS techniques for managing dimensionality lessening of IDS datasets, with the important features attained using the FS techniques after which the SVM ML classifier is then employed. Jaisankar et al. [38] proposed an Intellectual IDS by utilizing a fuzzy rough set over the C4.5 technique. Ganapathy et al. [39] suggested an Innovative Weighted Fuzzy C-Means clustering technique over Immune GA to enhance the IDS. Nancy et al. [40] used an active recursive FS approach to choose important and relevant features for enhancing the classification procedure. Table 1 shows the summary of related works reviewed.

3 Material and Method

The materials in respect to datasets and the ML classifier method employed in this study are discussed in this section.

Table 1 Summary of related works

Authors	Technique	Dataset	Weaknesses
Yin et al. [41]	RNN	NSL-KDD	Lower detection accuracy (DA) for minority classes like R2L and U2R, increased complexity thereby requiring more time for training
Shen et al. [42]	Ensemble learning method, BAT optimizer	KDD CUP, NSL-KDD, Kyoto	Low detection accuracy for the U2R invasion
Shone et al. [43]	Deep AE and RF classifier	KDDCUP, NSL-KDD	Low performance for DA for R2L and U2R
Ali et al. [44]	FLN, particle swarm optimization (PSO)	KDDCUP	DR is low for R2L and U2R
Jia et al. [45]	DNN	KDD CUP and NSL-KDD	DA is lower for the U2R
Wang et al. [46]	DNN	NSL-KDD	–
Yan et al. [47]	Stacked sparse autoencoder (SSAE) and SVM	NSL-KDD	Medium DR for U2R and R2L attacks
Xu et al. [48]	RNN with multilayer perception and a softmax classifier	KDD CUP and NSL-KDD	Lower DR for R2L and U2R
Al-Qatf et al. [49]	Sparse AE and SVM	NSL-KDD	No result presented for the performance of the approach contrary to R2L and U2R
Papamartizivanous et al. [50]	Sparse AE	KDD CUP and NSL-KDD	Incapable of identifying R2L and U2R
Khan et al. [51]	Deep stacked AE, SMOTE	KDDCUP, UNSW-NB15	DA of 89.14% achieved
Xiao et al. [52]	CNN, PCA, and AE	KDDCUP	Lower DR for U2R and R2L
Yao et al. [53]	RF	KDDCUP	–
Gao et al. [54]	DT, RF, KNN and DNN	NSL-KDD	Displays substandard outcomes for feebler attack classes
Wei et al. [55]	DBN, particle swarm, fish swarm, and genetic algorithm	NSL-KDD	The technique is multifaceted and necessitates further training period

(continued)

Table 1 (continued)

Authors	Technique	Dataset	Weaknesses
Zhang et al. [56]	CNN and gcForest, p-Zigzag	UNSW-NB15 and CIC-IDS2017	DA for attack classes with reduced training data is short
Malaiya et al. [57]	AE and seq2seq		The cost of training is high. The model that performs best in the detection of the minority attack classes are not shown
Karatas et al. [58]	SMOTE and six ML algorithms are employed	CSECIC-IDS2018	The execution time is done at a higher cost
Jiang et al. [59]	CNN and bi-directional long short-term memory (BiLSTM), SMOTE	NSL-KDD and UNSW-NB15	The postulated method is complex and DR of minority data classes advanced slightly
Yang et al. [60]	AE with regularization and DNN	NSL-KDD, UNSW-NB15	DR achieved was reasonable for U2R and R2L attacks
Yu et al. [46]	Few-shot learning (FSL), DNN, CNN	NSL-KDD, UNSW-NB15	DR achieved was reasonable for U2R and R2L attacks
Andresini et al. [61]	AE, ID convolution layer, two stacked fully connected layers	–	Model effectiveness in identifying the minority attack class is not given

3.1 Dataset

Datasets are a gathering of information annals that include a variety of qualities or features as well as relevant data about the cyber-security model. As a result, understanding the nature of cyber-security data, which contains several sorts of cyber-attacks and important aspects, is critical. The justification for this is that raw security data collected from pertinent cyber sources may be utilized to evaluate innumerable forms of security events or spiteful activity to generate a data-driven security method that would assist us to accomplish our objectives. Two typical network streams of traffic datasets, such as the KDD CUP dataset are utilized in this study. The following subsections provide a thorough explanation of these datasets. These datasets contain 4,898,431 instances with 41 attributes. All attacks from the KDD CUP dataset are listed in Table 2.

DoS is a type of invasion in which a genuine individual does not have entree to the organization and system resources. DoS attacks consist of worm, processtable, pod, land, Neptune, back, teardrop, apache2, mailbomb and udopstor.

Table 2 KDD CUP attack types

Datasets attacks	Attack types in KDDCUP
Dos	Worm, processtable, pod, land, Neptune, back, teardrop, apache2, mailbomb and udopstor
Probe	Satan, IPsweep, Nmap, Portsweep, Mscan, Sa int
R2L	Named, ftp_write, snmpgue, ss, sendmail, guess_password, imap, multihop, xlock, phf, xsnoop, sumpgeattack, httptunnel, warezmaster
U2R	ps, perl, buffer_overflow, xterm, rootkit, loadmodule, sqlattack

R2L is an invasion where an invader attempts to access an entree to the prey devices deprived of having a network in it. R2L consist of named, ftp_write, snmpgue, ss, sendmail, guess_password, imap, multihop, xlock, phf, xsnoop, sumpgeattack, httptunnel, warezmaster.

U2R is an invasion where an invader attempts to access a privilege by having a local entree in the prey devices. U2R consists of ps, perl, buffer_overflow, xterm, rootkit, loadmodule, sqlattack.

A probe is an attack where an invader aims at the host and attempts to lay hold on records about the host. The probe consists of Satan, IPsweep, Nmap, Portsweep, Mscan, Sa int.

They have three abnormalities, namely protocol types, services, and flags, which are all discussed. This method was created to turn these abnormalities into numerical characteristics. These characteristics have proven to enhance classification conversion using this method.

3.1.1   KDDCup'99

The KDDCup'99 dataset is a modified form of the DARPA dataset, which contains 7 weeks of network traffics and 4 Gb of binary Tcpdump data organized into 5 million entries. The KDDCup'99 training dataset contains 4,900,000 sole connection vectors with 41 attributes that are either attack or normal. The following security threats are included in this dataset: DoS, U2R, R2L, and Probing attack. In a U2R attack, the invader gets root entree to the host by gaining access to a user account. An attacker uses the R2L attack to obtain access to a remote host by sending data packets to it. In addition, in a probing assault, the attacker tries to gather knowledge concerning a computer network in preparation for future security attacks [62].

3.2   ML-Based Classifier

Classification is a supervised learning method that is commonly employed to simulate cyber incursions based on a variety of attack categories. Data is always tagged before being used in supervised learning. The classifier learns the labels in the training phase so that it can accurately forecast unknown data in the test phase. We employ prominent machine learning techniques in our study for a variety of objectives. Several approaches are summarized in the table.

3.2.1 Support Vector Machine

SVM is an ML supervised technique founded on the notion of a hyperplane with extreme sideline separation in n-dimensional feature space. It might be employed to unravel linear and non-linear challenges. Kernel roles are employed to unravel non-linear problems. The aim is to utilize the kernel role to translate a reduced-dimensional input vector into an extreme-dimensional feature space. The support vector is then used to find an ideal extreme marginal hyper-plane, which serves as a decision boundary [63–65]. By precisely forecasting the normal and anomalies classes, the SVM technique might be utilized to advance the competency and correctness of NIDS [47, 66].

3.2.2 Feature Selection Methods

Machine learning algorithms that use feature selection (FS) algorithms offer a high level of efficacy and efficiency. In general, the FS technique has employed inappropriate and duplicated features, whereas feature selection methods choose the most meaningful subsets of characteristics [67]. FS techniques are employed to eliminate immaterial and repetitious features, which will aid to intensify the rapidity and accuracy of classification in real-world data mining applications.

The dataset was first organized plus the attacks types and obtainable attributes for developing ML-based IDS methods. Table 3 shows the several categories of attacks in the KDDCUP dataset.

We talk about our machine-learning-based intrusion detection approach, which has four major components:

- Attack Class Label: To incorporate them into model intrusion detection systems, all the varied threats have been tallied as separate unique class labels. Different forms of attacks, such as DoS, U2R, R2L, and PROBE, are represented as discrete classes in Table 3: Class 1, Class 2, Class 3, and Class 4.
- Security Features or Attributes: These are utilized on their own to forecast the cyber risks listed above. Protocol type, service, length, and error rate are examples of features.
- Training and Testing Datasets: The dataset is divided into two categories: training and testing. The training data set is used to train the IDS model, while the testing dataset is used to assess the model's generalization. We employ a major portion of the above-mentioned cybersecurity data to create the IDS model, and the rest for testing.

Table 3 KDDCUP datasets innumerable categories of attacks

Types of attack	Attack name	Number of instances
DoS	Smurf	2,807,886
	Neptune	1,072,017
	Back	2203
	Pod	268
	Teardrop	979
U2R	Buffer overflow	30
	Load module	9
	Perl	3
	Rootkit	10
R2L	FTP write	8
	Guess password	53
	Imap	12
	MultiHop	7
	Phf	4
	Spy	2
	Warez client	1020
	Warez master	20
PROBE	Ipsweep	12,481
	Nmap	2316
	Portsweep	10,413
	Satan	15,892
Normal		972,781

4 Investigational Evaluation

This section contains the ID performance measures. The outcomes of tests on cyber-security datasets involving various sorts of assaults were described. All of the assessment measures are based on the various properties in the confusion matrix, a 2-dimensional matrix that contains information about the actual and projected class [68, 69]. This entails:

i.   True Positive (TP): The data instances are appropriately predictable as an Attack by the classifier.
ii. False Negative (FN): The data instances are erroneously predictable as Normal instances.
iii. False Positive (FP): The data instances erroneously classified as an Attack.
iv. True Negative (TN): The instances appropriately classified as Normal instances.

The accurate forecasts are represented by the diagonal of the confusion matrix, whereas the incorrect forecasts are represented by non-diagonal components. These confusion matrix properties are shown in Table 4. Furthermore, the many assessment measures that have been utilized in recent research are as follows [70]:

- Precision: It's the proportion of successfully predicted Attacks to all Attacks samples.

$$\text{Precision} = \frac{TP}{TP + FP} \tag{1}$$

- Recall: It's the proportion of all Attacks samples successfully categorized to all Attacks samples that are truly Attacks. It's also known as a Detection Rate.

$$\text{Recall} = \frac{TP}{TP + FN} \tag{2}$$

- False Alarm Rate (FAR): It's also known as the false positive rate, and it's calculated as the ratio of Attack samples that were incorrectly predicted to all Normal samples.

$$\text{False Alarm Rate} : \frac{FP}{FP + TN} \tag{3}$$

- True Negative Rate (TNR): It is defined as the number of correctly identified Normal samples divided by the total number of samples that are Normal.

$$\text{True Negative Rate} = \frac{TN}{TN + FP} \tag{4}$$

- Accuracy: It's the proportion of cases that have been successfully categorized into the total number of instances. It's also known as Detection Accuracy, and it's only relevant as a performance metric when a dataset is well-balanced.

$$\text{Accuracy} = \frac{TP + TN}{TP + TN + FP + FN} \tag{5}$$

- F-Measure: Precision and Recall are combined to form the harmonic mean. To put it another way, it's a statistical approach for evaluating a system's correctness by taking into account both precision and recall.

Table 4 Confusion matrix

Predicted class			
Actual class		Attack	Normal
	Attack	TP	FN
	Normal	FP	TN

$$\text{F Measure} \; = \; 2\left(\frac{\text{Precision} \; \times \; \text{Recall}}{\text{Precision} \; + \; \text{Recall}}\right) \tag{6}$$

The six discussed evaluation metrics are computed by testing the postulated methods using the KDDCUP benchmark dataset.

4.1 Investigational Findings and Discussion

The efficacy of ML classification approaches for detecting intrusions is demonstrated in this section. To assess and test the proposed system, two common network datasets are used. The system was built utilizing the Python programming language and 64-bit Windows 10 with a Core i5 CPU and 8 GB RAM. In the KDDCUP, a total of 21 assaults were examined. DoS, Probe, U2R, and R2L assaults are among the four primary attacks in the dataset. KDD stands for Knowledge Discovery and Development. The suggested system was tested using 117, 142 record packets for DoS, 299 record packets for U2R, 319 record packets for R2L, 1, 292 record packets for Probe, and 29, 154 record packets for normal in the KDD dataset. The IDS was enhanced using an upgraded ML approach (ICA + SVM). The KDDCUP dataset has 41 features, four of which are anomalies, and the rest features are numerical.

In our IDS model, we combine a popular classification approach SVM with an FS technique ICA to achieve this. To assess the intrusion detection model, we compared the efficacy of the popular classification approach to that of other current ML classification techniques. On the dataset, we utilize tenfold cross-validation to test the IDS model. Models are evaluated using tenfold cross-validation, which divides the data into ten distinct groups of samples. Nine partitioned sets are trained from them, while the remaining one is tested. It repeats the process ten times before calculating the average accuracy. Precision, recall, F1-score, and accuracy are measured as stated above to compare the potentiality of models. For assessment, we used the equivalent set of training and testing data in both the classification-based IDS models. From Table 5, we discovered that the SVM with ICA FS classifier-built IDS method

Table 5 Results of the proposed system for the KDDCUP dataset

Attack	Methods	Precision	Recall	F1-score
DoS	ICA + SVM	0.98	0.99	0.98
	SVM	0.99	0.92	0.97
U2R	ICA + SVM	1	0.75	0.83
	SVM	0.66	0.81	0.74
R2L	ICA + SVM	0.98	0.80	0.85
	SVM	0.74	0.82	0.76
Probe	ICA + SVM	0.98	0.93	0.95
	SVM	0.97	0.97	0.97

Table 6 Comparison with the existing systems

Authors	Method	Accuracy
Zhang et al. [71]	LSTM, ensemble, auto-encoder	95
Shone et al. [43]	Random forest (RF)	99.5
Khan et al. [51]	RF	89
Al-Qatf et al. [49]	SVM, PCA	97
Proposed system	ICA + SVM	100

executes better than the other classifier for noticing intrusion in terms of precision, recall, and F1-score. The system had a detection accuracy of 100%.

This proposed method was compared with other recent works as displayed in Table 6. The implementation results evidence that our proposed approach performs better in terms of detection for the KDD CUP dataset. Accuracy.

5   Conclusion and Future Work

For security reasons, the possibility and usefulness of ML-built ID modeling is a major issue for IT professionals, e-commerce designers, and application developers. In general, cyber-security data collection comprises various types of cyber-attacks together with important characteristics. As a result, depending on a range of attack categories and characteristics, certain classifiers might not execute well in relation to accuracy and real forecast rate. A large number of investigated studies have been conducted to identify and avert incursion, and the majority of these studies have relied only on machine learning algorithms categorization approaches. Machine learning techniques are highly effective in detecting intrusion. To identify intrusion, an enhanced ML classification approach is utilized in this study. The suggested approach for enhancing the IDS was tested using a standard dataset known as KDD CUP. The ICA method was used to identify the most important network data features for enhancing the performance of the suggested system while dealing with the dimensionality reduction feature selection approach. The SVM ML classification method is used to process these chosen characteristics. Precision, recall, F1-score, and total accuracy are some of the performance measures we looked at. As a consequence, the findings demonstrate and specify that the postulated system is better suited for ID. Deep learning algorithms will be the focus of future studies and we can repeat this experiment with multiple cases with FS techniques using DL algorithm.

References

1. Bouyeddou B, Harrou F, Kadri B, Sun Y (2021) Detecting network cyber-attacks using an integrated statistical approach. Clust Comput 24(2):1435–1453. https://doi.org/10.1007/s10 586-020-03203-1
2. Azeez NA, Ayemobola TJ, Misra S, Maskeliūnas R, Damaševičius R (2019) Network intrusion detection with a hashing based Apriori algorithm using hadoop mapreduce. Computers 8(4):86
3. Ring M, Wunderlich S, Scheuring D, Landes D, Hotho A (2019) A survey of network-based intrusion detection data sets. Comput Secur 86:147–167. https://doi.org/10.1016/j.cose.2019. 06.005
4. Shukla AK (2020) An efficient hybrid evolutionary approach for identification of zero-day attacks on wired/wireless network system. Wirel Pers Commun 1–29. https://doi.org/10.1007/ s11277-020-07808-y
5. Zakeri A, Hokmabadi A (2019) Efficient feature selection method using real-valued grasshopper optimization algorithm. Expert Syst Appl 119:61–72
6. Jin D, Lu Y, Qin J, Cheng Z, Mao Z (2020) SwiftIDS: real-time intrusion detection system based on LightGBM and parallel intrusion detection mechanism. Comput Secur 97:101984
7. Dwivedi S, Vardhan M, Tripathi S (2020) Distributed denial-of-service prediction on IoT framework by learning techniques. Open Comput Sci 10(1):220–230
8. Mohammadi S, Mirvaziri H, Ghazizadeh-Ahsaee M, Karimipour H (2019) Cyber intrusion detection by combined feature selection algorithm. J Inf Secur Appl 44:80–88
9. Mafarja M, Aljarah I, Faris H, Hammouri AI, Ala'M AZ, Mirjalili S (2019) Binary grasshopper optimization algorithm approaches for feature selection problems. Expert Syst Appl 117:267–286
10. Shukla AK, Pippal SK, Chauhan SS (2019) An empirical evaluation of teaching-learning-based optimization, genetic algorithm and particle swarm optimization. Int J Comput Appl 1–15. https://doi.org/10.1080/1206212X.2019.1686562
11. Saremi S, Mirjalili S, Lewis A (2017) Grasshopper optimisation algorithm: theory and application. Adv Eng Softw 105:30–47
12. Ibrahim HT, Mazher WJ, Ucan ON, Bayat O (2019) A grasshopper optimizer approach for feature selection and optimizing SVM parameters utilizing real biomedical data sets. Neural Comput Appl 31(10):5965–5974
13. Mirjalili SZ, Mirjalili S, Saremi S, Faris H, Aljarah I (2018) Grasshopper optimization algorithm for multi-objective optimization problems. Appl Intell 48(4):805–820
14. Singh I, Kumar N, Srinivasa KG, Sharma T, Kumar V, Singhal S (2020) Database intrusion detection using role and user behavior-based risk assessment. J Inf Secur Appl 55:102654
15. Dwivedi S, Vardhan M, Tripathi S (2020) Incorporating evolutionary computation for securing wireless network against cyberthreats. J Supercomput 1–38
16. Tidjon LN, Frappier M, Mammar A (2019) Intrusion detection systems: a cross-domain overview. IEEE Commun Surv Tutor 21(4):3639–3681
17. Saharkhizan M, Azmoodeh A, Pajouh HH, Dehghantanha A, Parizi RM, Srivastava G (2020) A hybrid deep generative local metric learning method for intrusion detection. In: Handbook of big data privacy. Springer, Cham, Switzerland, pp 343–357
18. Alsaadi HI, Almuttairi RM, Bayat O, Ucani ON (2020) Computational intelligence algorithms to handle dimensionality reduction for enhancing intrusion detection system. J Inf Sci Eng 36(2):293–308
19. Shao XL, Liu YW, Geng MJ, Han JB (2014) The parallel implementation of mapreduce for the Bayesian algorithm to detect botnets. CAAI Trans Intell Syst 1:26–33
20. Wang S, Zou H, Sun Q, Yang F (2012) Bayesian approach with maximum entropy principle for trusted quality of web service metric in E-commerce applications. Secur Commun Netw 5(10):1112–1120
21. Amor NB, Benferhat S, Elouedi Z (2004) Naive Bayes versus decision trees in intrusion detection systems. In: Proceedings of the 2004 ACM symposium on applied computing, pp 420–424

22. Liu H, Hou X, Yang Z (2016) Design of intrusion detection system based on improved k-means algorithm. Comput Technol Dev 1:101–105
23. Al-Yaseen WL, Othman ZA, Nazri MZA (2017) Multi-level hybrid support vector machine and extreme learning machine based on modified K-means for intrusion detection system. Expert Syst Appl 67:296–303
24. Shams EA, Rizaner A (2018) A novel support vector machine-based intrusion detection system for mobile ad hoc networks. Wirel Netw 24(5):1821–1829
25. Zargar ST, Joshi J, Tipper D (2013) A survey of defense mechanisms against distributed denial of service (DDoS) flooding attacks. IEEE Commun Surv Tutor 15(4):2046–2069
26. Horng SJ, Su MY, Chen YH, Kao TW, Chen RJ, Lai JL, Perkasa CD (2011) A novel intrusion detection system based on hierarchical clustering and support vector machines. Expert Syst Appl 38(1):306–313
27. Pandeeswari N, Kumar G (2016) Anomaly detection system in cloud environment using fuzzy clustering-based ANN. Mob Netw Appl 21(3):494–505
28. Shah SAR, Issac B (2018) Performance comparison of intrusion detection systems and application of machine learning to Snort system. Futur Gener Comput Syst 80:157–170
29. Zhang B, Liu Z, Yanguo J, Ren J, Zhao X (2018) Network intrusion detection method based on PCA and Bayes algorithm. Secur Commun Netw 2018:1–11. https://doi.org/10.1155/2018/1914980
30. Peng K, Leung V, Zheng L, Wang S, Huang C, Lin T (2018) Intrusion detection systembased on decision tree over big data in fog environment. Wirel Commun Mob Comput 2018:1–10. https://doi.org/10.1155/2018/4680867
31. Aldhyani T, Joshi MR (2014) Analysis of dimensionality reduction in intrusion detection. Int J Comput Intell Informat 4(3):199–206
32. Kamarudin MH, Maple C, Watson T, Safa NS (2017) A new unified intrusion anomaly detection in identifying unseen web attacks. Networks 2017(2539034):1–18
33. Salunkhe UR, Mali SN (2017) Security enrichment in intrusion detection system using classifier ensemble. J Electr Comput Eng 201(10):1–6
34. Zhang HR, Han ZZ (2003) An improved sequential minimal optimization learning algorithm for regression support vector machine. J Softw 14(12):2006–2013
35. Zhang W, Fan J (2015) Cloud architecture intrusion detection system based on KKT condition and hyper-sphere incremental SVM algorithm. J Comput Appl 35(10):2886–2890
36. Ganapathy S, Kulothungan K, Muthurajkumar S, Vijayalakshmi M, Yogesh P, Kannan A (2013) Intelligent feature selection and classification techniques for intrusion detection in networks: a survey. EURASIP J Wirel Commun Netw 2013(1):1–16
37. Jaisankar N, Ganapathy S, Yogesh P, Kannan A, Anand K (2012) An intelligent agent-based intrusion detection system using fuzzy rough set-based outlier detection. In: Soft computing techniques in vision science. Springer, Berlin, Heidelberg, pp 147–153
38. Jaisankar N, Ganapathy S, Kannan A (2012) Intelligent intrusion detection system using fuzzy rough set based C4. 5 algorithms. In: Proceedings of the international conference on advances in computing, communications and informatics, pp 596–601
39. Ganapathy S, Kulothungan K, Yogesh P, Kannan A (2012) A novel weighted fuzzy C-means clustering based on immune genetic algorithm for intrusion detection. Procedia Eng 38:1750–1757
40. Nancy P, Muthurajkumar S, Ganapathy S, Kumar SS, Selvi M, Arputharaj K (2020) Intrusion detection using dynamic feature selection and fuzzy temporal decision tree classification for wireless sensor networks. IET Commun 14(5):888–895. https://doi.org/10.1049/iet-com.2019.0172
41. Yin C, Zhu Y, Fei J, He X (2017) A deep learning approach for intrusion detection using recurrent neural networks. IEEE Access 5:21954–21961
42. Shen Y, Zheng K, Wu C, Zhang M, Niu X, Yang Y (2018) An ensemble method based on selection using bat algorithm for intrusion detection. Comput J 61(4):526–538
43. Shone N, Ngoc TN, Phai VD, Shi Q (2018) A deep learning approach to network intrusion detection. IEEE Trans Emerg Top Comput Intell 2(1):41–50

44. Ali MH, Al Mohammed BAD, Ismail A, Zolkipli MF (2018) A new intrusion detection system based on fast learning network and particle swarm optimization. IEEE Access 6:20255–20261
45. Jia Y, Wang M, Wang Y (2019) Network intrusion detection algorithm based on deep neural network. IET Inf Secur 13(1):48–53
46. Wu X, Jiang G, Wang X, Xie P, Li X (2019) A multi-level-denoising autoencoder approach for wind turbine fault detection. IEEE Access 7:59376–59387
47. Yan B, Han G (2018) Effective feature extraction via stacked sparse autoencoder to improve intrusion detection system. IEEE Access 6:41238–41248
48. Xu C, Shen J, Du X, Zhang F (2018) An intrusion detection system using a deep neural network with gated recurrent units. IEEE Access 6:48697–48707
49. Al-Qatf M, Lasheng Y, Al-Habib M, Al-Sabahi K (2018) Deep learning approach combining sparse autoencoder with SVM for network intrusion detection. IEEE Access 6:52843–52856
50. Papamartzivanos D, Mármol FG, Kambourakis G (2019) Introducing deep learning self-adaptive misuse network intrusion detection systems. IEEE Access 7:13546–13560
51. Khan FA, Gumaei A, Derhab A, Hussain A (2019) A novel two-stage deep learning model for efficient network intrusion detection. IEEE Access 7:30373–30385
52. Xiao Y, Xing C, Zhang T, Zhao Z (2019) An intrusion detection model based on feature reduction and convolutional neural networks. IEEE Access 7:42210–42219
53. Yao H, Fu D, Zhang P, Li M, Liu Y (2018) MSML: a novel multilevel semi-supervised machine learning framework for intrusion detection system. IEEE Internet Things J 6(2):1949–1959
54. Gao X, Shan C, Hu C, Niu Z, Liu Z (2019) An adaptive ensemble machine learning model for intrusion detection. IEEE Access 7:82512–82521
55. Wei P, Li Y, Zhang Z, Hu T, Li Z, Liu D (2019) An optimization method for intrusion detection classification model based on deep belief network. IEEE Access 7:87593–87605
56. Zhang X, Chen J, Zhou Y, Han L, Lin J (2019) A multiple-layer representation learning model for network-based attack detection. IEEE Access 7:91992–92008
57. Malaiya RK, Kwon D, Kim J, Suh SC, Kim H, Kim I (2018) An empirical evaluation of deep learning for network anomaly detection. In: 2018 international conference on computing, networking and communications (ICNC). IEEE, pp 893–898
58. Karatas G, Demir O, Sahingoz OK (2020) Increasing the performance of machine learning-based IDSs on an imbalanced and up-to-date dataset. IEEE Access 8:32150–32162
59. Jiang K, Wang W, Wang A, Wu H (2020) Network intrusion detection combined hybrid sampling with deep hierarchical network. IEEE Access 8:32464–32476
60. Yang Y, Zheng K, Wu B, Yang Y, Wang X (2020) Network intrusion detection based on supervised adversarial variational auto-encoder with regularization. IEEE Access 8:42169–42184
61. Andresini G, Appice A, Di Mauro N, Loglisci C, Malerba D (2020) Multi-channel deep feature learning for intrusion detection. IEEE Access 8:53346–53359
62. Tavallaee M, Bagheri E, Lu W, Ghorbani AA (2009) A detailed analysis of the KDD CUP 99 data set. In: 2009 IEEE symposium on computational intelligence for security and defense applications. IEEE, pp 1–6
63. Chen WH, Hsu SH, Shen HP (2005) Application of SVM and ANN for intrusion detection. Comput Oper Res 32(10):2617–2634. https://doi.org/10.1016/j.cor.2004.03.019
64. Roopa Devi EM, Suganthe RC (2020) Enhanced transudative support vector machine classification with grey wolf optimizer cuckoo search optimization for intrusion detection system. Concurr Comput: Pract Exp 32(4):e4999. https://doi.org/10.1002/cpe.4999
65. Abdulsalam SO, Mohammed AA, Ajao JF, Babatunde RS, Ogundokun RO, Nnodim CT, Arowolo MO (2020) Performance evaluation of ANOVA and RFE algorithms for classifying microarray dataset using SVM. Lect Notes Bus Inf Process 402:480–492
66. Ghanem K, Aparicio-Navarro FJ, Kyriakopoulos KG, Lambotharan S, Chambers JA (2017) Support vector machine for network intrusion and cyber-attack detection. In: 2017 sensor signal processing for defense conference (SSPD). IEEE pp 1–5. https://doi.org/10.1109/SSPD.2017.8233268

67. Alsaadi HIH, ALmuttari RM, Ucan ON, Bayat O (2021) An adapting soft computing model for the intrusion detection system. Comput Intell
68. Awotunde JB, Ogundokun RO, Jimoh RG, Misra S, Aro TO (2021) Machine learning algorithm for cryptocurrencies price prediction. Stud Comput Intell 972:421–447
69. Deng X, Liu Q, Deng Y, Mahadevan S (2016) An improved method to construct basic probability assignment based on the confusion matrix for classification problem. Inf Sci 340:250–261
70. Ogundokun RO, Awotunde JB, Sadiku P, Adeniyi EA, Abiodun M, Dauda OI (2021) An enhanced intrusion detection system using particle swarm optimization feature extraction technique. Procedia Comput Sci 193:504–512
71. Zhang G, Wang X, Li R, Song Y, He J, Lai J (2020) Network intrusion detection based on conditional Wasserstein generative adversarial network and cost-sensitive stacked autoencoder. IEEE Access 8:190431–190447

A Binary Firefly Algorithm Based Feature Selection Method on High Dimensional Intrusion Detection Data

Yakub Kayode Saheed ⓘ

Abstract Network intrusion detection system are significant features that contribute to the enterprises and organization network success. In the past decade, Intrusion Detection system (IDS) using several methods have been proposed and implemented to safeguard the networks within an organization to be reliable, secure, and readily available. Although, some of these methods are built utilizing machine learning concepts. IDS utilizing machine learning techniques are accurate, effective and efficient in spotting attacks. However, the machine learning model performance reduces with high dimensional intrusion dataset. Therefore, it is sacrosanct to implement an efficient feature selection technique that can generate a great impact on the classification stage. Also, the feature selection phase is the most serious phase in IDS based on machine learning. This phase is expensive both in time and efforts. Furthermore, most of the proposed machine learning intrusion detection system suffered from low detection accuracy and high false positive rate when the models are experimented on high dimensional dataset. The aim of this paper is to propose a binary firefly algorithm (BFFA) based feature selection for IDS. We first performed normalization in the first stage of the model. Subsequently, the BFFA algorithm was used for feature selection stage. We adopted random forest algorithm for the classification phase. The experiment was performed on high dimensional University of New South Wales-NB 2015 (UNSW-NB15) dataset with seventy five percent of the data used for training the model and twenty percent for testing. The findings showed an accuracy of 99.72%, detection rate of 99.84%, precision of 99.27%, recall of 99.84% and F-score of 99.56%. The results were gauge with the state-of-the-art results and our results were found outstanding.

Keywords Intrusion detection system · Binary firefly algorithm · Random forest · Feature selection · High dimensional data · UNSW-NB15 · Binary movement

Y. K. Saheed (✉)
School of IT & Computing, American University of Nigeria, Yola, Nigeria
e-mail: yakubu.saheed@aun.edu.ng

© The Author(s), under exclusive license to Springer Nature Switzerland AG 2022
S. Misra and C. Arumugam (eds.), *Illumination of Artificial Intelligence in Cybersecurity and Forensics*, Lecture Notes on Data Engineering and Communications Technologies 109, https://doi.org/10.1007/978-3-030-93453-8_12

1 Introduction

Today, the Internet's applications benefit society in a variety of ways, including entertainment, education, business, and electronic communication, and it has become an integral part of people's everyday lives. Cyber security, on the other hand, has become vulnerable as a result of the vast growth of computer nets and the quick appearance of interruption incidents [1]. The critical nature of establishing cyber safety has garnered widespread interest from commercial enterprise and academics worldwide. Given the large use of various safety programs such as malware protection, firewalls, data encryption, and the user verification, numerous establishments and corporations fall victim to modern cyber-assaults [2]. To get access to the scheme, hackers may purposefully maneuver the target system's weaknesses and unveiling a variety of assaults, that may result in the release of sensitive material. As technology advances, these attacks continue to put cyber systems' confidentiality, integrity, and availability at risk [3]. As a result, intrusion detection systems (IDSs) must be implemented [4–6], to safeguard systems against a range of threats. The IDS gives insight into the monitoring and reporting of user and system behavior, keeping an eye out for arithmetical irregularities, vulnerabilities, and social analysis of usage patterns [7, 8]. To be more precise, IDSs are extensively implemented in a variety of dispersed schemes, detecting unwanted interruptions and rapidly resolving them to avoid future infiltration and spreading [9].

Despite major advancements in computer security, current technologies are insufficient to fully protect computer networks from harmful assaults [10]. Due to the rapid growth of intrusion tactics, standard safety approaches such as data encryption, firewalls, and user verification are insufficient to adequately preserve network security [11, 12]. An IDS is a computer program that automates the procedure of spotting malicious activities on a system. It is critical for observing and analyzing everyday computer system activity in order to notice breaches and security intimidations. IDSs are often divided into two categories: anomaly-based schemes and misuse-based or signature-based schemes [13, 14]. Signature-based approaches identify irregularities by comparing the signatures of predefined attacks [15, 16]. While these methods have significant advantages in terms of little false positive-rates and simplicity, they are unable to detect novel mimicking attacks. Anomaly-based approaches are predicated on the premise that the intruder's behavior is dissimilar to that of the rest of the network [17]. These approaches analyze the network's typical traffic and classify any abnormal activity as malicious. This technique enables the detection of both unknown and recognized assaults. The primary drawback of this approach is its high of false-positives. The primary objective of an IDS is to identify as many assaults as feasible while generating as few false alarms as possible; to put it another way, the system must be capable of accurately noticing attacks. Nevertheless, an effective system that is incapable of handling a high volume of network influx and is sluggish to make decisions will fall short of the aim of a detection system. As a result, it is required to design a system capable of detecting most attacks, emitting few false alarms [9], processing vast amounts of data, and making real-time choices.

Regrettably, as attackers improve their sophistication, new risks and vulnerabil-ities arise at a rapid pace. Mostly on one side, the risk of vital infrastructure being compromised increases considerably in a short period of time. As a result, numerous ways have been investigated and established in order to enhance the detection accu-racy and effectiveness of IDSs. Machine learning (ML) is one of them [18–20], that can be used for both misuse and anomaly recognition models. By monitoring network traffic that passes via main nodes in the network, an IDS not just to distin-guishes benign from unwanted traffic and infers the exact type of attack that occurred in the storage network [7]. Additionally, the diversity of attack kinds and network traffic features complicates ML by expanding the problem's search area and resulting in significant time complication and computational issue [21]. Importantly, feature selection (FS) has been demonstrated to be an effective method for an IDS [22], detecting extremely interesting features and eliminating irrelevant ones with minimal performance degradation [23]. There are three primary techniques to feature selec-tion: filter, wrapper, and embedding. The Firefly method is a traditional wrapper algo-rithm that utilizes feature selection. Yang devised a unique nature-inspired method, dubbed the FFA, in recent years to address dynamic programming challenges [24]. Later, FFA was expanded to address issues involving multimodal optimization [25], discrete optimization [26, 27] uncertain optimization and dynamic [28], and multi-objective optimization [29]. However, FFA's applications to dual optimization issues are quite imperfect. Lately, a binary firefly method (BFFA) [30] was presented for solving FS problems, in which the probability estimated using the sigmoidal function is used to convert a real value to a binary integer. When it comes to addressing FS issues, the existing FFA has the following drawbacks:

- When a firefly's position is updated using the typical FFA algorithm, it goes in the directions of all other fireflies with an advanced illumination. If there are multiple brighter fireflies in the same place, the firefly will completely search the environment, wasting processing resources.
- Currently, when comparing a firefly's attraction to a sparkling one, just their remoteness is evaluated. However, the brighter one's return is ignored. Since attractiveness is inversely proportional to distance, a firefly with outstanding illu-mination is located far from the population will have a low chance of procreation; on the other hand, a firefly with substandard illumination but is located near to the populace would have a very good chance of procreation. Therefore, the existing technique will perform poorly while looking for ideal locations.
- Additionally, there are other control factors for FFA, such as attraction and light attenuation coefficient, that have a significant effect on the information search of a firefly [29, 31], and tuning them requires considerable work.

While the IDS has resulted in a number of useful network security measures [32–34], current solutions are confined to static data publication. That is, such methods presuppose the availability of the whole dataset during the release period. This defini-tion assumes a serious problem, as data are collected on a continual basis (and hence keep growing) and there exist constant desire for contemporary data. One strategy is to employ ID methods throughout the whole data every time it is supplemented with

novel entry. This ensures that researchers always have the most up-to-date information. While this is possible with currently available approaches, there are two big
downsides. To begin, it necessitates redundant work because the entire dataset must
be evaluated even if only a few entries are added. At times, ID algorithms may fail
to function correctly due to the constantly expanding size of the dataset. Second,
enormous storage capacity will be required to keep all past datasets, which may be
impractical in some cases. As a result, it is vital to build an IDS system that is capable
of handling both incremental and static data. As a result of the foregoing study, we
suggest a BFFA in this study to enhance FFA's potential for solving FS challenges,
the major contributions of this paper are summarized below.

- Rather than a distance-based indicator of attraction, a BFFA is established to
  quantify the firefly's appeal.
- We utilized an up-to-date UNSW-NB15 dataset that reflects contemporary attacks
  as against dataset that does not represent recent attacks.
- We present a one-of-a-kind technique that incorporates advantages of FS and
  ensemble RF with the goal of detecting intrusions efficiently and accurately.
- On an enormous testbed comprised of the UNSW-NB15 dataset, the approach is
  correlated to established methods. The proposed technique outperforms similar
  methods, in terms of F-score, detection rate, accuracy, and precision, while
  maintaining the FAR at a safe level, according to experimental data.

The rest of the paper is organized as follows. In Sect. 2, the associated work
was given. Section 3 then presents the recommended approach, followed by Sect. 4
which presents the evaluation results collected through experiments and comparative
analyses. Finally, the conclusion is presented in Sect. 5.

2 Related Work

The FS strategy [35, 36], which can be employed as a significant phase in ML classifiers to reduce computation cost, tries to delete unnecessary features while retaining
or even improving the IDS performance. In this review, we presented FS in IDS based
on bioinspired algorithms and non-inspired algorithms. As the term implies, algorithms that are inspired by nature are known as bio-inspired algorithms motivated by
the qualities of natural organisms. It's being used to develop and propose answers
to problems through the application of natural approaches. Additionally, these algorithms are classified into three types: ecology-based, evolutionary and swarm-based
[37]. Ahmad et al. [38] proffered an IDS that makes use of MLP-based feature
subset selection. They combine GA and PCA. They used GA to find the primary
feature maps with the best sensitivity for a subset of features, then PCA to outline
the features vectors to the primary feature vectors and pick attributes commensurate
with the largest eigenvalues. The KDD-Cup dataset was employed, and the suggested
technique optimized 12 out of 41 rows of data. The result indicated that the accuracy

exceeded 99% and that the features were optimal, enhanced precision, reduced preparation and computational costs simplified the intrusion analysis engine's engineering and increased detection rates. In Kuang et al. work [39], A novel SVM prototype for intrusion detection was proposed that combines KPCA with GA. The experimental findings demonstrated accurateness obtained from the KPCA + SVM classifier are more and more significant than those obtained from SVM with parameters chosen at random. Additionally, the research consequences demonstrated that KPCA outperforms PCA on the sets of ID data. This solution was chosen because KPCA, when used in conjunction with PCA, can do higher-level information exploration on initial inputs. Pham et al. [40] developed a hybrid model that combines tree-based classifiers using the gain ratio approach for FS with bagging for feature combination. The experimental findings indicate that the bagging is done using J48 as a base algorithm performed the greatest on a thirty-five-feature set of the NSL-KDD data. In Aslahi et al. [41] research, an IDS was proposed using a hybrid method of SVM, and GA. The proposed combination technique was used to lessen the attributes from 41 to 10. The features were prioritized using GA methods, with the most significant being assigned the highest level of significance, and the least significant being assigned the lowest level of significance. The results indicate that the predicted hybrid classifiers can attain a true positive assessment of 97.3%, whereas 0.017 was the false-positive rate. Additionally, Zhong et al. [42] introduced HELAD, a novel anomaly detection approach that relies on the incremental statistics that have been damped classifier for FS and natural incorporation of different DL approaches for classification. The study [43] proffered a wrapper-based element select that utilized the BA for subcategory generation and a SVM as the algorithm. KDD-Cup 99 has four arbitrary subsets that were used in the studies. Each subset consists of approximately 4000 records. The predicted method's performance is evaluated using typical IDS calculations. The result demonstrates that the detection accuracy 99% was attained, the attribute set was lessen to eight features, and the FAR was decreased with 0.004 FAR. Ref. [44] coupled an ACO algorithm with a part weighting SVM to choose the components. To begin, they developed a full fitness weighting index by combining SVM group accuracy with highlight subset evaluation. They then used the ant colony technique to do global optimization and numerous exploration abilities in order to find the ideal solution feature. The findings demonstrated that the presented technique is proficient of successfully reducing dimensionality of features while increasing the accuracy of network intrusion detection to 95.75%. Ref. [45] suggested a bagging approach based on REPTree that is faster to build and achieves the maximum accuracy with the fewest false-positives using the NSL-KDD data. Rani and Xavier [46] recommended a hybrid invasive detection system. Similarly, the system is built on the C5.0 DT, which also makes use of a 1-class SVM. The findings obtained on the NSL-KDD data demonstrated that the intended strategy outperformed existing methods in terms of discovery rate and reduced false alarm rate. An IDS, dubbed the LSSVM-IDS, was constructed utilizing the proposed FS approach. The effectiveness of LSSVM-IDS was evaluated using three different IDS evaluation datasets, including Kyoto 2006, KDD Cup99, and NSL-KDD data. Hota and Shrivas [23] created a method that eliminated extraneous features via the utilization of several

FS approaches. The results reveal that when only seventeen (17) characteristics are used in the NSL-KDD dataset, C4.5 together with IG achieves the maximum accuracy. Authors [47] suggested an enhanced approach for selecting the most pertinent information in order to improve detection accuracy precision and overall system performance. The KDD99 dataset was used in their technique, with the selected essential features including only twelve (12) attributes outside the forty-one (41)-full feature. This now decreases the extent of the KDD CUP99 bench data by almost 70%. The test results revealed that their upgraded prototype can achieve an advanced rate of detection, a higher performance, a lesser false alarm, and a more rapid and effective detection procedure. The Table 1 gave a summary of existing FS in IDS based on bioinspired and non-bioinspired algorithms.

3 Methodology

We describe the methodology used in this section by adopting the method of research paper methodology writing reported in [50]. To improve the detection capability of IDS and defend service providers from malicious attacks, we offer an effective ML-based IDS based on a metaheuristic BFFA algorithm and an ensemble of RF algorithm method. The experiments employed 25% of the UNSW-NB15 for training and 75% for testing. This approach is used to assess the model's performance and categorize benign activity and numerous forms of assaults. Figure 1 illustrates the recognition mechanism for the proposed ML-based IDS, which comprises of the list of four phases:

- Preprocessing of datasets: The first step is to change fresh data into an analysis-ready set-up by performing preprocessing on the acquired data.
- Feature Selection/dimensionality reduction method: To address the issue of high-dimension data, a FS strategy based on BFFA is utilized to lessening the dataset's dimension and choose the most important characteristics for each kind of assault.
- Training classifiers: To enhance the precision of the IDS, we train the RF algorithm.
- Attack recognition: The detecting model is validated by dividing the data in half, using 75% for training and 25% of the data for testing and validation, and making classification judgments using RF.

3.1 Features Selection

The purpose of FS is to identify a subset of the features set of characteristics that is sufficiently representational of the data and contains extremely interesting attributes for prediction [51]. The approaches to FS can be broadly classified as filter, wrapper and embedding methods [52, 53]. Although, filter methods examine the significance of the attributes first from data and the choice of the attributes is dependent on the

Table 1 Summary of the existing FS approaches in IDS

Authors	Feature selection methodology	Utilized Datasets	Number of features selected	Classification algorithms	Performance evaluation
Ahmad et al. [38]	G.A + PCA	KDD Cup 99	12	MLP	Accuracy; 99%
Kuang and Zhang [39]	Hybrid kernel PCA + G.A	KDD Cup 99	12	Multilayer, SVM	Detection rate: 99.22
[41]	Hybrid method of G.A and SVM	KDD Cup 99	10	SVM	TPR: 0.97 TNR: 0.01
Zhong et al. [42]	DIS-DBN	CIC-IDS2017	50	Autoencoder-LSTM	Accuracy:99.58
Alomari and Othman [43]	Bees algorithm	KDD Cup 99	8	SVM	DR: 98.38 FAR: 0.004
Pham et al. [40]	IGR	KDD Cup 99	35	C4.5	84.25
Xingzhu [44]	ACO + FWSVM	KDD Cup 99	–	SVM	Detection rate: 95.75
Gao et al. [48]	PCA	KDDTest+	20	Ensemble CART	84.54
Rani and Xavier [46]	Cuttlefish algorithm	NSL-KDD	–	C5.0 + one class SVM	Accuracy; 98.20 FAR: 1.405
Parker et al. [49]	Auto encoder-MI	AWID-CLS-R	7	RBFC	98
Madbouly [47]	Enhanced correlation-based mode	NSL-KDD	12	Ensemble of classifiers	Probe accuracy: 99.98 DoS accuracy: 99.3 R2L accuracy: 98.10 U2R accuracy: 86.91

metrics, the accuracy is employed in wrapper methods as a component of the attribute groups analysis and selection procedures. In opposed to wrapper techniques, embedding methods are substantially less costly since they involve an interplay between selection of features and learning procedure. Although, embedding methods combine a regularized framework approach to improve the attributes labelling variables and the forecaster characteristics [54], making a change in the classification task is not easy to gain better attainment [55]. Current ID data always comprise a abundantly

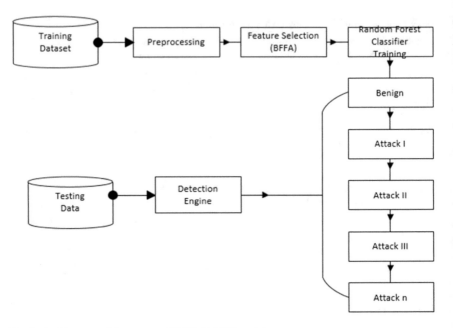

Fig. 1 Architecture of the proposed BFFA-RF IDS model

surplus and unrelated feature [56], which diminish the efficiency of ML methods and produce uninterpretable findings [57]. Hence, the initial step in this research is to minimize the dimensions and pick the subset of features of the employed dataset [58]. In this research, a BFFA is presented to enhance the design of the FS and improve the classification accuracy.

3.2 Firefly Algorithm

The firefly (FFA) is a heuristic optimization process inspired by nature and modeled after the brightness and attracting behavior of fireflies [59, 60]. Algorithm FFA is a population-based algorithm [61], which emit brief, rhythmic lights and exhibit a range of flashing behaviors [62]. The brightness number is used in FFA to assess a firefly's goodness, which is influenced by the terrain of an optimization issue [31]. In the maximizing issue, for instance, the illumination value is equal to the objective function's value. A firefly's position is updated according to its attraction, which is often proportionate to the range among fireflies [63, 64]. When two fireflies are present, the less brilliant one will gravitate towards other. To be more precise, a firefly, symbolized by $Z_i = (Z_i, 1, Z_2, \ldots, Z_i, F)$, changes its position in the following manner [25]:

$$Z_{i,j} = Z_{i,j} + \beta(I, m) \left(X_{m,j} - Z_{i,j}\right) + \partial(\text{rand-0.5}) \tag{1}$$

where Z_m denotes a more attractive version of Z_i, which can be any fireflies with a higher luminosity. The light permeability is used to regulate the light intensity decrease. The notation ri, m symbolizes the range of Zi and Zm(i, m) denotes the desirability of Zi in comparison to Zm. At ri, m = 0, 0 is the desirable value. The variable, ∂ denotes a randomization variable, whereas rand represents a random number among 0 and 1.

3.3 Binary Firefly Algorithm

The fundamental firefly approach is presented to solve problems involving continuous optimization, and the algorithm cannot be easily extended to discrete optimization problems [65]. To address the permutations flow-shop challenges, Sayadi et al. [66] suggested a BFFA, which was modified to solve discrete issues. There are two ways to obtain the discrete algorithm from the original firefly algorithm. The first way is to keep updating the location of fireflies and afterwards discretize, and the second technique updates the firefly's position in discrete space directly. For a comprehensive review of BFFA methods, read Tilahun et al.'s [67] recent paper. The BFFA is discussed in this research as part of the second technique. Zhang et al. suggested a BFFA by translating the other two terms in Eq. (1) with the sigmoid into the following probability vector [30].

$$M_{0,j} = \begin{cases} 1, & \frac{1}{1+e} \\ 0, & \text{otherwise} \end{cases} > \text{rand} \tag{2}$$

3.4 Random Forest

Breimanis introduce RF in [68], which is yet another DT method that works by generating several DT. It takes hundreds of response variable without variables deletion and categorize them using their relevance [69]. RF can be regarded as an ensembles of DT in which every tree participates with a vote cast for the task of its most common class to the data input [70]. In comparison to other ML techniques (e.g., ANN and SVM), there are less factors to be supplied while executing RF.

$$\{a (i, \theta_u), u = 1, 2, \ldots v \ldots\} \tag{3}$$

where a denotes RF classifier, $\{\theta_u\}$ represents for random vectors dispersed separately similar, so each of the tree gets a majority for better renowned class at inputs variable i. The form and dimension of θ depend on the use in tree structures. The establishment of each DT that comprises the forest is vital to the success of RF. RF has a low computing cost and is unaffected by parameters or outliers. Furthermore, in comparison to individual DT, over-fitting is less of a concern, so there is no need to trim the trees, which is a time-consuming job [68].

4   Results and Discussion

As stated previously, the purpose of this work is to design an effective IDS with a high degree of accuracy and a low rate of false alarms. For this aim, an FS approach, called BFFA is performed to find a subset of the features in order to exclude the irrelevant attributes, and increase the predictive performance.

4.1   Description of the Dataset

One of the difficulties researchers confront while evaluating IDS is locating a relevant dataset. Obtaining a real-world dataset that accurately portrays network traffic without any form of anonymization or alteration is an issue that the cybersecurity community has repeatedly confronted [71]. Even if the data is disseminated or distributed for general use, it'll be deeply obfuscated or significantly altered. This will result in the loss or degradation of many crucial collected data that are regarded critical by researchers. As a result, numerous authors have chosen to work with simulated datasets, including the famous KDDCup'99 data [72] and one of its successors, the NSL-KDD data [73]. Lately, a substantial effort has been made to build data sets that are representative of real-world data. The University of New South Wales released an IDS dataset dubbed UNSW-NB15 in 2015. The UNSW-NB 15 data set was developed by extracting a combination of current contemporary attack and normal activities from network traffic using the IXIA PerfectStorm tool [74]. This data collection comprises 2,540,044 records in four (4) csv format. Additionally, a subset of this data was separated into training and testing subset. The training data consisted of 175,341 data, whereas the testing set consisted of 82,332 data from across all attack kinds and normal records [73]. The assaults in the UNSW-NB15 data were classified into (9) nine categories as: fuzzers, backdoors, analysis, DoS, generic, exploits, shellcode, reconnaissance, and worm.

Table 2 Predictive performance of the proposed model

Method/metrics	Accuracy	Detection rate	Precision	Recall	F-score	False alarm
BFFA-RF (%)	99.72	99.84	99.27	99.84	99.56	0.16

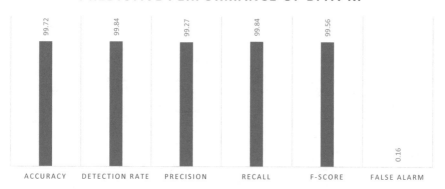

Fig. 2 Performance of the proposed BFFA-RF

4.2 Performance Classification of the Proposed BFFA-RF

The predictive performance of the proposed BFFA-RF model is given in Table 2 and Fig. 2. The accuracy of our model achieved 99.72%, detection rate of 99.84%, precision of 99.27%, F-score of 99.56 and false alarm of 0.16.

The predictive performance of our proposed model BFFA-RF gave outstanding performance in terms of the detection rate, precision, accuracy, recall, F-score and false alarm as shown in Fig. 2. This is as a result of the efficient performance of the BFFA in eliminating the redundant and irrelevant features.

4.3 Comparison with the State-of-the-Art

We compared the findings of proposed BFFA-RF with other existing methods in this section. We compared our BFFA-RF performance to that of earlier research that analyzed IDS using the datasets KDDCup'99, KDDtest, CIC-IDS2017, NSLKDD, and AWID-CLS-R. The results of the comparison with several known techniques are provided in Table 3. The proposed approach, which is based on experimental results on UNSW-NB15, achieves the best detection accuracy and surpasses other current IDS strategies. Apart from outperforming previous approaches in terms of detection accuracy, the suggested method also outperforms them significantly in terms of detection rate and false alarm metrics.

Table 3 Comparison with the state-of-the-art approaches

Authors	Utilized datasets	#Features selected	Classification algorithms	Accuracy	DR	FAR
Ahmad et al. [38]	KDD Cup 99	12	MLP	99	N/A	N/A
Kuang and Zhang [39]	KDD Cup 99	12	Multilayer SVM	N/A	99.22	N/A
Aslahi-Shahri et al. [41]	KDD Cup 99	10	SVM	N/A	97.00	0.01
Zhong et al. [42]	CIC-IDS2017	50	Autoencoder-LSTM	99.58	N/A	N/A
Alomari and Othman [43]	KDD Cup 99	8	SVM	N/A	98.38	0.004
Pham et al. [40]	KDDTest+	35	C4.5	84.25	N/A	N/A
Xingzhu [44]	KDD Cup 99	–	SVM	N/A	95.75	N/A
Gao et al. [48]	KDDTest+	20	Ensemble CART	84.54	N/A	N/A
Rani and Xavier [46]	NSL-KDD	–	C5.0 + one class SVM	98.20	N/A	1.405
Parker et al. [49]	AWID-CLS-R	7	RBFC	98	N/A	N/A
Madbouly [47]	NSL-KDD	12	Ensemble of classifiers	86.91	N/A	N/A
Proposed approach	UNSW-NB15	19	RF	99.72	99.84	0.160

5 Conclusion and Future Work

Although numerous ML approaches have been suggested to cost-effectiveness of IDSs, current ID algorithms continue to struggle to attain acceptable performance. In this research, we suggest a novel ID methodology for dealing with high-dimensional traffic that is based on BFFA for feature selection and ensemble RF approaches. To begin, we present a BFFA algorithm for finding the best subset features based on feature correlation. After then, the RF is utilized to construct the model. Eventually, the presented IDS is validated by utilizing 75% of the data for training and 25% of the testing data from the UNSW-NB15 ID dataset. The experimental findings are encouraging, with a classification accuracy of 99.72%, a DR of 99.84%, and a FAR of 0.16% using a subset of 19 attributes from the UNSW-NB15 dataset. Additionally, our method exceeds other techniques to feature selection in terms of precision, recall, and F-Measure. Comparing the proposed BFFA-RF method to the state-of-the-art approaches demonstrates that it has the potential to deliver a significant competitive edge in the intrusion detection area. Although, the suggested BFFA-RF approach demonstrated better performance, in future work, its potential to handle rare attacks from enormous network traffic should be enhanced.

References

1. Zhou Y, Cheng G, Jiang S, Dai M (2020) Building an efficient intrusion detection system based on feature selection and ensemble classifier. Comput Netw 174. https://doi.org/10.1016/j.com net.2020.107247
2. Al-Jarrah OY, Alhussein O, Yoo PD, Muhaidat S, Taha K, Kim K (2016) Data randomization and cluster-based partitioning for botnet intrusion detection. IEEE Trans Cybern 46(8):1796–1806. https://doi.org/10.1109/TCYB.2015.2490802
3. Balogun BF, Gbolagade KA, Arowolo MO, Saheed YK (2021) A hybrid metaheuristic algorithm for features dimensionality reduction in network intrusion, vol 3. Springer International Publishing
4. Elhag S, Fernández A, Bawakid A, Alshomrani S, Herrera F (2015) On the combination of genetic fuzzy systems and pairwise learning for improving detection rates on intrusion detection systems. Expert Syst Appl 42(1):193–202. https://doi.org/10.1016/j.eswa.2014.08.002
5. Wang K, Du M, Maharjan S, Sun Y (2017) Strategic honeypot game model for distributed denial of service attacks in the smart grid. IEEE Trans Smart Grid 8(5):2474–2482. https://doi. org/10.1109/TSG.2017.2670144
6. Wang K, Du M, Sun Y, Vinel A, Zhang Y (2016) Attack detection and distributed forensics in machine-to-machine networks. IEEE Netw 30(6):49–55. https://doi.org/10.1109/MNET.2016. 1600113NM
7. Azeez NA, Ayemobola TJ, Misra S, Maskeliūnas R, Damaševičius R (2019) Network intrusion detection with a hashing based apriori algorithm using hadoop mapreduce, Computers 8(4). https://doi.org/10.3390/computers8040086
8. Odusami M, Misra S, Adetiba E, Abayomi-Alli O, Damasevicius R, Ahuja R (2019) An improved model for alleviating layer seven distributed denial of service intrusion on webserver. J Phys Conf Ser 1235(1). https://doi.org/10.1088/1742-6596/1235/1/012020
9. Awujoola OJ, Ogwueleka FN, Irhebhude ME (2021) Wrapper based approach for network intrusion detection model with combination of dual filtering technique of resample and SMOTE. Springer
10. Mohammadi S, Mirvaziri H, Ghazizadeh-Ahsaee M, Karimipour H (2019) Cyber intrusion detection by combined feature selection algorithm. J Inf Secur Appl 44:80–88. https://doi.org/ 10.1016/j.jisa.2018.11.007
11. Tavallaee M, Stakhanova N, Ghorbani AA (2010) Toward credible evaluation of anomaly-based intrusion-detection methods. IEEE Trans Syst Man Cybern Part C Appl Rev 40(5):516–524. https://doi.org/10.1109/TSMCC.2010.2048428
12. Tapiador JE, Orfila A, Ribagorda A, Ramos B (2015) Key-recovery attacks on KIDS, a keyed anomaly detection system. IEEE Trans Dependable Secur Comput 12(3):312–325. https://doi. org/10.1109/TDSC.2013.39
13. Chen X, Zhang F, Susilo W, Tian H, Li J, Kim K (2014) Identity-based chameleon hashing and signatures without key exposure. Inf Sci (NY) 265:198–210. https://doi.org/10.1016/j.ins. 2013.12.020
14. Wang D, Zhang Z, Wang P, Yan J, Huang X (2016) Targeted online password guessing: an underestimated threat. In: Proceedings of the ACM conference on computer and communication security, vol 24–28, pp 1242–1254. https://doi.org/10.1145/2976749.2978339
15. Kabir E, Hu J, Wang H, Zhuo G (2018) A novel statistical technique for intrusion detection systems. Futur Gener Comput Syst 79:303–318. https://doi.org/10.1016/j.future.2017.01.029
16. Maggi F, Matteucci M, Zanero S (2010) Detecting intrusions through system call sequence and argument analysis. IEEE Trans Dependable Secur Comput 7(4):381–395. https://doi.org/ 10.1109/TDSC.2008.69
17. Karimipour H, Dinavahi V (2017) Robust massively parallel dynamic state estimation of power systems against cyber-attack. IEEE Access 6:2984–2995. https://doi.org/10.1109/ACCESS. 2017.2786584

18. Du M, Wang K, Chen Y, Wang X, Sun Y (2018) Big data privacy preserving in multi-access edge computing for heterogeneous internet of things. IEEE Commun Mag 56(8):62–67. https:// doi.org/10.1109/MCOM.2018.1701148
19. Du M, Wang K, Xia Z, Zhang Y (2018) Differential privacy preserving of training model in wireless big data with edge computing. IEEE Trans. Big Data 6(2):283–295. https://doi.org/ 10.1109/tbdata.2018.2829886
20. Mishra P, Varadharajan V, Tupakula U, Pilli ES (2019) A detailed investigation and analysis of using machine learning techniques for intrusion detection. IEEE Commun Surv Tutor 21(1):686–728. https://doi.org/10.1109/COMST.2018.2847722
21. Aljawarneh S, Aldwairi M, Bani M (2018) Anomaly-based intrusion detection system through feature selection analysis and building hybrid efficient model. J Comput Sci 25:152–160. https:// doi.org/10.1016/j.jocs.2017.03.006
22. Shamshirband S et al (2014) Co-FAIS: Cooperative fuzzy artificial immune system for detecting intrusion in wireless sensor networks. J Netw Comput Appl 42(2008):102–117. https://doi.org/ 10.1016/j.jnca.2014.03.012
23. Hota HS, Shrivas AK (2014) Decision tree techniques applied on NSL-KDD data and its comparison with various feature selection techniques. Smart Innov Syst Technol 27(1). https:// doi.org/10.1007/978-3-319-07353-8
24. Yang XS (2010) Nature-inspired metaheuristic algorithms. Luniver press
25. Yang X-S (2009) Furefly algorithms for multimodal optimization. In: SAGA 2009, LNCS, pp 169–178
26. Marichelvam MK, Prabaharan T, Yang XS (2014) A discrete firefly algorithm for the multi-objective hybrid flowshop scheduling problems. IEEE Trans Evol Comput 18(2):301–305. https://doi.org/10.1109/TEVC.2013.2240304
27. Rahmani A, Mirhassani SA (2014) A hybrid firefly-genetic algorithm for the capacitated facility location problem. Inf Sci (NY) 283(June):70–78. https://doi.org/10.1016/j.ins.2014.06.002
28. Nasiri B, Meybodi MR (2016) Improved speciation-based firefly algorithm in dynamic and uncertain environments. J Inf Sci Eng 32(3):661–676. https://doi.org/10.6688/JISE.2016.32.3.9
29. Yang XS (2013) Multiobjective firefly algorithm for continuous optimization. Eng Comput 29(2):175–184. https://doi.org/10.1007/s00366-012-0254-1
30. Zhang L, Shan L, Wang J (2017) Optimal feature selection using distance-based discrete firefly algorithm with mutual information criterion. Neural Comput Appl 28(9):2795–2808. https:// doi.org/10.1007/s00521-016-2204-0
31. Zhang Y, Song X, Gong D (2017) A return-cost-based binary firefly algorithm for feature selection. Inf Sci (NY) 418–419:561–574. https://doi.org/10.1016/j.ins.2017.08.047
32. Hwang K, Cai M, Chen Y, Qin M (2007) Hybrid intrusion detection with weighted signature generation over anomalous internet episodes. IEEE Trans Dependable Secur Comput 4(1):41–55. https://doi.org/10.1109/TDSC.2007.9
33. Dartigue C, Jang HI, Zeng W (2009) A new data-mining based approach for network intrusion detection. In: Proceedings, seventh annual communication networks and services research conference CNSR 2009, pp 372–377. https://doi.org/10. 1109/CNSR.2009.64
34. Gupta KK, Nath B, Member S (2010) Random fields for intrusion detection 7(1):35–49
35. Maza S, Touahria M (2018) Feature selection algorithms in intrusion detection system: a survey. KSII Trans Internet Inf Syst 12(10):5079–5099. https://doi.org/10.3837/tiis.2018.10.024
36. Mi J, Wang K, Li P, Guo S, Sun Y (2018) Software-defined green 5G system for big data. IEEE Commun Mag 56(11):116–123. https://doi.org/10.1109/MCOM.2017.1700048
37. Tu Q, Li H, Wang X, Chen C (2015) Ant colony optimization for the design of small-scale irrigation systems. Water Resour Manag 29(7):2323–2339. https://doi.org/10.1007/s11269-015-0943-9
38. Ahmad I, Abdullah A, Alghamdi A, Alnfajan K, Hussain M (2011) Intrusion detection using feature subset selection based on MLP. Sci Res Essays 6(34):6804–6810. https://doi.org/10. 5897/SRE11.142

39. Kuang F, Xu W, Zhang S (2014) A novel hybrid KPCA and SVM with GA model for intrusion detection. Appl Soft Comput J 18:178–184. https://doi.org/10.1016/j.asoc.2014.01.028
40. Pham NT, Foo E, Suriadi S, Jeffrey H, Lahza HFM (2018) Improving performance of intrusion detection system using ensemble methods and feature selection. ACM Int Conf Proc Ser. https://doi.org/10.1145/3167918.3167951
41. Aslahi-Shahri BM et al (2016) A hybrid method consisting of GA and SVM for intrusion detection system. Neural Comput Appl 27(6):1669–1676. https://doi.org/10.1007/s00521-015-1964-2
42. Zhong Y et al (2020) HELAD: a novel network anomaly detection model based on heterogeneous ensemble learning. Comput Netw 169:107049. https://doi.org/10.1016/j.comnet.2019.107049
43. Alomari O, Othman ZA (2012) Bees algorithm for feature selection in network anomaly detection. J Appl Sci Res 8(3):1748–1756
44. Xingzhu W (2015) ACO and SVM selection feature weighting of network intrusion detection method. Int J Secur Appl 9(4):259–270. https://doi.org/10.14257/ijsia.2015.9.4.24
45. Gaikwad DP, Thool RC (2015) Intrusion detection system using bagging ensemble method of machine learning. In: The first international conference on computing, communication, control and automation ICCUBEA 2015, pp 291–295. https://doi.org/10.1109/ICCUBEA.2015.61
46. Rani MS, Xavier SB (2015) A hybrid intrusion detection system based on C5.0 decision tree and one-class SVM. Int J Curr Eng Technol 5(3):2001–2007
47. Madbouly AI, Barakat TM (2016) Enhanced relevant feature selection model for intrusion detection systems 4(1):21–45
48. Gao Y, Liu Y, Jin Y, Chen J, Wu H (2018) A novel semi-supervised learning approach for network intrusion detection on cloud-based robotic system. IEEE Access 6©:50927–50938. https://doi.org/10.1109/ACCESS.2018.2868171
49. Parker LR, Yoo PD, Asyhari TA, Chermak L, Jhi Y, Taha K (2019) Demise: interpretable deep extraction and mutual information selection techniques for IoT intrusion detection. In: ACM international conference proceeding series.https://doi.org/10.1145/3339252.3340497
50. Misra S (2021) A step by step guide for choosing project topics and writing research papers in ICT related disciplines, vol 1350. Springer International Publishing
51. Saheed YK, Akanni AO, Alimi MO (2018) Influence of discretization in classification of breast cancer disease. Univ PITESTI Sci Bull Electron Comput Sci 18(2):13–20
52. Hajisalem V, Babaie S (2018) A hybrid intrusion detection system based on ABC-AFS algorithm for misuse and anomaly detection. Comput Netw 136:37–50. https://doi.org/10.1016/j.comnet.2018.02.028
53. Saheed YK, Hamza-Usman FE (2020) Feature selection with IG-R for improving performance of intrusion detection system. Int J Commun Netw Inf Secur 12(3):338–344
54. Bolón-Canedo V, Sánchez-Maroño N, Alonso-Betanzos A (2016) Feature selection for high-dimensional data. Prog Artif Intell 5(2):65–75. https://doi.org/10.1007/s13748-015-0080-y
55. Liu H, Member S, Yu L, Member S (2005) Algorithms for classification and clustering, vol 17, no 4, pp 491–502
56. Acharya N, Singh S (2018) An IWD-based feature selection method for intrusion detection system. Soft Comput 22(13):4407–4416. https://doi.org/10.1007/s00500-017-2635-2
57. Chen XY, Ma LZ, Chu N, Zhou M, Hu Y (2013) Classification and progression based on CFS-GA and C5.0 boost decision tree of TCM Zheng in chronic hepatitis B. Evid-Based Complement Altern Med 2013. https://doi.org/10.1155/2013/695937
58. Salo F, Nassif AB, Essex A (2018) Dimensionality reduction with IG-PCA and ensemble classifier for network intrusion detection. Comput Netw. https://doi.org/10.1016/j.comnet.2018.11.010
59. Peng H, Zhu W, Deng C, Wu Z (2020) Enhancing firefly algorithm with courtship learning. Inf Sci (NY) 543:18–42. https://doi.org/10.1016/j.ins.2020.05.111
60. Hassan BA (2021) CSCF: a chaotic sine cosine firefly algorithm for practical application problems. Neural Comput Appl 33(12):7011–7030. https://doi.org/10.1007/s00521-020-05474-6

61. Kumar V, Kumar D (2021) A systematic review on firefly algorithm: past, present, and future. Arch Comput Methods Eng 28(4):3269–3291. https://doi.org/10.1007/s11831-020-09498-y
62. Karthikeyan S, Asokan P, Nickolas S, Page T (2015) A hybrid discrete firefly algorithm for solving multi-objective flexible job shop scheduling problems. Int J Bio-Inspir Comput 7(6):386–401. https://doi.org/10.1504/IJBIC.2015.073165
63. Xue X, Chen J (2020) Optimizing sensor ontology alignment through compact co-firefly algorithm. Sensors (Switzerland) 20(7):1–15. https://doi.org/10.3390/s20072056
64. Wu J, Wang YG, Burrage K, Tian YC, Lawson B, Ding Z (2020) An improved firefly algorithm for global continuous optimization problems. Expert Syst Appl 149:113340. https://doi.org/10.1016/j.eswa.2020.113340
65. Lin M, Liu F, Zhao H, Chen J (2020) A novel binary firefly algorithm for the minimum labeling spanning tree problem. C Comput Model Eng Sci 125(1):197–214. https://doi.org/10.32604/cmes.2020.09502
66. Sayadi MK, Ramezanian R, Ghaffari-Nasab N (2010) A discrete firefly meta-heuristic with local search for makespan minimization in permutation flow shop scheduling problems. Int J Ind Eng Comput 1(1):1–10. https://doi.org/10.5267/j.ijiec.2010.01.001
67. Tilahun SL, Ngnotchouye JMT (2017) Firefly algorithm for discrete optimization problems: a survey. KSCE J Civ Eng 21(2):535–545. https://doi.org/10.1007/s12205-017-1501-1
68. Jin Z, Shang J, Zhu Q, Ling C, Xie W, Qiang B (2020) RFRSF: employee turnover prediction based on random forests and survival analysis. In: Lecture notes in computer science (including Subseries lecture notes in artificial intelligence and lecture notes in bioinformatics), vol 12343. LNCS, pp 503–515. https://doi.org/10.1007/978-3-030-62008-0_35
69. Saheed YK, Hambali MA, Arowolo MO, Olasupo YA (2020) Application of GA feature selection on Naïve Bayes, random forest and SVM for credit card fraud detection. In: 2020 international conference on decision aid sciences and applications DASA 2020, pp 1091–1097. https://doi.org/10.1109/DASA51403.2020.9317228
70. Adnan MN, Islam MZ (2017) Forest PA: constructing a decision forest by penalizing attributes used in previous trees. Expert Syst Appl 89:389–403. https://doi.org/10.1016/j.eswa.2017.08.002
71. Aldwairi T, Perera D, Novotny MA (2018) An evaluation of the performance of restricted boltzmann machines as a model for anomaly network intrusion detection. Comput Netw 144:111–119. https://doi.org/10.1016/j.comnet.2018.07.025
72. Rosset S, Inger A (2000) Knowledge discovery in a charitable organizations donor database. SIGKDD Explor 1(2):85–90
73. Moustafa N, Slay J (2015) UNSW-NB15: a comprehensive data set for network intrusion detection systems (UNSW-NB15 network data set). In: Proceedings of the 2015 military communications and information systems conference, MilCIS 2015.https://doi.org/10.1109/MilCIS.2015.7348942
74. Moustafa N, Slay J (2017) The significant features of the UNSW-NB15 and the KDD99 data sets for network intrusion detection systems. In: Proceedings of the 2015 4th international workshop on building analysis datasets and gathering experience returns for security BADGERS 2015, pp 25–31. https://doi.org/10.1109/BADGERS.2015.14

Graphical Based Authentication Method Combined with City Block Distance for Electronic Payment System

Suleman Isah Atsu Sani⬤, John Kolo Alhassan⬤, and Abubakar Saddiq Mohammed

Abstract User authentication is an essential component in nearly all electronic payment systems. Authentication provides the foundation for a user, legal access control, and user accountability. The most foremost used authentication methods are textual and alphanumeric passwords. However, the alphanumeric password suffers some drawbacks such as easy guessing and hard to remember the password, which can make it vulnerable to attacks such as social engineering and dictionary attacks thereby exposing sensitive information and thus compromising the security of a system. To overcome these challenges associated with an alphanumeric password, we proposed a graphical-based authentication method combined with City block distance to measure the similarity score between the image passwords at the points of registration and login using the KNN algorithm to compute the distance. In this paper, similarity measure was found as an important task for text matching, image processing, and retrieval of images from the database. To achieve optimal performance of the system and make it robust, an experiment was conducted during the login session using the city block distance and other different distance measures that include Euclidean, Cosine similarity, and Jaccard utilizing the KNN algorithm. The experimental results show that the proposed city block distance method has the fastest execution time of 0.0318 ms, minimal matching error of 1.55231, and an acceptable login success rate of 64%, compared to when the graphical-based password is combined with other similarity score distance measures. This paper concludes that the proposed method would be the most reliable authentication for e-payment systems.

S. I. A. Sani (✉) · J. K. Alhassan · A. S. Mohammed
Department of Computer Science, Federal University of Technology, Minna, Nigeria
e-mail: sani.pg918915@st.futminna.edu.ng

J. K. Alhassan
e-mail: jkalhassan@futminna.edu.ng

A. S. Mohammed
e-mail: abu.saddiq@futminna.edu.ng

© The Author(s), under exclusive license to Springer Nature Switzerland AG 2022
S. Misra and C. Arumugam (eds.), *Illumination of Artificial Intelligence in Cybersecurity and Forensics*, Lecture Notes on Data Engineering and Communications Technologies 109, https://doi.org/10.1007/978-3-030-93453-8_13

Keywords Graphical password authentication · Similarity measures · e-payment ·
Matching error · City block distance · K-nearest neighbor algorithm · Evaluation

1 Introduction

Authentication is at the core of many security systems [1]. It is the process of verification of a user's identity to validate a legitimate user. Authentication provides users access to their unique resources. It answers the question such as, is the user genuine or not? Unauthorized access to user information always puts the security of an entire system at risk. Hence, it is very important to protect users' resources and information. User authentications have found usage in several different domains, including electronic payment systems. The most common authentication method is the text-based password, also known as an alphanumeric password. It is the traditional form of an authentication technique that is commonly used in knowledge-based authentication [2]. An alphanumeric password is a combination of upper and lower-case letters, numbers, special symbols like a punctuation mark, and asterisk.

Many users make use of textual passwords while logging into any authentication-based platform. This password should be memorable to the user, easy, and secure. However, due to its simplicity, it is sometimes hard to remember. The alphanumeric password is vulnerable to various kinds of attacks [3] including brute force, shoulder surfing, social engineering, and dictionary attacks. These vulnerabilities and other limitations associated with an alphanumeric password have led to the development of graphical-based authentication schemes as possible alternatives to the text-based scheme.

The idea of graphical password authentication was pioneered and developed by Blonder in 1996. Graphical passwords are classified into two categories namely: recognition-based and recall-based graphical passwords [4]. In the recognition-based graphical password, a user interacts with a set of images or visual objects to select images or objects. For example, during the registration phase, a user has to enter a username, textual password and thereafter select images or objects as a graphical password and at the authentication phase, a user will recognize the same objects or picture images, and the system will identify the user. While in the case of recall-based graphical passwords, the user has to draw an image during registration, and this same process will be repeated at authentication time.

In Pure Recall Based techniques, a user is not provided a clue to recall a password, while in Cued Recall Based techniques; the user will choose some memorable points from the image and recall them at authentication time. In order words, image recognition has been demonstrated to be better in assisting users in recalling information than relying solely on memory. Graphical password is a knowledge-based protocol for user authentication. It enables the user to remember the password in the form of graphical objects or images [5].

Graphical password provides users the advantage of memorability and a very user-friendly interface compared to alphanumeric passwords while keeping security in check. Hence, this paper presents a graphical-based authentication method combined with a city block distance measure for an electronic payment system. The benefit of combining this is to make the scheme more robust against any form of attack.

In recent years, researchers have emerged with different algorithms as touching or clicking point graphical passwords authentication and different approaches have been implored on various platforms. Some of the researchers are aimed at developing graphical password schemes that would be suitable to the users/clients while keeping security in check.

Electronic payments, despite their benefits, pose challenges around the world due to advancements in technology in use. The challenges faced by traditional authentication (alphanumeric or textual) and fingerprint authentication in electronic payment systems are prone to hacking or rather vulnerable to attacks because they are easy to guess, hard to remember passwords, and slow to log in.

Hence in this study, to overcome or address the challenges associated with traditional authentication (alphanumeric and text-based) and fingerprint authentication, a graphical-based authentication method combined with city block distance measure is proposed. The graphical-based authentication method provides the users/clients of electronic payment systems with the choice to select a stronger password and memorable images, compared to less and vulnerable alphanumeric and fingerprint authentication. In the graphical authentication method, a user is provided with a user-friendly interface that has fast login access and fast execution time, when compared with alphanumeric and fingerprint authentication while keeping security, preserved [3, 6].

This paper aims to develop a Graphical Based Authentication Method combined with city block distance for an Electronic Payment System. This was achieved by developing both a research framework and a mathematical equation for the proposed method which was tested and evaluated its performance with other existing methods. While the significance of the paper is focused on providing security, intelligence, protection and the benefits of graphical-based authentication mechanisms for users.

The scope of this research work is focused on developing a Graphical Based Authentication combined with city block distance for E-payment systems for users of online transactions to provide authentication mechanisms for better login, usability, security, a robust and effective system. In our proposed method a similarity measure was used to compute the distance between two vector points using different distance measures such as Euclidian, Cosine Similarity, City block and Jaccard.

This study employs similarity distance measures such as City block distance, Euclidean distance, Cosine similarity, and Jaccard which utilize the K-Nearest Neighbor (KNN) algorithm to compute the distance between two vector points at register coordinate $(x1, y1)$ and vector points at login coordinate $(x2, y2)$. The similarity distance measures use the KNN algorithm to compute the distance of the image data saved in the database and at the login phase. K-Nearest Neighbor (KNN) is a common categorization approach. KNN is known as a simple and easy algorithm.

The KNN method is used as a research algorithm by many researchers, this is due to KNN's ability to handle data computation and classification [7].

In the KNN method, distance measures are key parameters that depend on data collection. The distance matrix is used to calculate the distance between two data points. The most commonly used distance matrix functions in calculating distance matrices are city block distance and Euclidean distance. The calculation of similarity distance utilizing numerous matrix values is commonly used to extract the similarity of data objects and is aided in the classification process by efficient algorithms [7].

In this paper, the following contributions to knowledge are achieved: a research framework for a graphical-based authentication System (GBAS) was developed, a mathematical equation for graphical-based authentication method combined with city block distance for electronic payment was developed and similarity scores between the points of registration and login are defined using different distance measures such as Euclidean, city block, cosine similarity, and Jaccard, and the adoption of city block distance as a similarity measure and the use of matching error as a metric in this graphical-based authentication scheme justifies the robustness and effective performances of the scheme.

The rest of this research paper is organized as follows: Sect. 2 is the background and literature review, types of attacks on passwords, categories of graphical-based user authentication techniques and related studies of the existing research works. Section 3 research designed framework, mathematical equation for the proposed graphical-based authentication method combined with city block distance for e-payment system, implementation, and performance evaluation metrics of the proposed graphical method are highlighted in Sects. 3.5 and 4.4 respectively. Section 4 experimental results and analysis of results, Sect. 5 conclusion and future works.

2   Background and Literature Review

2.1   Types of Attacks on Passwords

The advancement of technology has created challenges, such as the necessity to secure highly valuable data. This is due to the activity of attackers, who continue to develop and deploy increasingly sophisticated tools and methods to conduct various assaults against user data. Data alteration, deletion, denial of service, impersonation, eavesdropping, and identity theft are examples of attacks [8]. The common types of attacks on passwords are as follows.

Brute force attacks. In this attack, the attacker program impersonates a real user and attempts to log into the system by selecting the correct password from the graphics. A brute force attack differs from a dictionary attack in that it does not use dictionaries of alternative passwords. Instead, to gain access to the system, the attacker tries every possible password. Graphical passwords, on the other hand, are more resistant to Brute Force attacks than text passwords because they have a larger password space [9].

Dictionary Attacks. This attack uses an exhaustive list of words, such as a dictionary, to break the password. This dictionary includes terms that are likely to be used as passwords by the user. Unlike a brute force attack, a dictionary attack cracks passwords using a structured key search, taking into account only those possibilities that are most likely to succeed, but it cannot crack the password every time, as a brute force attack would. This type of attack is uncommon with graphical passwords [10, 11].

Guessing Attacks. It is the most common type of attack on alphanumeric passwords where the attacker tries to defeat the authentication system by merely inputting words that he feels an average user can use as a password. It is very likely to be successful if the attacker has little knowledge of the user.

Shoulder Surfing Attacks. An attacker may often discover a user's password by peering over their shoulder, as the name suggests. This kind of intrusion is popular in crowded places where people are unaware of their surroundings.

Social Engineering Attacks. A description assault is another name for this. It refers to psychologically persuading people to conduct acts or reveal sensitive information. To gain people's trust and expose sensitive information, it uses several deceptions, which lead to a variety of scams and fraud [12].

2.2 Graphical Password Authentication Schemes (GPAS)

A password is a secret code that is used for authentication. Passwords are the most widely used means of distinguishing users of computer and communication systems. It should only be understood by the user. A graphical password is an authentication scheme that works by making the user choose from images displayed in a particular order in a graphical user interface [13]. For this reason, the graphical-password approach is sometimes referred to as graphical user authentication (GUA) [1, 13].

2.3 Categories of Graphical Password Authentication Techniques

Various techniques have been identified over time. The knowledge-based scheme is the most common among them, as it is regarded as the most important technique in terms of protection and usability. This method has been suggested to address

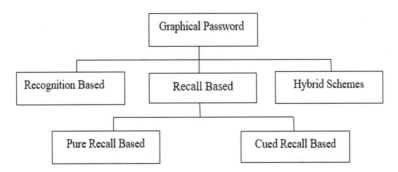

Fig. 1 Categorization of graphical password authentication techniques

some flaws in standard password techniques. The reason for this is that, as seen in Fig. 1, the text is much more difficult to identify, memorize, and recall than pictures [2]. Currently, certain existing graphical password authentication methods can be classified into four categories as follows.

Recognition Technique. Users pick pictures and symbols from a list of picture images in a recognition technique. During the authentication point, it is important to remember the pictures or signs from the collection of picture images that were selected earlier during the registration process [14].

Recall Based Technique. Clients have a hard time remembering passwords, so recall-based is very easy and fun to use. It is, however, more reliable than the recognition method [14].

Cued Recall Technique. Clients are given a hint or clue by the Cued Recall scheme. This hint or clue usually aids clients in quickly, accurately, and conveniently reproducing their password. Its operation is similar to that of recall schemes, but it combines recall and cueing [2].

Hybrid Schemes. This scheme is known as the hybrid strategy because it incorporates two techniques to form a new scheme to correct bugs or setbacks in one scheme, such as shoulder surfing and spyware attacks. The most common drawbacks are usually addressed by hybrid authentication techniques [2].

2.4 Performance Metrics

System evaluation metrics are regarded as an important phase in research work, in which standard goals are measured to compare experimental results with the existing graphical scheme [15]. This also clarifies evaluation as a systematic procedure for evaluating a designed scheme for its architecture, framework, and benefit. Evaluation is a vital process in which a thorough examination, consideration, or attentiveness, and judgment of the system yield accurate results. These include; evaluation method, security evaluation, usability evaluation, usefulness, and utility evaluation [16].

In this study, the Similarity Metrics, Distance measure metric, login success rate metrics, the Execution time of algorithms or time speed in milliseconds metric, and percentage matching errors of the distance measures are used to determine the robustness, efficiency, and performance of the system [17, 18]. In computing-related research works, evaluation is the yardstick for assessing and validating the degree of achievement of a technique [19].

Distance Metrics Overview. Distance metrics are essential for determining the similarity or regularity of data images. It is important to know how image data are associated with each other, how different data are from each other, and what measures are considered to compare them. The first goal of metric calculation in a particular problem is to get an appropriate distance and similarity function. Metric learning has emerged as a well-liked issue in many learning tasks, and it can also be utilized in a wide variety of settings.

A metric also referred to as a distance function, is a function that defines the distance or space between two or more elements/objects in a group. A group with a metric is known as metric space. This distance metric is extremely important in clustering techniques. The main contribution of this work is the investigation of performances of the similarity metric [18, 20].

Cosine Distance and Cosine Similarity. The cosine distance and cosine similarity metrics are primarily used to discover similarities between two data points. The cosine similarity, or the number of similarities, decreases as the cosine distance between the data points increases, and vice versa. As a result, points that are close to each other are more similar than points that are far apart. Cosine similarity is given by $\cos\theta$, and cosine distance is $1 - \cos\theta$ [20].

Cosine Similarity is introduced as a method of reducing the illegal user's login time, which is thought to be crucial to a password scheme's usability. It aims to motivate the user by providing a fun, friendly interface that improves the user experience and provides an acceptable login time. The use of the Cosine Similarity to Login is a promising technique [21].

2.5 Related Works

The usefulness of any research work cannot be proven until it is compared to other similar current works as well as the proposed study, in order to show how superior, the proposed studies are [19]. Many systems have been proposed to improve password authentication, in order to enhanced data and information security. However, each of these techniques has some drawbacks [8].

There are several different graphical-based authentication approaches proposed in the literature. Many of these proposed approaches are implemented using either touch or click point-based techniques. Even though, their implementations vary, the aim is to develop a graphical password authentication that would be user-friendly, secure and robust against any form of attack. For example, the work of [6] proposed a graphical password technique based on Persuasive Cued Click Points. Here, a user is authenticated based on a group of some clicked images as well as the approximate pixel of the user's click. This approach increases the application's security. Similar work was proposed by [3]. The results of both proposals suggested that graphical password authentication is more effective in overcoming the challenges such as fraud, hacking data, or steal data that are facing the traditional alphanumeric password authentication [9] presented a graphical password scheme based on colors and numbers, which is purely recognition-based. In the work, a user has to first enter a username followed by a request to rate colors from 1 to 8 randomly. Subsequently, in the login phase, the user after entering the correct username and selecting the right color is granted access to the system. To provide a robust system, the password format changes in each login session.

The works of [22, 23], proposed a combination of two factors for graphical password authentication (that helps in generating strong passwords). This work is similar to the pass faces system where one image is used with several faces on the screen and the user clicks 2 or 3 faces as the password but here, they implored the second layer of security to ensure strong authentication (which will make the login process slow). They also implored the use of login indicators that are generated once, and they allowed for all the images selected by the users to be displayed on a single web page. In their algorithm, at the authentication phase, a login indicator would be generated and given to the user through various ways such as audio, visual, or text.

In the work of [24], a pure recognition-based graphical passwords technique is proposed and referred to as the image sequence technique. In this technique, users are required to upload images during registration from their directory into the scheme in a particular sequence and during login; the user will have to remember the sequence in which the images were uploaded in the first instance during registration. Thereafter, those images are added into a group of random images in which the user will select the ones that were uploaded.

Awodele et al. [25] proposed a shoulder-surfing resistant graphical authentication scheme to address the major issues with the graphical authentication schemes that have been developed. The proposed scheme provides a high level of resistance to shoulder surfing attacks, reduces the need to upload pictures, and aids in the scheme's selection of objects. The proposed scheme utilizes a set of colored rows and columns which will assist users in identifying their chosen cell. The interface design elaborates on the cued recall graphical technique being utilized.

Sun et al. [26] proposed PassMatrix, a novel authentication system based on graphical passwords, to combat shoulder surfing attacks. PassMatrix, with a one-time valid login indicator and circularize horizontal and vertical bars covering the entire scope of pass-images, provides no hint for attackers to figure out or narrow down the password, even when multiple camera-based attacks are conducted. The results showed that the proposed system is more resistant to shoulder surfing threats while still being usable.

The work of [27] presented a new password scheme that employs a graphical user interface for password entry. The password consists of multiple graphical objects that are integrated to form one picture. The main advantage of this approach is that it makes user authentication more user-friendly where it is often easier to remember a scene than an alphanumeric password. The user creates the password scene by selecting from the available shapes where the selection process is combined with the selected objects to create the actual password. The scene created by the user is transformed into an alphanumeric password where the number of combinations used in creating this alphanumeric password from the given objects of the scene prevents brute-force attacks. These combinations include the choice of objects to use, the number of times each object is selected, the order of object selection and object sizes. The results and analysis of the proposed scheme show it to be secure and easy to use.

In the work of [11], a new graphical user authentication scheme called the Tri-Pass is proposed. In this scheme to create a password user has to choose one image from the pool of images and then select any three points by clicking on the image called the password point. To log in, the user has to repeat the same sequence of activities carried out in the first stage. The proposed new algorithm is based on two techniques, namely the PassPoint and Triangle algorithms [28] proposed a combination of the DAS and Story algorithms for graphical password authentication.

3 Methodology

3.1 Research Design Framework

The schematic block diagram shown in Fig. 2 depicts the research design framework of the graphical-based authentication method combined with city block distance. This framework involves steps from the Registration Interface to the Login Interface and finally, E-Payment Login interfaces where a user can make electronic payments (e-transactions) either by Credit Card, Smart Card, and Debit Card respectively. Clients can also make a transfer of funds. The algorithms follow the steps in Fig. 2 to produce and evaluate the results.

The interface components for the Proposed Graphical Based Authentication (GBA) (Fig. 2) are explained as follows.

Registration Interface. Step 1: User enters personal information: In every system, the user must register as a new user. The proposed graphical scheme's block diagram (Fig. 2) begins with the registration process, in which the user enters personal information such as a username, email identification, phone number, textual password, and saves it to the system's database. Here the registered user identification, password, email address and mobile number are used to create a textual password and save it in the system's database.

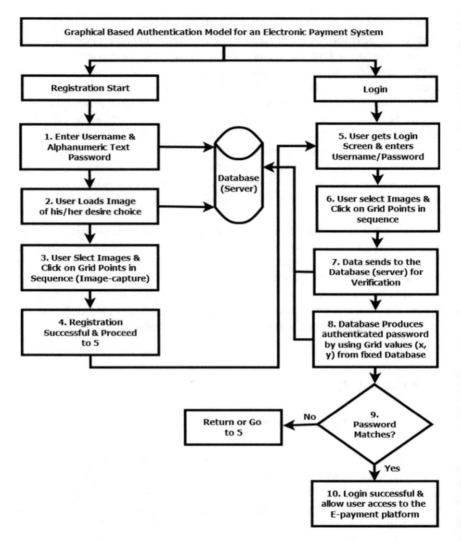

Fig. 2 The schematic block diagram for the proposed graphical-based authentication method combined with city block distance

Step 2: User loads images of their own choice: After saving the personal identification/details, the user loads their own desired images and presses the next button to now move on to a 3 × 3 graphical image grid display.

Step 3: User Selects images and clicks on the grid points in sequence: The image grid contains a set of images, and the user must select images in sequential order and click once on a selected point on each of the three images to generate a graphical password. The image is captured by the model using the generated registered coordinates $x1$, $y1$ and stored in the database using the user ID generated at the end of the registration process.

Step 4: Registration successful: After the password entry, the user clicks on the register button. The system sends a registration successful message. If not, successful it sends an error message.

Login Interface. Step 5: User gets login screen: After successful completion of registration procedures with the scheme, the user then proceeds to step 5 to login into the scheme as a new user by providing correctly username, textual password and graphical password which the user entered during the registration phase. Here the user selects a distance measure from the pulldown menu to login into the scheme.

Verification and Validation. Step 6: Select images and click grid points in sequence: Once a user login correctly with the username and textual password used during the registration, a 3 × 3 images grid display is shown. When the user clicks an image in the grid display, it zooms in to give the user a better look. Here again, a user picks a set of images in sequential order and clicks on the same 3 different images on the clicked point chosen as a password during the registration.

Step 7: The data or login coordinates values $x2$, $y2$ generated by the scheme are sent to the server for password matching or against the registered coordinates for verification and validation.

Step 8: Server produces encrypted password by accessing grid values from the fixed database: Here the server produces the encrypted images on a request by the user for validation.

Authentication Interface. Step 9: If the graphical images entered by the user in the correct order are correct, the user is legitimately granted access to the electronic payment interface to conduct electronic transactions (e-payment transactions).

Step 10: If the password does not match, it moves to step 5: If a user is denied access to the model the user simply switch to step 5, alternatively, the user can simply select the Forgot Password option, and a 7-digit random alphanumeric code will be sent to the email address used during the registration process, allowing the user to reset their password.

3.2 The Equation for the E-Payment Authentication

Let Eq. (1) represents the equation for the system.

$$EP = \left(\sum_{k=1}^{n} T + T_A \right) \tag{1}$$

where: EP is Electronic Payment (E-Payment), $\mathbf{T}$ is Transaction, $\mathbf{T_A}$ represents the type of distance measure authentication of users, $\mathbf{n}$ denotes the number of transactions, and $\mathbf{K} = 1$ is constant for authentication for one transaction.

Therefore, substituting T_A with the various distance measures in Eq. (1) gives the following equation:

$$EP = \left(\sum_{k=1}^{n} T + (ED, JD, CBD, CS) \right) \tag{2}$$

where: ED = Euclidean Distance, JD = Jaccard Distance, CBD = City Block, Distance (Manhattan Distance), CS = Cosine Similarity.

For simplicity and clarity, let's substitute T_A with a formula that denotes each distance measure, starting with ED, JD, and followed by CDB and CS in that order.

That is, for Euclidean distance (ED), the following is obtained,

$$EP = \left(\sum_{k=1}^{n} T + \sqrt{\sum_{i=0}^{n} (X_i + Y_i)^2} \right) \tag{3}$$

where: EP = Electronic Payment (E-Payment), (Y_i, X_i) are Registered coordinates, (Y_j, X_j) are Login coordinates.

In the case of Jaccard Distance (JD), the following is given:

$$EP = \left(\sum_{k=1}^{n} T + X_i * X_j \Big/ \left(|X_i|^2 + |X_j|^2 - X_i * X_j \right) \right) \tag{4}$$

where: (Y_i, X_i) are Registered coordinates, (Y_j, X_j) are Login coordinates.

For City Block Distance CBD, the equation is as follows;

$$EP = \left(\sum_{k=1}^{n} T + \sum_{j=1}^{n} |X_{sj} - X_{tj}| \right) \tag{5}$$

where: (Y_i, X_i) are Registered coordinates, (Y_j, X_j) are Login coordinates.

And finally, for the Cosine Similarity, the equation is given as follows.

$$EP = \left(\sum_{k=1}^{n} T + \left(\sum_{i=1}^{n} A_i * B_i \right) \Big/ \sqrt{\sum_{i=1}^{n} A2i} * \sqrt{\sum_{i=1}^{n} B2i} \right) \tag{6}$$

where: A_i^2 and B_i^2 are components of vector points A and B respectively [20], (Y_i, X_i) are Registered coordinates, (Y_j, X_j) are Login coordinates.

Therefore, in this study the Mathematical Equation for the E-Payment scheme is given as:

$$EP = \left(\sum_{k=1}^{n} T + T_A \right) \qquad (7)$$

where: T = Transaction (it includes registration, login, verification, e-payment) and T_A = Type of authentication.

3.3 Proposed Graphical Based Authentication Method Combined with City Block Distance

This research paper aims to design a graphical-based authentication method combined with city block distance for Electronic Payment System and to implement the system in four phases: Home page, Registration phase, Login Phase (Image password creation) and Image password authentication as shown in Fig. 2. The following system research tools were also used in implementing this graphical scheme: The scheme was hosted on 4 GB RAM, 800 MHz, Intel, Core i5, 15.6 Inches screen display with a resolution of 1366×768 pixels running on windows 7, keyboard and mouse.

Storage Space. Graphical passwords are a better alternative to textual passwords since they are more secure. Graphical authentication techniques require significantly more storage space than text-based passwords. Because a significant number of picture images must be retrieved and displayed for each round of verification in the authentication process, a huge number of images will have to be retained in a centralized storage database. Having a sufficiently large password space is the primary defense measure against brute force search. The following formulas can be used to compute text-based and graphical password space respectively [25, 29]:

$$SPACE = M^N \qquad (8)$$

where M denotes the number of printable characters i.e. (94) excluding SPACE, while N denotes the length of the password.

$$\text{Graphical Password Space} = \sum_{l=1}^{k} \binom{j+l-1}{j-1} = \sum_{l=1}^{k} \frac{(j+l-1)!}{l!(j-1)!} \qquad (9)$$

where *j* is the total number of pictures, *l* is password length, while *k* is the maximum password length.

The proposed system's performance of these distance-measuring algorithms is that they allow for password matching with minimal error and a sequence of image selection, making it non-vulnerable to brute-force search.

3.4   The Requirements for the Authentication Method

The two essential requirements used for this graphical scheme are Software and Hardware system.

Application Software. The implementation of these graphical schemes was developed using system research tools such as HTML, CSS, XAMPP and JavaScript all at frontend and backend, and Web Browser was used (Google Chrome) and a Server (Database).

Hardware. The following hardware components were required and used to implement the graphical scheme in this study: a database server and a 4 GB memory device, such as a laptop/desktop.

The Adaptability of the Proposed System for Mobile Devices. In this paper, a graphical authentication scheme was proposed as an alternative to traditional text-based password techniques, driven in part by the fact that humans recall picture images better than text-based passwords. Graphical passwords are more difficult to guess or crack with brute force [9]. If the number of possible pictures is large enough, the possible password space of a graphical password scheme may exceed that of text-based schemes, implying that it is more resistant to brute force and spyware attacks [9]. Because of these advantages, graphical passwords are gaining popularity. The proposed system is scalable in terms of RAM and storage. Therefore, it can be implemented on mobile devices such as personal digital assistants (PDAs), Smartphones, iPods, iPhones, and tablets. Similarly [30], has demonstrated that combining mobile banking and blockchain improves the scalability and security of online banking. The advancement of smartphone processing power and features has made it simple for banks to enter the mobile technology arena to provide personalized and customer-oriented financial and non-financial services in a convenient, cost-effective, ubiquitous, and accessible manner [31].

3.4.1   Data Collection

The secondary data collection method was used for this study. This was done by data capturing of different picture images. The picture images collected as data are obtained from [32–34]. The images collected were used to determine their x and y coordinates during the registration and login processes.

3.5   Performance Evaluation

This research study conducted an experiment involving five (5) participants (users) to compare different distance measures and their various times taken for execution, login success rate and matching error. The registration and login time were utilized to test the reliability, efficiency and robustness of the graphical scheme.

The experiment that was conducted in this research work, involved critical and quantitative analyses, and performance evaluation of the proposed method (graphical-based combined with City block distance). In addition, the performance evaluation of the proposed method was quantitatively compared with other existing methods (Euclidian, Cosine Similarity and Jaccard) using evaluation metrics such as Time of Execution and Matching error.

Login Success Rate. Five 5 participants (users) were selected from the Computer Science Department, Federal University of Technology, Minna. Five (5) participants were requested to take part in the registration procedure by entering their username, text password, email and phone number. Also, they have to choose an image password in sequential order and click on the three different images one at a time. Each attempt was reported as either successful or failure. This means that the success rate can be calculated based on the successful login of a user as.

$$\text{Login Success Rate (LSR)} \ = \ \frac{N_S}{N_A} * 100\% \tag{10}$$

where N_S is the Number of Successful Logins by a user, N_A is the Number of Attempts to Login by a user.

Execution Time for the Distance Measures. In this experiment, the same 5 participants used the four different distance measures: Euclidean distance, Cosine similarity, City Block distance and Jaccard distance. Here the time taken for execution differs from one distance measure to the other. Every participant was permitted to register and create an image password in sequential order. After this, participants login into their accounts respectively by utilizing one distance measure at a given time. The execution time was taken from the point where each participant (user) clicks on the distance similarity measure from the dropdown menu to submit a password for authentication until the participant views E-payment interface files on successful login. The execution time is computed as:

$$\text{Execution Time (ET)} = T_2 - T_1 \tag{11}$$

where T_2 is the End Time of Successful Login by a user, T_1 is the Start Time of login by a user.

Matching Errors for the Distance Measures. The idea of the Matching error stems from the computation of an average angular error [35], it is an important metric that allows for an effective evaluation of the performances of the different distance measures (Euclidean distance, Cosine similarity, City block distance and Jaccard distance) for similarity measure between different vectors points. It describes the angle between the vector points at the login phase and the vector points at the user registration, all in 2D space. The matching error between two vector points is computed as the inverse cosine of the ratio of the dot product of the vectors and the product of their lengths: The general equation is given by:

$$\mathbf{ME} = \arccos \frac{U * U_e + V * V_e + 1}{\sqrt{(U)^2 + (V)^2 + 1}\sqrt{(U_e)^2 + (V_e)^2 + 1}} \tag{12}$$

where: **U, V** denote the vector point (x, y) obtained during the login phase $(x2, y2)$.

Ue, Ve − (**e**) represents the vector points obtained during user registrations and save in the database $(x1, y1)$.

Arccos. Represents the inverse cosine and **ME** is the matching error computed between the vector points obtained during the user login and the vector points during the registration phase. To avoid division by zero, one is added to both the nominator and denominator as shown in Eq. (12). Now Eq. (12) can also be expressed using variables x, y as expressed in Eq. (13):

$$\mathbf{ME} = \arccos \frac{(X2 * X1) + (Y2 + Y1)}{\sqrt{(X2)^2 + (Y2)^2 + 1}\sqrt{(X1)^2 + (Y1)^2 + 1}} \tag{13}$$

Equation (13) was used in this paper to calculate the matching errors involving matching of two vector points during the register and login sessions of coordinates $(x1, y1)$ and $(x2, y2)$.

In this research work, the matching error gave the error rate of the graphical scheme in the process of matching two points at the registration and login phase. A minimum matching error means that the graphical scheme is robust and reliable, ensuring the security and usability of the entire system.

3.6 Implementation

System implementation completely involves the assembling and articulating of various components to form the new graphical method and the experimental performance test is carried out on it to obtain its result. The threshold in this context is a range of value (s) to accommodate bits of error in coordinate selection. In this proposed graphical method, each user logs on to the home page and then clicks on register to fill in their demographic data to create an account with the system.

Home Page. Figure 3 shows the interface that was used by users to interact with the system while making use of the graphical scheme. The new users are expected to

Fig. 3 Home page interface of the graphical scheme

Fig. 3 Home page interface of the graphical scheme

make registration while existing users are expected to input in their accurate personal information details for authentication.

Registration Phase. This phase requires the user to enter personal details into the system, which is the first crucial step to utilize the graphical scheme. For authentication to be possible in a web application, each user needs to create and save the account details in the database.

During the registration phase, the user must click on the registration button as shown in Fig. 3, which leads the user to a page where to fill in personal unique identification details as shown in Fig. 4. On clicking the user registration button, the system will generate a unique user ID known as coordinates. Immediately the user submits personal details and the text password, it will be redirected to the next phase, where the user is required to create a graphical password and save it in the database.

Image Graphical Password Creation. During this phase, the user will be shown a series of images from which to select a password. The images from which the user will select click points for accurate login will be generated at random and presented to the user in an image grid format, as shown in Fig. 5. During the password creation process, the user must select one-click point per image.

After the user has entered a username and password, the user will then choose the image as a graphical password. This phase involves the following:

(a) The user selects preferred images as a password from the images displayed in Fig. 5. In this scheme three (3) different images are considered from Figs. 5 to 8.

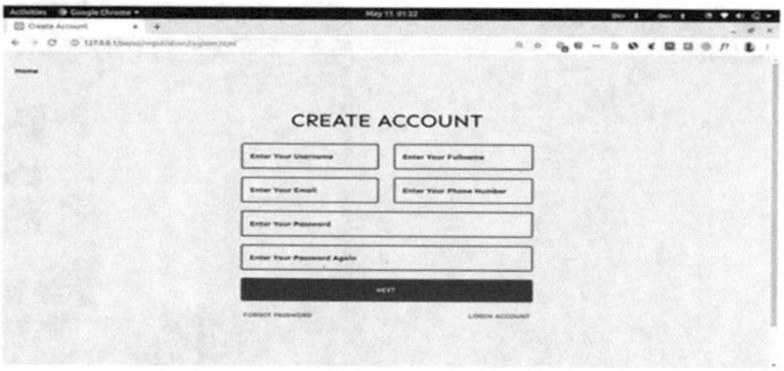

Fig. 4 Registration phase

Fig. 5 A 3 × 3 grid images
display

(b)   The user clicks on the images in the same sequence as selected.

Login Phase. In the Login phase as shown in Fig. 6, the user enters a username and textual password to check if the ID is valid or not. If it is valid then the matching images will be displayed. On these images, the user will have to choose click points by using the single click technique (Fig. 7).

After a user has successfully registered with the system and the personal details provided by the user are validated or text matched with that in the database to ensure a valid user has been given access. This entirely involves the username and the graphical password. The user will either use Jaccard, City Block, Cosine similarity and Euclidean distance measure to login into the system.

The registration form in (Fig. 5) is a set of 3 × 3 grid images which serve as the graphical password, where each user selects three images in a sequence that would be entered during login time in the same sequence to access the E-payment interface.

Fig. 6 The first image used
to create a graphical
password

Fig. 7 Login phase
(interface)

Fig. 8 A 3 × 3 graphical
grid image display during the
login phase

Each image provides a coordinate $(x1, y1)$ on account creation and is stored in the database.

The user logs in using their respective username and text password followed by their graphical password based on recognition, the users are provided with four distance measures namely, Euclidean, Cosine Similarity, City Block, and Jaccard distance to which they will select to login. These distance similarity measures compute the recognition rate of the scheme used with the containment of minimal error in its threshold.

In this phase, the user must log in with a Username and Text password to view the image grid (Fig. 8), to be able to go through the exact sequence of click points.

The user must click on the images chosen as the password during the graphical password creation. The images must be in the same order as they were when the password was created. When the user clicks an image in the 3 × 3 image grid displayed in Fig. 8, the image enlarges to provide the user with a larger view of the image. The user can now click on the point chosen as a password when creating the password. The user must repeat the process until all three images are selected with their respective click-points.

Every image's click-point refers to x, y coordinates. These coordinates are saved in the database with a link to the images selected for each user ID. The user is given the option to select any of the distance measures after clicking on the final image and its click-point. Users are provided with four distance measures in the implemented system.

After clicking on the final image (third image) and its click-point, the user is given the freedom to choose any of the distance measure. In this scheme, the user is provided with four distance measures namely; Euclidean, Cosine similarity, City Block and Jaccard.

Applications for the Electronic Payment System. The Electronic Payment System applications access and process sensitive data for user. The generated coordinates, user ID, all transaction information and demographic data of the users of the e-payment are back up on the XAMP application in an encrypted format.

The implementation of the e-payment was developed using the following applications HTML, MySQL, PHP, CSS, XAMPP and JavaScript, all at frontend and

Fig. 9 E-payment phase and
various card types/payment
procedures

backend, and Web Browser was used to access the e-payment platform database
server.

Electronic Payment Interface. This phase display to the user an Electronic Payment
login screen where user/client can use their Automated Teller Machine (ATM) cards,
Credit card, Smart card and Debit card for transactions. For example, a visa card,
verve card can be used to make electronic payments or fund transfers. In this platform,
some Card details such as Card number, Card Verification Value (CVV), the valid
date will be required to allow effective E-payments or fund transfer (see Fig. 9).

The Database (Server). Database refers to a relationship that provides tables and
other images or objects in an index format. In a database, tables comprise Rows
and Columns. All the user detail data information such as username, text password,
picture images, graphical password, registered and login coordinates are all saved
and stored in the database. Database (Server) is being utilized for storing data information purposes. In this graphical scheme, the XAMPP server is used. This database
phase displays all login data and user's activities, such as login distance measure
and matching errors, user unique ID and threshold values, Fig. 10 displays the regis-
tered/login coordinates of graphical images and demographic details all in rows and
columns.

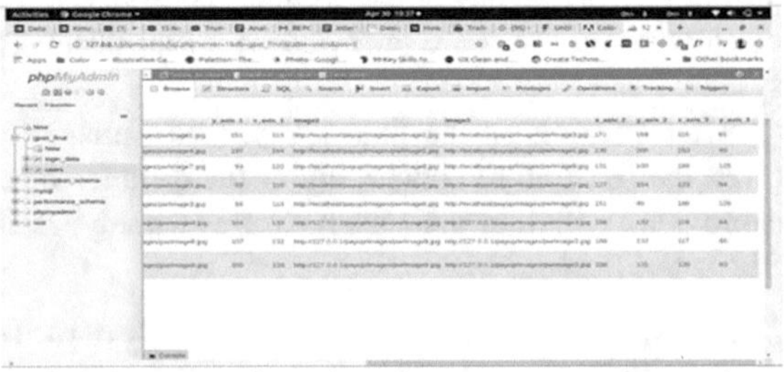

Fig. 10 Database showing
user's registered and login
coordinates

Algorithm 1: Pseudo Code for the algorithm in registration phase

---

Input: Email, Full name, Phone Number, Text Password
Output: user's Password
Step 1: If Email and Phone number is not in Database show first image to user click on a point of that.
Please Enter E: Email, F: Full name, P: Phone number PS: Password
If E or PS is not in Database
register (E, F, P, PS)
else Show Error Message, break;

Step 2: Display a 3x3 Image Grid, select Image and Get Coordinate
[Img1] ← select image from the grid;
Display Enlarged Image of [Img1]
[X1, Y1]←Get coordination of user's click;
For (posX = 0, posY = 0; oElement; oElement = oElement.offsetParent)
{
PosX += oElement.offsetLeft;
PosY += oElement.offsetTop;
}
Return [posX, posY];
}
Else
{
Return [oElement.x, oElement.y];
}
[X1] ← PosX
[Y1] ← PosY
[X1, Y1] ← [Img1]

Step 3: Repeat **step 2** for 2 more images to selected
Step 4: Assign Coordinates
[H1, P1] ← [X1, Y1] for the first Image
[H2, P2] ← [X1, Y1] for the second Image
[H3, P3] ← [X1, Y1] for the third Image
Step 5: Clean and Format data extracted
Step 6: Create Database Table
Step 7: Save (Email, Full name, Phone number,Img1,H1,P1,Img2,H1,P2,Img3, H3,P3,TimeTaken)
Database← Insert Record into Database;

Algorithm 2: Pseudo Code for the algorithm in login phase

Input: Username (U), Text Password (TP)
Output: Authorization
Step 1: If username (U) and Text password (TP) is in Database then GOTO step 2 else show error message "Invalid login credential"
Step 2: Display an array of 3x3 image grid and select a valid image.
Step 3: click on a point on the selected image and extract coordinates.
Step 4: If user click wrong coordinate point, Display error message. This is just for the selected image.
Step 5: Repeat Step 2, Step 3 and step 4 for two more valid images sequentially
Step 6: Apply KNN machine learning algorithm to compute distance measure
[(H1, P1), (H2, P2), (H3, P3)] ← Get coordinates of 3 valid images from database
[(X1, Y1), (X2, Y2), (X3, Y3)] ← Get coordinates of 3 images from Login phase
[R] ← Compute distance measure for the 3 images using City Block Distance;
Where R1 = Image1 City block Distance Threshold
    R2 = Image2 City block Distance Threshold
    R3 = Image3 City block Distance Threshold

[R1] ← Sqrt (abs ((H1 - X1) + (P1 - Y1)));
[R2] ← Sqrt (abs ((H2 - X2) + (P2 - Y2)));
[R3] ← Sqrt (abs ((H3 - X3) + (P3 - Y3)));
[E] ← Compute Matching Error
Where E1 = Image1 Machine Error
    E2 = Image2 Machine Error
    E3 = Image3 Machine Error

[E1] = ACOS((($X1$ + H1) + (Y1 + P1) + 1) / (SQRT(POW((Y1),2) + POW((P1),2)+1)
* SQRT(POW((X1),2) + POW((H1),2)+1)));

[E2] = ACOS((($X2$ + H2) + (Y2 + P2) + 1) / (SQRT(POW((Y2),2) + POW((P2),2)+1)
* SQRT(POW((X2),2) + POW((H2),2)+1)));

[E3] = ACOS((($X3$ + H3) + (Y3 + P3) + 1) / (SQRT(POW((Y3),2) + POW((P3),2)+1)
* SQRT(POW((X3),2) + POW((H3),2)+1)));

[T] ← Compute time Taken
[T] ← End time – Start Time
Step 7: Authenticate Image chosen and coordinate points as graphical Password (GP)
Step 8: If graphical password (GP) is correct then GOTO step 9 else Display Error Message "Invalid Graphical Password"
Step 9: save (Coordinates points, Username, R1, R2, R3, E1, E2, E3, Time taken) as Logs
Step 10: If the logs is saved user can login successfully to access E-payment Interface

---

3.7 Similarity Measure

The similarity measure is an important task for document retrieval, image processing, text matching and retrieval of images from the database that is similar to the query image. To achieve an optimal performance of the system and make it robust in the face of many challenges, an experiment was conducted during the login session using city block distance and other different distance measures that include Euclidean distance, Cosine similarity, and Jaccard distance (Fig. 11).

3.7.1 K-Nearest Neighbor Algorithm

K-Nearest Neighbor algorithm (KNN) is a distance measure machine learning algorithm that uses distance measures at its core [7]. Distance measure provides a foundation for effective machine learning algorithms like K-nearest Neighbor [7].

The similarity distance measures utilize the KNN algorithm to get the accurate matching of password clicking points while the login success and matching error are used as metric methods of evaluating the performance of the similarity distance measures.

Similarity distance measures such as city block distance, Euclidean distance, cosine similarity, and Jaccard distance utilize the KNN algorithm to compute distance between two vector points at register coordinate $(x1, y1)$ and vector points at login coordinate $(x2, y2)$. The similarity distance measures use the KNN algorithm to compute and measure the distance of the image data saved in the database and at the login phase.

Euclidean Distance. The Euclidean distance is the most widely used similarity measure. It has vast applications in image and document retrieval, text matching. The Euclidian distance measures are used to compute the distance between a given vector point and some other vector points save in the database. It determines the root of square differences between the coordinates of a pair of objects [18]. Euclidean is the distance between two points in a plane or 3D space that measures the length of a segment connecting the two points. For vectors x and y, distance $d\ (x, y)$ is given by the general equation for Euclidean distance:

Fig. 11 Dropdown menu of the various algorithms used during the login sessions

$$d = \sqrt{\sum_{i=1}^{n} (x_i + y_i)^2} \tag{14}$$

where: x and y are n-dimensional vectors.

In this paper, mathematical Eq. (15) was used to calculate Euclidean distance involving two pair matching points at register and login of $(x1, y1)$ and $(x2, y2)$:

$$d = \sqrt{(x1 - x2)^2 + (y1 - y2)^2} \tag{15}$$

where: $x1, y1$—registered coordinates saved in the database during registration $x2$, $y2$—login coordinates. Here the threshold value for Euclidean distance is ≤ 5.

Cosine Similarity. The Cosine Similarity begins by finding the cosine of the two non-zero vectors. The introduction of cosine similarity in this work is to eliminate ambiguous matching results between different vector points such as those obtained during user registration and those from the user login phase. The idea of cosine similarity is to measure the orientation of two vectors. The cosine similarity has been used in many applications including data mining, text matching and document retrieval.

Given two vectors of points, A and B, the cosine similarity, $\cos(\theta)$, is represented by using a dot product and magnitude as defined in Eq. (16) as the general equation [18].

$$A \cdot B = \|A\| \cdot \|B\| \cos\theta$$
$$\cos\theta = \frac{A \cdot B}{\|A\|\|B\|} = \frac{\sum_{i=1}^{n} A_i B_i}{\sqrt{\sum_{i=1}^{n} A_i^2} \sqrt{\sum_{i=1}^{n} B_i^2}} \tag{16}$$

where A_i^2 and B_i^2 are components of vector points A and B respectively [24], and the value for cosine similarity is less than or equal to one.

In this research work, Cosine similarity was computed using Eq. (17) as it involves two vector points as register and login of coordinates $(x1, y1)$ and $(x2, y2)$.

$$\text{Cosine Similarity} = \frac{(x1 * y1) + (x2 * y2)}{\sqrt{x1^2} * \sqrt{x2^2} + \sqrt{y1^2} * \sqrt{y2^2}} \tag{17}$$

where parameter: $x1, y1$—registered coordinates saved in the database during registration, while $x2, y2$—login coordinates.

City Block Distance (Manhattan). The City Block distance (CBD) is also called the Manhattan distance in an n-dimensional vector space with fixed Cartesian coordinates between two vectors X_{sj} and X_{tj} is the sum of the lengths of the line segment projections between the points onto the coordinate axes. The mathematical formula of the city block or the Manhattan distance is given by (18):

$$\text{CBD}_{\text{st}} = \sum_{j=1}^{n} |x_{\text{sj}} - x_{\text{tj}}| \tag{18}$$

where **n** is the number of variables, and X_{sj} and X_{tj} are the values of the *j*th variables at points *x* and *y* respectively.

The City block or Manhattan distance is the simple sum of the horizontal and vertical components, while the diagonal distance is computed by using the Pythagorean Theorem. For example, if two points $u = (x1, y1)$ and $v = (x2, y2)$ are two points, then the city block distance between u and v is given as:

$$\text{CBD} = \|x1 - x2\| + \|y1 - y2\| \tag{19}$$

In most cases, the City Block distance is greater than or equal to zero. For identical points, the measurement would be zero, while for points with little similarity, the measurement would be high. City block distance is used in image processing, visual image tracking, as face recognition, and many other image processing to measure both horizontal and vertical distances.

Where CB ranges between 0 and 1 but not greater than one. In this paper, City block distance was calculated by using Eq. (20)—Pythagorean Theorem:

$$\textbf{CBD} = \sqrt{(x1 - x2)^2 + (y1 - y2)^2} \tag{20}$$

where: $x1, y1$—registered coordinates saved in the database during registration, while $x2, y2$—login coordinates.

Jaccard Distance (JD). The Jaccard distance calculates similarity by dividing the intersection by the union of the vector points. The Jaccard distance measure can be used to compute the similarity between two data sets. The general formula could be stated as:

$$\text{JD} = \frac{(x_i \cdot x_j)}{(|x_i|^2 + |x_j|^2 - x_i \cdot x_j)} \tag{21}$$

where: JD refers to the Jaccard distance, x_i, x_j are *n*-dimensional vectors. The range value for the Jaccard distance (JD) is between 0 and -1.

Like other distance measures, the Mathematical Eq. (22) was used in this paper to calculate Jaccard distance involving two points at register and login sessions of coordinates $(x1, y1)$ and $(x2, y2)$:

$$\mathbf{JD} = (x1 * y1)\frac{(x1 * y1) + (x2 * y2)}{(x1 + x2) + (y1 + y2) - (x1 * y1) + (x2 * y2)}$$

$$(22)$$

JD ranges between 0 and 1 but not greater than one, where: $x1$, $y1$—registered coordinates saved in the database during registration, while $x2$, $y2$—login coordinates.

JD—threshold value ≤ 1.

Each user has a login coordinate which is cross-referenced with the initial coordinate on registration.

4 Experimental Results

This section analyses the performances of all the different distance measures used for the proposed graphical-based authentication methods to identify and recommend the most efficient and robust amongst them. To provide a fair and accurate evaluation for all the distance measures, metrics such as the login success rate, execution time and the average matching error are used. In particular, the average matching error in this aspect is very crucial for achieving a high login success rate. While the experiment has no limit as to the number of participants, here an experiment was conducted using five (5) participants (users). Each of these participants was allowed to interact with the system and for every interaction, the performance of each distance measure based on how successful a user can log in into the system, the duration of time it takes to log in and the average matching error between the click points at registration and login stages are evaluated and recorded.

4.1 Login Success Rate

Five (5) participants (users) were selected from the Computer Science Department, Federal University of Technology, Minna. The participants took part in the registration procedure by entering their username, email, phone number and text password. They have to choose an image password in sequential order and click on the three different images one at a time. Each attempt is recorded as either successful or failure. This means that the success rate can be calculated based on the successful login of a user. The result obtained by the participants on successful login is shown in Tables 1 and 2.

Figure 12 shows a 2D graphical bar chart representation of results in Table 1 as the login success. The bar chart shows that different users are login with distance measures one at a time and both individual login success with their corresponding average success login are recorded. It also shows how each different user

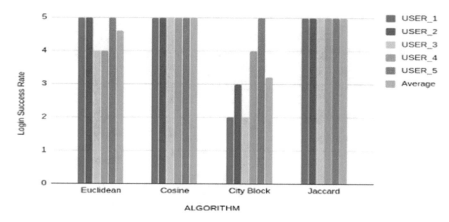

Fig. 12 Bar chart for the login success

Table 1 Login success

Distance measure	Users					
	User_1	User_2	User_3	User_4	User_5	Average
Euclidean distance	5	5	4	4	5	4.6
Cosine similarity	5	5	5	5	5	5
City block distance	2	3	2	4	5	3.2
Jaccard distance	5	5	5	5	5	5

is login successfully with different distance measure and the average login success is determined.

In Table 2, the participants' login success rate in percentage is shown. Here the Cosine Similarity and Jaccard distance measures gave a 100% login success rate which indicates that they have a higher recognition rate.

Table 2 Average login success rate in percentage

Distance measure	Users					
	User_1 (%)	User_2 (%)	User_3 (%)	User_4 (%)	User_5 (%)	Average (%)
Euclidean distance	100	100	80	80	100	92
Cosine similarity	100	100	100	100	100	100
City block distance	40	60	40	80	100	64
Jaccard distance	100	100	100	100	100	100

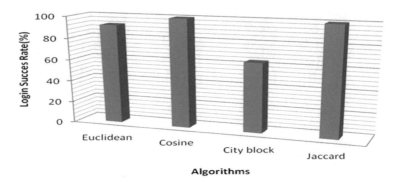

Fig. 13 The average login success rate in percentage (%) for each user on all the distance measures

Figure 13 shows a 3D graphical bar chart representation of results in Table 2 as the average login success rate in percentage (%) in the Y-axis. The bar chart shows the success login rate of 5 different users that used 4 different distance measures one at a time to log in successfully and their login success rate and average login success rate are determined in percentage (%) in the X-axis.

From the results shown (see Tables 1 and 2, Figs. 12 and 13), it is observed that both Cosine Similarity and Jaccard Distance recorded a 100% average login success rate, indicating that they both have a higher recognition rate than Euclidean Distance and City Block Distance.

4.2 Execution Time

In this experiment, the same 5 participants used the four different distance measures: Euclidean, Cosine similarity, City Block distance and Jaccard distance. Here the time taken for execution differs from one distance measure to the other. Every participant is permitted to register and create an image password in sequential order. After this, they can log in to their accounts respectively by utilizing one distance measure at a given time. The execution time is taken from the point each participant (user) clicks on the distance measure from the dropdown menu to submit a password for authentication until the participant views E-payment interface files on successful login. The time taken by the four distance measures allows the data to be retrieved from the database (server) gave the following results.

Figure 14 shows a 2D graphical bar chart representation of the duration of execution time results in Table 3 as the execution time for distance measures in milliseconds (Y-axis). The graphical bar chart shows how each of the 5 different login users using 4 different distance measures one at a time during the login procedure, to obtain both individual and average execution time of each distance measure in (X-axis).

Figure 15 shows a 3D graphical bar chart representation of the duration of execution in Table 3 as the average execution time in milliseconds (Y-axis) by each distance

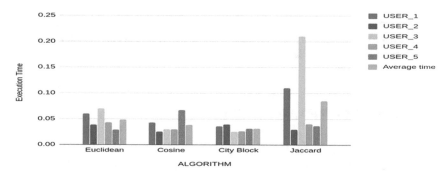

Fig. 14 Shows a 2D graphical bar chart representation of the execution time

Table 3 Execution time for distance measures

Distance measure	Users					
	User_1	User_2	User_3	User_4	User_5	Average time
Euclidean distance	0.06	0.039	0.07	0.043	0.029	0.0482
Cosine similarity	0.043	0.025	0.03	0.03	0.067	0.039
City block distance	0.036	0.04	0.025	0.026	0.032	0.0318
Jaccard distance	0.11	0.03	0.21	0.041	0.037	0.0856

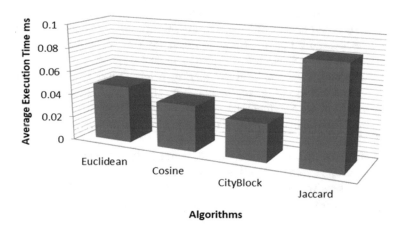

Fig. 15 Bar chart of the average execution time in milliseconds

measure. The bar chart shows how 4 different distance measures are used to the obtained average execution time of each distance measure (X-axis). The graphical bar chart shows outstanding performance by the City Block distance measure with the best execution time of 0.0318 ms.

In this experiment, it is observed that City Block Distance has the best execution time followed by Cosine Similarity, Euclidian Distance and Jaccard Distance in that order respectively (see Table 3 and Fig. 15).

4.3   Matching Errors for the Distance Measures

Accuracy is always of the utmost importance in a high-standard performance measure. Compliance and input errors are factors that lead to mechanism matching and positional errors, and in this scheme, registration and login (input) errors are considered the error source. To keep the output error within the desired limits, the login distance measure use tolerance or threshold allocation, which affects the system's dynamic performance.

In this research paper, the matching error is taken as one of the standard performance evaluation metrics that gave good accounts of the errors that occurred during the registration process and the login procedures. The matching error gave the error rate of the graphical scheme due to the problem of matching points at the registration and login phase. During the point clicking on the picture images, there are minimal errors, and, in this case, the threshold of each distance measure is considered to minimize the system's matching error. If the matching error is at the minimum point at registration and login time during the matching of the coordinates, then the graphical scheme is robust and reliable. In this scheme, the average matching errors were determined by 5 participants that use four distance measures to register and log in. The distance measures used are; Euclidean distance, Cosine similarity, City Block distance (Manhattan) and Jaccard as shown in Table 4.

Figure 16 shows a 2D graphical bar chart representation of results of matching errors in Table 4 and as in Fig. 16, the matching error is in (Y-axis). The bar chart shows how each of the 5 different users and 4 different distance measures are used to obtain both individual and average matching errors of each distance measure in (X-axis).

Figure 17 shows a 3D graphical bar chart representation of results in Table 4 as the average matching error for each distance measure (Y-axis). The bar chart shows how 4 different distance measures are used to obtain the average matching error for each distance measure (X-axis). The bar chart shows the best performance by the

Table 4 Average matching error of the images clicking point per user login

Distance measure	Users					Average matching error
	User_1	User_2	User_3	User_4	User_5	
Euclidean distance	1.55341	1.55404	1.55406	1.55162	1.54851	1.55233
Cosine similarity	1.55342	1.55429	1.55406	1.55137	1.54844	1.55232
City block dist	1.55334	1.55408	1.55403	1.55162	1.54850	1.55231
Jaccard distance	1.55330	1.55407	1.55418	1.55166	1.54887	1.55242

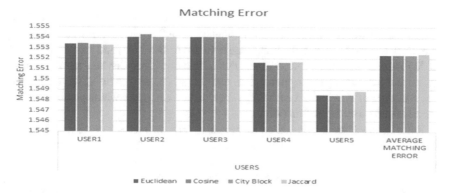

Fig. 16 Matching error obtained for each distance measure during users' interaction with the system

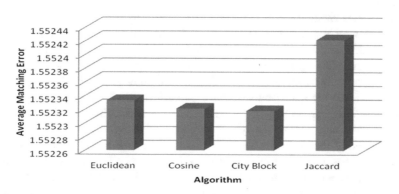

Fig. 17 Average matching error obtained for each distance measure during users' interaction with the system

City Block distance measure with an average matching error of 1.55231 during the image click point.

As can be seen from Table 4 and Fig. 17, the City Block distance has the lowest matching error (minimum) in comparison to Cosine similarity distance, Euclidean distance and Jaccard distance.

4.4 Analysis of Results

One of the most crucial reasons for introducing similarity measures such as City block distance (CBD), Cosine Similarity (CS) and Jaccard Distance (JD) in this scheme is to analyze results and justify the effective performances of the individual distance measures used in the scheme based on their execution time, login success rate and matching errors. In Table 3 and Fig. 14 it is observed that City Block distance has the best average execution time than other distance measures. City block distance has an average execution time value of 0.0318 ms followed by Cosine similarity distance (0.039 ms), Euclidean distance (0.0482 ms) and Jaccard distance (0.0856 ms) respectively. This analysis means that city block distance has the fastest execution time during the login session, and this indicates that the system usability and security could be robust with City block distance in place which conforms with the aims and objectives of this study.

It is also observed that based on login success rate in Tables 1 and 2 and Figs. 12 and 13, Cosine similarity distance and Jaccard distance have better average login success with values of 5 successive logins each and also average login success values of 5 rated as 100% each, followed by Euclidean distance (4.6) as 92% and City block distance (3.2) rated as 64%. This analysis also means that Cosine Similarity and Jaccard distance performs better than other distance measures during the login session.

As for the matching errors, from Table 4 and Figs. 16 and 17, City Block Distance gave the lowest average matching error of 1.55231, while Cosine similarity distance has 1.55232, Euclidean distance 1.55233 and Jaccard distance 1.55242 having higher matching error during their interactions with the model (scheme). The results of the matching error obtained for all the distance measures as shown in Figs. 16 and 17 showed that the City Block distance outperformed all the remaining distance measures. This outstanding performance by City Block distance with minimum average error indicates that the system is robust and user-friendly.

Considering the general analysis of the performance results of the individual distance measures used in this graphical scheme, it is concluded that the utmost requirement in any computing arena is the consideration for the execution time and minimum error of the system which are paramount for a user-friendly and accurate system. Not only does the city block distance measure perform better in terms of the execution time, but it has also outperformed other distance measures with low matching error in the scheme. Furthermore, the city block distance has also performed above average with a login success rate of 64%, which is within the range of an acceptable result for a system. In addition, the low login success rate recorded by the city block distance is due to its high sensitivity to password matching (matching of vector points) during login. As such it has a high restriction to maintaining the threshold and password clicking points hence this gives low matching error as required by a system.

The fast execution time and minimum error performances achieved by City Block distance gave it an upper hand in terms of security, usability and robustness of the scheme. This means that if a system is slow in execution time, the users will have to stay much longer time on the system and this can attract attention or could create avenues for hackers to compromise the system or expose users to Spyware, Internet phishing and Internet surfing attack.

Finally, the analysis of results for the similarity distance measures used in the proposed scheme concludes that City Block distance has an outstanding performance in terms of execution time, minimum matching error and average performance in

Table 5 Comparison of results for the evaluation performance of the distance measures

Distance measure	Login success rate (ISR) (%)	Average execution time (AET) (ms)	Average matching error (AME)
Euclidean distance	92	0.0482	1.55233
Cosine similarity	100	0.039	1.55232
City block distance	64	0.0318	1.55231
Jaccard distance	100	0.0856	1.55242

login success rate and so it is recommended as the best distance measure in the scheme (Table 5).

5 Conclusion and Future Work

In this paper, a graphical password combined with city block distance is presented. A mathematical equation for graphical-based authentication method combined with city block distance for electronic payment system was developed and similarity scores between the points of registration and login are defined using different distance measures utilizing the K-NN algorithm. The experimental result shows that the proposed graphical-based password combined with city block distance has the fastest execution time with minimal matching error compared to when the graphical-based password is combined with other similarity score distance measures such as Euclidean, Jaccard and Cosine similarity.

This system's performance measures are highlighted in Sects. 3.5 and 4 respectively. All similarity distance measures were subjected to an evaluation process using standard metrics such as execution time, login success, and matching error to test and evaluate the efficiency, and performance of the mathematical equation developed for the electronic payment system with other existing methods and also to identify and recommend the best performing distance measure. To select the best successful login, the four distance measures were utilized one at a time to log in users into the proposed graphical scheme and each distance measure's login success rate was recorded.

During the experiment, the execution time of each distance measure was also recorded, and the city block was determined to be the fastest distance measure with an execution time of 0.0318 ms, matching error of 1.55231 and an acceptance of 64%. The matching errors that occurred during the registration and login stages were recorded and computed in the final evaluation, and the optimal error was also determined. The combination of city block distance in the proposed graphical passwords scheme is crucial in mitigating attacks such as shoulder surfing and social engineering.

In conclusion, from the experimental results obtained it can be deduced that city block distance is preferable for high dimensional authentication systems in order to achieve robustness in terms of speed and accurate matching of passwords.

In this study, the proposed graphical-based authentication method combined with city block distance gave a fair level of security, usability, and robustness. However, future advanced research and broad studies in graphical authentication schemes are recommended, in order to achieve higher levels of superiority and stronger security techniques for authentication and usefulness, including training on an operational basis and better ways to select strong, and accurate image click points as graphical passwords.

References

1. Akram T, Ahmad V, Haq I, Nazir M (2017) Graphical password authentication. Int J Comput Sci Mob Comput 6(6):394–400
2. Kadu D, Therese S (2017) Different graphical password authentication techniques. In: International conference on emanation in modern technology and engineering (ICEMTE-2017), pp 56–58
3. Razvi SA (2017) Implementation of graphical passwords in internet banking for enhanced security. In: International conference on intelligent computing and control systems (ICICCS), pp 35–41
4. Alsaiari H, Papadaki M, Dowland PS, Furnell SM (2014) Alternative graphical authentication for online banking environments. In: Proceedings of the eighth international symposium on human aspects of information security & assurance (HAISA 2014) (Haisa). Haisa, pp 122–136
5. Khan A (2015) Secure recognition-based graphical authentication scheme using captcha and visual objects. Eastern Mediterranea University, Gazimağusa, North Cypru
6. Veerasekaran S, Khade A (2015) Using persuasive technology in click based graphical passwords. Int J Tech Res Appl 31(31):29–36
7. Pulungan AF, Zarlis M, Suwilo S (2020) Performance analysis of distance measures in K-nearest neighbor. https://doi.org/10.4108/eai.3-8-2019.2290748
8. Osho O, Musa FA, Misra S, Uduimoh AA, Adewunmi A, Ahuja R (2019) AbsoluteSecure: a tri-layered data security system. In: Damaševičius R, Vasiljevienė G (eds) Information and software technologies. ICIST 2019. Communications in computer and information science, vol 1078. Springer, Cham, pp 243–255. https://doi.org/10.1007/978-3-030-30275-7_19
9. Shah M, Naik R, Mullakodi S, Chaudhari S (2018) Comparative analysis of different graphical password techniques for security. Int Res J Eng Technol (IRJET) 5(4):1873–1877
10. Thirunavukkarasu M (2017) An improving method of grid graphical password authentication system. Int J Eng Res Appl 07(05):40–43. https://doi.org/10.9790/9622-0705044043
11. Yesseyeva E, Yesseyev K, Abdulrazaq MM, Lashkari AH, Sadeghi M (2016) Tri-pass : a new graphical user authentication scheme. Int J Circuits Syst Signal Process
12. Sharifi E, Shamsi M (2014) Evaluate the security and usability of graphical passwords. Int J Adv Res Comput Sci Electron Eng (IJARCSEE) 3(8)
13. Computing M (2014) Comparative study of graphical. Int J Comput Sci Mob Comput 3(9):361–375
14. Istyaq S, Saifullah K (2016) A new hybrid graphical user authentication technique based on drag and drop method. Int J Innov Res Comput Commun Eng. https://doi.org/10.15680/IJIRCCE
15. Wazir W, Khattak HA, Almogren A, Khan MA, Ud Di I (2020) Doodle-based authentication technique using augmented reality. IEEE Access J 8:1–13. https://doi.org/10.1109/ACCESS.2019.2963543

16. Mihajlovic M (2019) Finding the most similar textual documents using case-based reasoning. Cornell University. arXiv:1911.00262v1
17. Khan MA, Ud Din I, Jadoon SU, Khan MK, Guizani M, Awan KA (2019) G-RAT: a novel graphical randomized authentication technique for consumer smart devices. IEEE Trans Consum Electron 65(2):215–223. https://doi.org/10.1109/TCE.2019.2895715
18. Bora DJ, Gupta AK (2014) Effect of different distance measures on the performance of K-means algorithm: an experimental study in matlab. Int J Comput Sci Inf Technol 5(2):2501–2506
19. Misra S (2020) A step by step guide for choosing project topics and writing research papers in ICT Related disciplines. In: Misra S, Muhammad-Bello B (eds) Information and communication technology and applications. ICTA 2020. Communications in computer and information science, vol 1350. Springer, Cham, pp 727–744. https://doi.org/10.1007/978-3-030-69143-1_55
20. Similarity J, Algorithm C, Vector D, Distance M, Measure P, Score S, Han J (2020) Cosine similarity 1:1–22
21. Taghva K, Veni R (2010) Effects of similarity metrics on document clustering. In: ITNG2010—7th international conference on information technology new generations, pp 222–226
22. Deorankar AV (2017) Secure graphical authentication system for web. Paripexindian J Res 6(5):412–413
23. Mahore TR (2017) Secure graphical password scheme. Int J Res Publ Eng Technol (IJRPET) 3(3):144–147
24. Ahsan M, Li Y (2017) Graphical password authentication using images sequence. Int Res J Eng Technol (IRJET) 4(11):1824–1832
25. Awodele O, Olamide K, Remo I (2017) Shoulder surfing resistant graphical authentication scheme for web-based applications. Am J Comput Sci Appl 1(7). https://doi.org/10.28933/ajcsa-2017-09-1801
26. Sun HM, Chen ST, Yeh JH, Cheng CY (2018) A shoulder surfing resistant graphical authentication system. IEEE Trans Dependable Secur Comput 15(2):180–193. https://doi.org/10.1109/TDSC.2016.2539942
27. Mohammad E, Maria A (2018) An improved authentication scheme based on graphical passwords. Int Conf Innov Comput Inf Control 2(8):775–783
28. Osunade O, Oloyede IA, Azeez TO (2019) Graphical user authentication system resistant to shoulder surfing attack. Adv Res 19(4):1–8. https://doi.org/10.9734/AIR/2019/v19i430126
29. Khodadadi T, Alizadeh M (2015) Security analysis method of recognition-based graphical password. J Teknol (Sci & Eng) 75(5):57–62. https://doi.org/10.11113/jt.v72.3941
30. Awotunde JB, Ogundokun RO, Misra S, Adeniyi EA, Sharma MM (2020) Blockchain-based framework for secure transaction in mobile banking platform. In: Abraham A, Hanne T, Castillo O, Gandhi N, Nogueira Rios T, Hong TP (eds) Hybrid intelligent systems. HIS 2020. Advances in intelligent systems and computing, vol 1375. Springer, Cham, pp 525–534. https://doi.org/10.1007/978-3-030-73050-5_53
31. Osho O, Mohammed UL, Nimzing NN, Uduimoh AA, Misra S (2019) Forensic analysis of mobile banking apps. In: Misra S et al (eds) Computational science and its applications—ICCSA. Lecture notes in computer science, vol 11623. Springer, Cham (2019). https://doi.org/10.1007/978-3-030-24308-1_49
32. Google (2020) Animal images. Google. https://www.google.com/search/animalimages. Accessed 25 Aug 2020
33. Shutterstock (2020) Animal images. Shutterstock. https://www.shutterstock.com/images. Accessed 25 Aug 2020
34. Pexels (2020) Animals. Pexels. https://www.pexels.com/search/animals. Accessed 25 Aug 2020
35. Baker S, Scharstein D, Roth S, Black MJ, Szeliski R (2011) A database and evaluation methodology for optical flow. Int J Comput Vision 92:1–31. https://doi.org/10.1007/s11263-010-0390-2

Authenticated Encryption to Prevent Cyber-Attacks in Images

S. Hanis, N. Edna Elizabeth, R. Kishore, and Ala Khalifeh

Abstract The increased usage and availability of multimedia-based applications have led authentication and encryption to gain considerable importance. Cryptography plays an important role in protecting images from theft and alteration. Digital images are used in a large number of applications such as education, defense, medicine, space and industry. This chapter aims at providing a secure authenticated encryption algorithm for the storage and transmission of digital images to avoid cyber threats and attacks. The designed algorithm makes use of the deep convolutional generative adversarial network to test if the image is a fake image originated by the intruder. If found fake exclusive OR operations are performed with the random matrices to confuse the intruder. If the image is not fake, then encryption operations are directly performed on the image. The image is split into two four-bit images and a permutation operation using a logistic map is performed and finally the split images are merged together. Finally, exclusive OR operations are performed on the merged image using the convolution- based round keys generated to generate the concealed image. In addition, authentication is also achieved by calculating the mean of the actual image. The performance analysis shows that the designed technique offers excellent security and also helps in testing the authenticity of the stored images.

Keywords Intelligent authentication · Encryption · Authentication · Cyber-attacks · Artificial intelligence

S. Hanis (✉) · N. E. Elizabeth · R. Kishore
Sri Sivasubramaniya Nadar College of Engineering, Chennai, India
e-mail: haniss@ssn.edu.in

N. E. Elizabeth
e-mail: ednaelizabethn@ssn.edu.in

R. Kishore
e-mail: kishorer@ssn.edu.in

A. Khalifeh
German Jordanian University, Amman, Jordan
e-mail: Ala.Khalifeh@gju.edu.jo

© The Author(s), under exclusive license to Springer Nature Switzerland AG 2022   325
S. Misra and C. Arumugam (eds.), *Illumination of Artificial Intelligence in Cybersecurity and Forensics*, Lecture Notes on Data Engineering and Communications Technologies 109, https://doi.org/10.1007/978-3-030-93453-8_14

1 Introduction

Reliable transmission and storage pose greater challenge in a multimedia based cyberspace [1, 2]. There is an exponential growth seen in the requirement for cyber-secure storage and transmission of videos and images online [3]. Image encryption should possess the following attributes. It should provide a computationally strong encryption system, faster encryption and perfect reconstruction after decryption and should be flexible. Images contain a large amount of redundant information and also they are bulky. Therefore, traditional cyber security algorithms that are used for concealing the data cannot be applied directly to images. Also, the images occupy a greater amount of bandwidth and memory space. This, in turn, affects the cost of computation and speed of the encryption algorithm. Data augmentation techniques were discussed in the literature to enhance the performance of palsy face detection systems [4] and there are techniques to assess the image quality before passing through facial recognition systems that improves the performance [5].

There are different encryption techniques [6–16], authentication algorithms [17–22] and data hiding techniques [23] available in the literature. For example, in [24] the authors have used two-dimensional Lorentz and logistics in concealing the intelligible information present in the image and inferred that the algorithm performs good. In [25] the authors have proposed an encryption system using bulban map and the execution speed of the algorithm is enhanced. In [26] the authors have designed an encryption algorithm using coupled map lattice and wavelet transform but it offers only limited accuracy. In [27] an image encryption technique using alternate logistic maps and bit-level permutations was proposed by the authors and the performance analysis shows that it offers good security. However, the designed algorithms are not capable of detecting any modifications done by intruders.

In a real-time scenario, combining separate authentication and encryption algorithms compromises the overall effectiveness of the cyber security algorithm [28]. Authenticated image encryption offers both confidentiality and authenticity by combining both authentication and encryption. Complex image encryption and authentication techniques have been proposed in the literature to secure images during storage and transmission. In the recent past, the motivation of emerging cryptographic algorithms is on reducing the computational time of the encryption algorithm without compromising the security. These algorithms aid in reducing cyber-attacks. The overview of the encryption process is as follows (Fig. 1).

This chapter discusses the image encryption along with authentication and key generation technique utilizing a logistic map [29] and a set of chaotic equations. The performance of the designed technique is also addressed. The rest of the chapter is structured [30] as shown. In Sect. 2, methodology for key generation and round key generation are presented. Section 3 gives a detailed framework of encryption algorithm and decryption. Section 4 elaborates the performance analyses. Section 5 summarizes the discussions and possibilities for further developments.

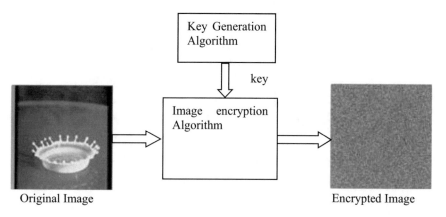

Original Image   Encrypted Image

Fig. 1 Basic image encryption process

2 Key Generation

It is a very important factor for concealing images. Strong key leads to secure storage and transmission of images whereas a weak key allows the intruder or the attacker to view the intelligible information. Here, logistic map based discrete modified convolution technique is used for generating keys [31].

2.1 Nonlinear Convolution Using Logistic Map

The logistic map was used basically in population dynamics. It is a chaotic equation that demonstrate dynamic behavior and is highly sensitive to changes in the initial conditions. It is presented mathematically as,

$$R_{n+1} = \mu R_n (1 - R_n) \tag{1}$$

where $0 < R_n < 1$ along with the parameter μ acts as a control element. The logistic map acts chaotic while $3.569 < \mu \leq 4$.

Discrete-time signals are generally processed using the primitive technique called discrete convolution When the two sets of discrete signals are described using binary values, the convolution is termed as binary convolution [32]. The convolution [33] of two discrete signals h and g of lengths l_1 and l_2 is a discrete sequence x of length $l_1 + l_2 - 1$ and can be represented as,

$$x = h * g = \begin{bmatrix} h_1 & 0 & \cdots & 0 & 0 \\ h_2 & h_1 & \cdots & \vdots & \vdots \\ h_3 & h_2 & \cdots & 0 & 0 \\ \vdots & h_3 & \cdots & h_1 & 0 \\ h_{n-1} & \vdots & \cdots & h_2 & h_1 \\ h_n & h_{n-1} & \cdots & \vdots & h_2 \\ 0 & h_n & \cdots & h_{n-2} & \vdots \\ 0 & 0 & \cdots & h_{n-1} & h_{n-2} \\ \vdots & \vdots & \cdots & h_n & h_{n-1} \\ 0 & 0 & \cdots & \cdots & h_n \end{bmatrix} \begin{bmatrix} g_1 \\ g_2 \\ g_3 \\ \vdots \\ g_l \end{bmatrix} \tag{2}$$

Moreover, the non-linearity is introduced by changing the order of h_1 randomly using the permutation based on the chaotic map in each column of the Toeplitz matrix h thereby modifying the matrix to a new updated matrix H. Mathematically, the process is represented as:

$$x_1 = Hg \bmod 2 \tag{3}$$

The random matrix x_1 generated as per the Eq. (3) is used as round key in the second stage of the diffusion process.

2.2 Round Key Generation

Round key generation plays a vital role in encryption schemes considering its effect on the security of the concealed output image. A 256 bits key was selected randomly and enlarged to a required size during the encryption and decryption process. Let the dimension of the original image be $M \times M$. The round key generation technique is as follows:

Step 1:   A random W bit key is chosen as an initial seed and it is assigned to the sequences h and g. The initial key is represented as $K_i = [k_1, k_2, k_3, \ldots, k_{256}]$ and $h = g = K_i$ where $K_i \in (0, 1)$ and $W = 256$ bits.

Step 2:   Consider the length of the convolved output as $l = l_1 + l_2 - 1$. A Toeplitz matrix of size $l \times l$ is then generated using the sequence h, with the size of 511×511.

Step 3:   The order of the elements in Toeplitz matrix are shifted to random positions, using the technique explained in Sect. 2.1. Here, the row index $i = [1, 2, 3, \ldots, l]$, $f_{row}(i) = i/l - c_l$ where c_l is equal to $0.1/l$ and $\mu\_row = 3.712381$.

Step 4:   Discrete convolution operation is performed using (3), the output sequence x_1 of length 511 bits is produced. The length of the sequence is made equal to 512 bits by inserting one bit randomly.

Step 5:   In the next round 256 bits from the output are forwarded as seed keys.

Step 6:   Step 4 and Step 5 are repeated until the key size becomes $M \times M \times 8 \times 2$.

3   Encryption and Decryption

In this system, the image to be encrypted is tested using Deep Convolutional Generative Adversarial Network (DCGAN) for genuineness and is chosen as I_0 if found genuine. The brightness of the image is considered as the information in this context. The LSB and the MSB bits of the image are extracted. Chaotic permutation technique is used for reducing the correlation between pixel values by shuffling the LSB and MSB pixel positions. The shuffled LSB and the MSB bits are represented as I_1 and I_2. The shuffled bits are then combined together to form a single 8-bit grayscale image. The 8-bit image is diffused using nonlinear convolution as keys. The encrypted image obtained is represented as EI. Figure 2 depicts the image encryption algorithm. The following subsections presents a detailed description of each block of the image encryption system.

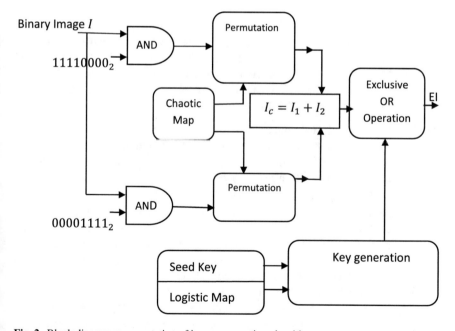

Fig. 2   Block diagram representation of image encryption algorithm

3.1 Deep Convolutional Generative Adversarial Network Based Test for Special Images

Deep Convolutional Generative Adversarial Network (DCGAN) was used as the first step in the image encryption process. The objective of DCGAN was to detect the special images generated by the intruder from the original image to be encrypted. Here, the images were tested for special images such as full black, full white, half black and half white and so on. The network undergoes training, validation and testing. The main aim of the training is to develop a model which classifies the image into a fake and intelligible image. Ten thousand patches of 512×512 were generated from the training set. The patches and the ground truth were given to estimate the model parameters from the trained model. The DCGAN was optimized by using a gradient function with a learning rate of 0.01. In order to conceal the key, if the image was tested to be a fake image, then a random matrix was added to it to diffuse the special image and then the encryption algorithm was allowed to run on the image. This step provides additional protection to safeguard the key from the intruder.

3.2 Permutations, Merging and Diffusion

Consider an intelligible grayscale image as I_0 of dimension $M \times M$. Take the mean of the image and select the fractional portion of the mean of the image and perform an XOR operation with the original image to obtain I. The four LSB bits are sliced by doing AND operation with (111100002). The four MSB bits are sliced by doing AND operation with (000011112). Then, the pixel locations of both the MSB and LSB bits were disordered and added using the method presented below:

Step 1: Normalize the row and column indexes to f_r and f_c by dividing the index by their maximum values. Let the row and column indexes of the original image be $x = [x_1, x_2, x_3, \ldots, x_M]$, $j = [y_1, y_2, y_3, \ldots, y_M]$, the row and column indices that are normalized are,
$$f_r = \left[\frac{x_1}{i_M}, \frac{x_2}{i_M}, \frac{x_3}{i_M}, \ldots, \frac{x_M}{i_M}\right] \ \& \ f_c = \left[\frac{y_1}{j_M}, \frac{y_2}{j_M}, \frac{y_3}{j_M}, \ldots, \frac{y_M}{j_M}\right].$$

Step 2: Subtract a constant value from f_r and f_c to shatter the symmetric values obtained by using chaotic map. Let, c_x be the constant value for f_r and c_y be the constant value for f_c chosen at random between 0 and 1 then, $f_r = f_r - c_x$ and $f_c = f_c - c_y$, $c_1 = 0.0071$ and $c_2 = 0.0033$ are the values of constant used in this algorithm.

Step 3: The control parameters for image I are assumed to be $\mu r_1 = 3.722381$ and $\mu c_1 = 3.599703$. The assumed values produce a chaotic behavior between the next state and the current state of a pixel location. The control parameter can be any value within the chaotic range of the logistic map.

Step 4: Iterating the row (f_r) and column (f_c) index using the rule in (1) to produce new sequences f_r' and f_c'.

Step 5: Arrange the sequences f_r' and f_c' in ascending order store the new index values. Shuffle the pixels using the new index values for the MSB portion of the image and obtain the permuted image I_1. Repeat step 4 and step 5 to obtain permuted LSB image I_2.

Step 6: Adding the MSB and LSB portion of the image to generate a single 8-bit unintelligible image I_c of size $M \times M$.

Finally, the permuted and merged image is further diffused by performing an XOR operation with the round keys generated to form the encrypted image.

Step 1: The convolution based key generated is divided into two keys K_A and K_B of size $M \times M \times 8$.

Step 2: XOR operation is performed with the key K_A generated and the permuted-merged image.

Step 3: XOR operation is performed with key K_B and the resultant obtained from step 2 to obtain the final encrypted image.

3.3 Authentication

The mean of the image I was calculated and the last three digits were extracted from the fractional portion of the mean. Let the mean of the image be represented as Mean $= b_2 b_1 b_0 \cdot b_{-1} b_{-2} b_{-3} \ldots$. Thus, the derived mean can be stated as $m = b_{-1} b_{-2} b_{-3}$. This is used to test the authenticity of the image. Any changes in the encrypted image made by the adversary would lead to a mismatch in the key. In such cases, the received image is discarded and a request for retransmission is generated.

3.4 Decryption

Decryption operation was done by first performing the reverse of diffusion operation and then reversing the confusion operation. The two portions of the image were extracted by performing AND operation and were combined to get a single image. The XOR operation was performed on the image to get the original intelligible image. Figure 3 displays the example of encryption, permutation and decryption operation using a single house image of dimension 512×512. The intelligible information relating to the plain image was completely lost in the permutation/confusion stage itself. Further, diffusion was performed on the permuted image for getting the encrypted image. The decrypted image was obtained by carrying out the encryption procedure in reverse order. Images of good quality were obtained in the decryption process.

(a) House image   (b) Permuted house image

(c) Encrypted house image   (d) Decrypted house image

Fig. 3 Example showing image encryption and decryption process

4 Performance Analyses

Performance-based on the security level of the algorithm was conducted on various test images to substantiate the image encryption scheme. The images used for testing were 8-bit grayscale of dimension 512×512. The images were tested using standard test images and also using the SIPI database. The other experiments were carried out using the MATLAB platform. The efficacy of the algorithm against statistical attacks was tested using global entropy, local entropy, histogram and correlation coefficient analysis. The key space and key sensitivity analysis were performed to test the strength of the algorithm against brute force attacks. The quality of the decrypted image was assessed using the robustness against occlusion attacks and noise attacks.

4.1   Entropy

The entropy of the encrypted image was computed using global and local entropy measures. The global entropy was calculated directly from the image and the local entropy was calculated by first splitting the encrypted images into blocks. The entropy corresponding to each block was calculated and averaged over some random blocks. Entropy is a measure to quantify randomness. For a gray scale image with 256 Gy levels, the Shannon entropy is represented as,

$$H(I) = -\sum_{i=0}^{255} p(I_i)\log_2 p(I_i) \tag{4}$$

where $p(I_i)$ depicts the probability of symbol I. The maximum value of entropy was seen as 8. Entropy is computed for a different combination of images, and the measured entropy of cipher image was found to be almost close to 8 (7.9994) as shown in Fig. 4. Therefore, the proposed algorithm is highly resilient towards Entropy attacks.

The global Shannon entropy did provide accurate results about the randomness of the image when tested on some special images. Therefore, the local entropy measure was also used for testing the randomness measure. A random choice of c disjoint chunks of image $B_1, B_2, \ldots B_c$ each with T_b pixels was made for an encrypted image EI. The local Shannon entropy can be mathematically described as,

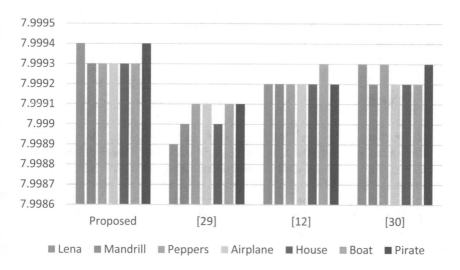

Fig. 4 The measure of entropy for different images in comparison with existing techniques

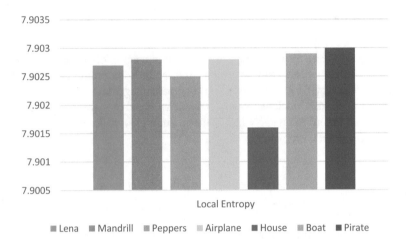

Fig. 5 The measure of local entropy for different images using the proposed technique

$$\overline{H_{k,T_b}}(EI) = \sum_{i=1}^{c} \frac{H(B_i)}{c} \tag{5}$$

where $H(B_i)$ is the entropy corresponding to each chunk of the encrypted image chosen at random and can be calculated via Eq. 5. The parameters c and T_b are chosen to be 30 and 1936 [34]. The local entropy scores were found to be greater than 7.90 as shown in Fig. 5 for images of dimension 512×512. Thus, the proposed encryption algorithm provides high randomness which in turn is capable of resisting cyber-attacks.

4.2 Histogram

The distributions of grey levels in an image were studied using image histograms. The attacker cannot succeed extracting useful statistical information if the distribution is uniform. The test images used in the simulation proposed by the researcher were of dimension 512×512. Figure 6 is the histogram plot of the plain and ciphered images. The encrypted image histogram shows a relatively uniform distribution with the implication of the inability to extract plain text information from statistical data. Thus, the encryption scheme is highly resilient towards statistical attacks and can be protected from cyber-attacks.

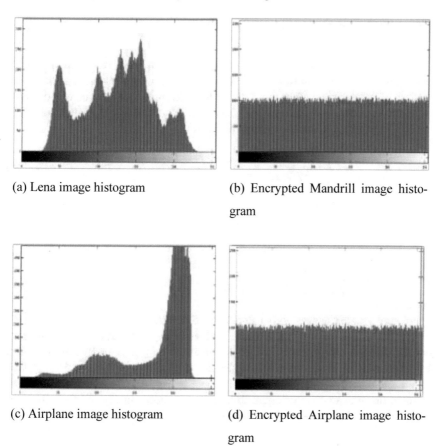

(a) Lena image histogram

(b) Encrypted Mandrill image histogram

(c) Airplane image histogram

(d) Encrypted Airplane image histogram

Fig. 6 Example showing histogram of encrypted and decrypted image

4.3 Keyspace Analysis

The set of all possible keys used for the generation of cipher image represent the key space. The encryption scheme comprises a key space of $\approx 10^{12} \times 10^{12} \times 10^{12} \times 10^{12}$ relating to the four logistic control parameters, the secret keys $(c_1, c_2, c_3) \approx 10^4 \times 10^4 \times 10^4$ and for diffusion process using chaotic equations the key space used is $2^{256 \times 256 \times 16}$. Hence, the key space adds up to $\approx 2^{199} \times 2^{256 \times 4096}$ and is large enough to avoid exhaustive search attacks and can withstand cyber search attacks.

4.4   UACI and NPCR Analysis

Randomness tests, namely, unified averaged changed intensity ($UACI$) and the
number of changing pixel rate ($NPCR$) were used for the evaluation of the effectiveness of the designed algorithm relating to differential attacks. The two ciphered
images EI_1 and EI_2 that differ by one pixel may be considered. The values of UACI
and NPCR were calculated using Eqs. 6 and 7.

$$\text{UACI} = \frac{1}{W \times H} \sum_{i,j} \frac{|EI_1(i,j) - EI_2(i,j)|}{255} \times 100\% \tag{6}$$

$$\text{NPCR} = \sum_{i,j} \frac{S(i,j)}{W \times H} \times 100\% \tag{7}$$

where $S(i,j) = \begin{cases} 1, EI_1(i,j) \neq EI_2(i,j) \\ 0, EI_1(i,j) = EI_2(i,j) \end{cases}$ and $W \times H$ is the dimension of images
EI_1 and EI_2.

For a grayscale image of dimension 512×512 that is represented by 8-bits, with
the significance level 0.05, the expected values of UACI should lie between 33.37
and 33.55% and NPCR should be greater than the critical value 99.59% [35]. For
the proposed algorithms, the UACI values lie in the range of 33.39–33.50% and are
within the expected range of values as shown in Fig. 8. Also, the $NPCR$ values
obtained are greater than or equal to 99.60% as shown in Fig. 7. Therefore, the
proposed algorithm is resistant to cyber-security based differential attacks.

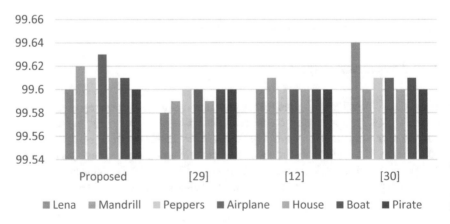

Fig. 7 NPCR % for proposed and existing techniques

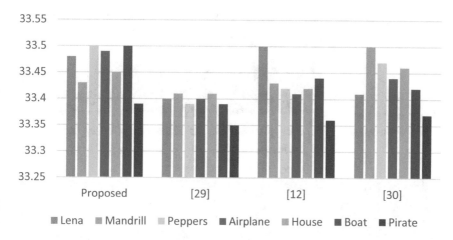

Fig. 8 UACI % for proposed and existing techniques

4.5 Known and Chosen Plain Text Attack

Key generation depends on the initial seed key, chaotic convolution and also on the dimension of the plain image. Therefore, it is very difficult for an invader to extract the key using special plain images. Figure 9 shows the encrypted images and the histogram of special images. The figure shows a good sign of randomness. Therefore, the designed algorithm is robust to plain and cipher-text attacks.

4.6 Correlation

The adjacent pixels of plain intelligible images have a high correlation. Therefore, the cipher image should be highly de-correlated for preventing statistical attacks. Let, u and k represent gray values of two nearby pixels in an image. The correlations between nearby pixels are calculated using the Eqs. (8)–(11).

$$E(u) = \frac{1}{M} \sum_{i=1}^{M} u_i \tag{8}$$

$$V(u) = \frac{1}{M} \sum_{i=1}^{M} (u_i - E(u))^2 \tag{9}$$

$$\text{covariance}(u, k) = \frac{1}{M} \sum_{i=1}^{M} (u_i - E(u))(k_i - E(k)) \tag{10}$$

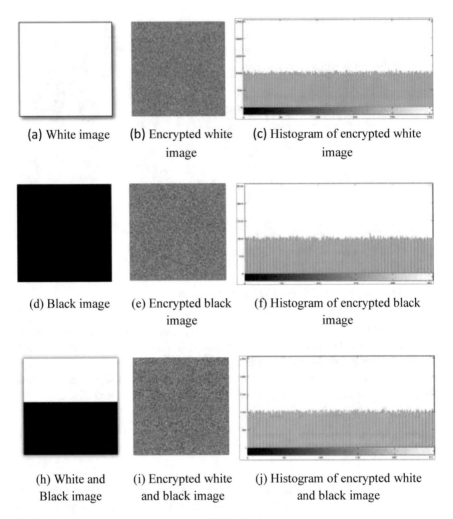

Fig. 9 Special images, encrypted output and its histogram

$$\gamma_{uk} = \frac{\text{covariance}(u, k)}{\sqrt{V(u)}\sqrt{V(k)}} \qquad (11)$$

The correlation coefficient values of the original images of dimension 512×512, permuted and combined image and the cipher image are verified. The correlation coefficient values are found to be very low and is close to 0, indicating de-correlation between all the pixel values of the encrypted image. Figure 10 represent correlation plot for the plain, permuted-combined, and cipher images. The graphs depict the coefficients of cipher image as scattered throughout the pixel range for the encrypted image. Therefore, the cipher image remained uncorrelated and protected from statistical based cyber-attacks.

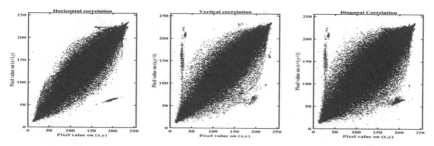

(a) Correlation coefficient plot of house image (horizontal, vertical, diagonal)

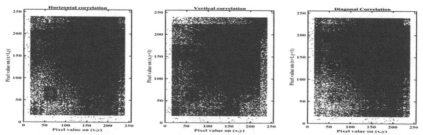

(b) Correlation coefficient plot of permuted house image (horizontal, vertical, diagonal)

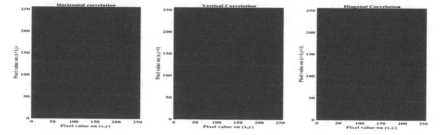

(c) Correlation coefficient of encrypted house image (horizontal, vertical, diagonal)

Fig. 10 Plot of correlation for original, permuted and encrypted house image

4.7 Key Sensitivity Analysis

Efficient encryption schemes have high sensitivity to slight variations in the key. The encrypted dual image was decrypted for the estimation of key sensitivity changing a single bit of the seed key. The results depict the recovery of original images only with a correct key, and an unintelligible image obtained with a wrong key as represented in Fig. 11.

(a) Decrypted Lena Image   (b) Decrypted House image

Fig. 11 Key sensitivity tests with key changed by 1-bit

4.8   Occlusion Attack Analysis

Some portions of the encrypted image pixels were replaced with zeros and the correct
decryption keys for reconstruction of original images was used for the purpose of
testing the robustness against occlusion attack for image encryption scheme. PSNR
values of the occluded decrypted images for image encryption scheme is as greater
than 30 dB even after occlusion, implying that the information is not completely lost
and the images obtained were with lesser data loss. The reason for obtaining good
quality decrypted images is that the pixels were shuffled well.

The ½ occluded and ¼ occluded cipher and deciphered images are shown in
Fig. 12. The figure shows the content of the images as clearly visible after decryption,
even though some portions of the encrypted images are lost. The quality of the
decrypted images obtained was visually good and the PSNR values calculated were
all greater than 30 dB [36]. Therefore, the proposed technique was considered reliable
in transmission losses.

4.9   Noise Attack Analysis

When the encrypted images are transmitted via a transmission channel there is a
possibility of the encrypted images facing noise attacks. There are many kinds of
noises of which impulse noise was considered for the test. The impulse noise can be
added intentionally or accidentally. It appears accidentally during image transmission
due to channel interference and atmospheric disturbances. The impulse noise of noise
density 0.01 was added to the encrypted images for the purpose of testing and was
decrypted using the correct keys and the PSNR values were calculated. The PSNR
values obtained are all seen as greater than 30 dB and the visual quality of the images
was maintained. Figure 13 shows the impulse noise attacked encrypted image and
the decrypted image. The information present in the image could be seen despite

(a) ½ Occluded Encrypted Image   (b) Decrypted Lena Image

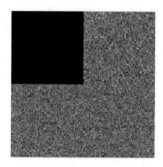

(c) ¼ Occluded Encrypted house   (d) Decrypted house image

Fig. 12 Example of ½ and ¼ occluded encrypted and decrypted image

(a) Encrypted house image   (b) Decrypted house image
added with noise

Fig. 13 Example of noise attack analysis

being affected by noise. This analysis also helps in the verification of the robustness of the encryption algorithms proposed.

5 Conclusions and Future Work

In this chapter, a key generation method and a secure image encryption algorithm used for the development of cyber security has been presented. The performance of the encryption algorithm evaluated in terms of entropy, key space, occlusion and sensitivity depicts the strength of the algorithm. Besides, the algorithm presents excellent robustness to noise and occlusion attacks making it more suitable for real-time applications. Hence, the proposed technique is useful in cyber security based storage, transmission and also military applications where transmission losses occur due to signal coverage problems in the recovery of the information without compromising on the security level and robustness against attacks. Implementation of suitable encryption algorithms to prevent cyber-attacks in various services such as voice, video, and 3D images related to Artificial Intelligence applications can be explored as future work.

References

1. Nikheta Reddy G, Ugander GJ (2018) A study of cyber security challenges and its emerging trends on latest technologies. Int J Eng Res 5:1–5
2. Jasper SE (2017) US cyber threat intelligence sharing frameworks. Int J Intell Count Intell 30:53–65
3. Kizza JM (2005) Computer network security. Springer, New York
4. Abayomi-Alli OO, Damaševičius R, Maskeliūnas R, Misra S (2021) Few-shot learning with a novel Voronoi tessellation-based image augmentation method for facial palsy detection. Electronics 10(8):978
5. Abayomi-Alli A, Atinuke O, Onashoga SA, Misra S, Abayomi-Alli O (2020) Facial image quality assessment using an en-semble of pre-trained deep learning models (EFQnet). In: 2020 20th international conference on computational science and its applications (ICCSA), pp 1–8
6. Yang C, Sun Y, Wu Q (2015) Batch attribute-based encryption for secure clouds. Information 6:704–718
7. El Assad S, Farajallah M (2016) A new chaos—based image encryption system. Signal Process Image Commun 41:144–157
8. Emmanuel S, Thomas T, Vijayaraghavan AP (2020) Machine learning and cybersecurity. In: Machine learning approaches in cyber security analytics. Springer, Singapore, pp 37–47
9. Hua Z, Zhou Y (2016) Image encryption using 2D Logistic-adjusted—Sine map. Inf Sci 339:237–253
10. Jin J (2012) An image encryption based on elementary cellular automata. Opt Lasers Eng 50:1836–1843
11. Li C, Luo G, Qin K (2017) An image encryption scheme based on chaotic tent map. Nonlinear Dyn 87:127–133
12. Souyah A, Faraoun KM (2016) Fast and efficient randomized encryption scheme for digital images based on quadtree decomposition and reversible memory cellular automata. Nonlinear Dyn 715–732

13. Xiao D, Fu Q, Xiang T, Zhang Y (2016) Chaotic image encryption of regions of interest. Int J Bifurc Chaos 26(11):1650193
14. Ye G, Huang X (2016) A secure image encryption algorithm based on chaotic maps and SHA-3. Secur Commun Netw 9:2015–2023
15. Ejbali R, Zaied M (2017) Image encryption based on new Beta chaotic maps. Opt Lasers Eng 96:39–49
16. Fang D, Sun S (2020) A new secure image encryption algorithm based on a 5D hyperchaotic map. Plos One 15:e0242110
17. Huang X, Liu J, Ma J, Xiang Y, Zhou W (2019) Data authentication with privacy protection. In: Advances in cyber security: principles, techniques, and applications. Springer, Singapore, pp 115–142
18. Chan CS (2011) An image authentication method by applying Hamming code on rearranged bits. Pattern Recogn Lett 32:1679–1690
19. Lo CC, Hu YC (2014) A novel reversible image authentication scheme for digital images. Signal Process 174–185
20. Skraparlis D (2003) Design of an efficient authentication method for modern image and video. IEEE Trans Consum Electron 49(2):417–426
21. Tabatabaei SA, Ur-Rehman O, Zivic N, Ruland C (2015) Secure and robust two-phase image authentication. IEEE Trans Multimed 17(7):945–956
22. Wu WC (2017) Quantization-based image authentication scheme using QR error correction. EURASIP J Image Video Process 1–12
23. Rachael O, Misra S, Ahuja R, Adewumi A, Ayeni F, Mmaskeliunas R (2020) Image steganography and steganalysis based on least significant bit (LSB). In: Proceedings of ICETIT. Springer, Cham, pp 1100–1111
24. Tao L, Baoxiang D, Xiaowen L (2020) Image encryption algorithm based on logistic and two-dimensional Lorenz emerging approaches to cyber security. IEEE Access 8:13792–13805
25. Mohamed ZT, Xingyuan W, Midoun MA (2021) Fast image encryption algorithm with high security level using the Bülban chaotic map. J Real-Time Image Proc 18:85–98
26. Yaghoub P, Ranjbarzadeh R, Mardani A (2021) A new algorithm for digital image encryption based on chaos theory. Entropy 23:341
27. Bisht A, Dua M, Dua S, Jaroli P (2020) A color image encryption technique based on bit-level permutation and alternate logistic maps. J Intell Syst 29:1246–1260
28. Maimut D, Reyhanitabar R (2014) Authenticated encryption: toward next-generation algorithms. IEEE Secur Priv 12(2):70–72
29. Hanis S, Amutha R (2018) Double image compression and encryption scheme using logistic mapped convolution and cellular automata. Multimed Tools Appl 77:6897–6912
30. Misra S (2021) A step by step guide for choosing project topics and writing research papers in ICT related disciplines. In: Misra S., Muhammad-Bello B (eds) Information and communication technology and applications. ICTA 2020. Communications in computer and in-formation science. Springer, vol 1350, pp 727–744. https://doi.org/10.1007/978-3-030-69143-1_55
31. Hanis S, Amutha R (2019) A fast double-keyed authenticated image encryption scheme using an improved chaotic map and a butterfly-like structure. Nonlinear Dyn 95:421–432
32. Mount DM, Kanungo T, Nathan SN, Piatko C, Silverman R, Ange-la YW (2001) Approximating large convolutions in digital images. IEEE Trans Image Process 10:1826–1835
33. Hunt BR (1971) A matrix theory proof of the discrete convolution theorem. IEEE Trans. Audio Electro acoust 19:285–288
34. Yue W, Yicong Z, George S, Sos A, Joseph P, Premkumar N (2013) Local Shannon entropy measure with statistical tests for image randomness. Inf Sci 222:323–342
35. Wu Y, Joseph PN, Agaian S (2011) NPCR and UACI randomness test for image encryption. Cyber J: Multidiscip J Sci Technol J Select Areas Telecommun 31–38
36. Timothy Shih K (2002) Distributed multimedia databases: techniques and applications. Idea group, USA

Machine Learning in Automated Detection of Ransomware: Scope, Benefits and Challenges

Vani Thangapandian ⓘ

Abstract **Background** Ransomware is a special kind of malware which is rapidly blooming around the world in different forms. In recent times, Ransomware plays havoc in individual and corporate systems heavily and claimed abundant amount of money as ransom in the form of crypto currency. And it's growth is galloping in fast pace due to the Ransomware-as-a-service facility. So it is imperative to mitigate ransomware and its attacks on an emergency basis. **Aim** The objective of this work is to study about the research works exclusively done for ransomware attacks and to analyze the scope and challenges of Machine Learning methods in ransomware detection. **Methodology** The research works exclusively aimed at the mitigation of ransomware are collected from various renowned research databases and a systematic literature study is performed based on the traits of ransomware, data sets and methods, various performance measures used in the implementation of detection models. **Results** Many detection models that are developed with high accuracy have been discussed. Out of them, most of the models employ Machine Learning techniques for detection of ransomware as it facilitates automated detection. The proportion of the count (37.5%) of Machine Learning based models is considerably higher than that of other models (3% each).The vital role of Machine Learning in developing automated detection tool is reviewed from different perspectives and the limitations of Machine Language based model are also discussed. **Conclusion** Based on the survey, Machine Learning methods can be applied to develop automated detection tool if the challenges are properly addressed. This will be helpful to the researchers to build a comprehensive and efficient model for ransomware detection, based on Machine Learning.

Keywords Ransomware detection · Automated detection · Ransomware attacks · Detection parameter · Anti ransomware tool · Malicious code · Malware · Machine Learning · Random forest · Support vector machine.

V. Thangapandian (✉)
Rajeswari Vedachalam Government Arts College, University of Madras,
Chennai, TamilNadu, India

© The Author(s), under exclusive license to Springer Nature Switzerland AG 2022
S. Misra and C. Arumugam (eds.), *Illumination of Artificial Intelligence in Cybersecurity and Forensics*, Lecture Notes on Data Engineering and Communications Technologies 109, https://doi.org/10.1007/978-3-030-93453-8_15

345

1 Introduction

Ransomware is a dreadfully enhanced malware, and of late, it spawns into multiple families [33, 76]. The strength of Ransomware is the use of powerful cryptographic techniques, which accelerates the sudden growth of ransomware families. Though Ransomware is highly dreadful [34, 71], it hails from the malware family, and it shares many features from the other Malware discussed here. The software which is written with malicious intent is termed Malware. Malware is an umbrella term covering a set of different malicious codes capable of causing destruction uniquely. The malware family includes Virus, Worm, Trojan Horse, Logic Bomb, Hacktool, Rabbit, Rootkit, Backdoor, Spyware, and Ransomware. Each of the above has a different mechanism and unique style of destruction on the target. The target can be a single system, a network of systems or the entire corporate network.

The Ransomware has all the typical malware characteristics, including a backdoor approach, spyware, stealthy virus, and dreadful Rootkit [67]. The uniqueness of Ransomware which differentiates it from the family of Malware, is extortion. There are two broad categories of Ransomware, namely Locky Ransomware and Crypto-Ransomware. It first traces the vulnerabilities of the victim system stealthily. Once it gets the vulnerabilities successfully, it enters into the system through the backdoor approach. Eventually, it starts to take the system under its control. So it gets into the rootkits and locks the system by disrupting the accessibility to the user. Then, it will start to encrypt the files using a solid cryptographic technique if it is a crypto type. As a final step, it displays the warning message to the user regarding the ransom to be paid to claim the access control back by the user. In some ransomware, the ransom payment is scheduled to a particular time, beyond which the files/ data will be inaccessible to the user if payment has defaulted. In some extreme cases, the data will be deleted depending on the brutality of the Ransomware.

To aim ransomware attack, the attackers first identify the victim system and the list of vulnerabilities in that system. Generally, these vulnerabilities are the potential traits [84, 98] for executing a successful ransomware attack in a system. There are two types of traits [76] namely Static traits and Dynamic traits, based on the changing nature of these vulnerabilities. The static traits are classified by the unchanging values such as file names, file extensions, etc. On the other hand, the dynamic traits are classified by the behavioral patterns of the components such as API calls, Registry values, etc. Using any one of these traits, the attackers aim to compromise the security of the system. Hence these traits pose severe threats to the system. Figure 1 depicts the set of traits usually existing in a system.

The disruptions of Ransomware is appalling, and it poses severe threats to individuals, networks, the corporate world and society. Due to the pandemic situation, almost all the corporate and other organizations shifted their work base from offices to home to disconnect the chain of the spread of disease. Furthermore, everything has to be shifted suddenly without any prior plans, arrangements and schedules. It creates more vulnerability in the system, which is the inviting point for all such dreadful Ransomware. Apart from this, many other ransomware enablers also exist, including

Fig. 1 Potential traits

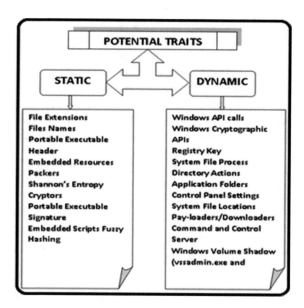

the availability of ransomware development kits, simple cryptographic techniques [10, 61], RaaS cloud services, untraceable payment methods, etc. Many research works have been carried out around the world to mitigate ransomware attacks. Many detection models have been designed, developed and tested with different data sets. These models are deployed using many detection mechanisms like Software Defined Network [11, 31, 40, 89], situational awareness model [91], sensor-based model [98], API logs [29, 95, 98, 103], web logs [91], pattern matching based model [43], Finite state machine [50, 85] and many Machine Learning-based models etc. Among these models, Machine Learning based models are comparatively extensively developed.

This paper is proposed as a Systematic Literature Review, and it discusses the different ransomware detection tools developed so far and highlights the strengths and weaknesses of Machine Learning-based detection tools. Past research works include several systematic literature reviews, but their focus of the study is different. Maigida et al. [76] presented a systematic literature review and metadata analysis on ransomware attacks. The main focus of this study is on how ransomware attacks are executed and the possible solutions to mitigate such attacks. Fernando et al. [20, 46, 48, 89, 110] described a literature study work on Ransomware, particularly in IoT platforms. The main focus of this work is the application of various machine learning and deep learning methods [5, 12, 15, 51, 78, 90, 91, 107] in detecting Ransomware. The authors described the working of each method in detail.

In the research work [44], the authors presented an evaluation work for ransomware detection in which they designed a platform for conducting experiments. They chose supervised learning methods to be tested against Ransomware in this platform and compared their performance. Their model accepts only a specific set of files only. Berrueta et al. [26] described an evaluation work in which the authors presented an open repository of ransomware datasets. The repository contains updated ransomware samples that can be utilized for comparing the performances of the existing ransomware detection tools. The main focus of the work is to analyze the need and importance of this datasets repository. The proposed work is produced as a systematic literature review [66] and discusses the inevitable role of Machine Learning methods in detecting Ransomware and the challenges or limitations of these models. This work does not include the research on generalized malware detection models and exclusively analyzes the ransomware detection models.

Section 2 discusses the relevant research works in the topic, Sect. 3 describes the methodology implemented in this paper, Sect. 4 discusses the results, Sect. 5 discusses the limitations, and Sect. 6 provides the conclusion and future scope of the study.

2 Background Study

Since last decade, many number of research works have been carried out around the world. Before the existence of ransomware attacks, millions of research works have been performed to develop models for malware detection and mitigation. As many features are common between ransomware and other malware, those models can be utilized for ransomware detection also. But ransomware is unique and different from other malware as the one and only motto of ransomware is extortion. The ransomware attacks are more destructive due to the vital features [7, 43, 108] of a ransomware (See Table 1).

The ransomware attack problem needs to be immediately addressed as the attack is time bound. Beyond that, the resources under attack may become unavailable or deleted permanently. To avoid such risks, the ransomware mitigation models have to be designed with main focus on ransomware features and these models have to be validated using ransomware attack related datasets. Hence more research works are focused on exclusive ransomware attacks.

The Fig. 2 shows that the trend of research work proposal on ransomware is exponentially growing in the past 5 years as the attacks are increasing proportionately.For the mitigation and analysis of ransomware attacks, many research works have been carried out in the past. Many authors focused on devising a ransomware detection model while others have tried to review the methodologies and techniques. Among the works, many are based on Machine Learning based models. The brief summary of the Machine Learning research works have been given in Table 2.

3 Methodology

To conduct the systematic literature study, Kichenham [66] model is followed and the systematic survey is conducted in various phases as follows.

Table 1 Vitality of a ransomware

S/N	Feature	Description
1	Inviolable encryption	Ransomware applies strong encryption techniques which is almost impossible to be decoded by any decryption tools
2	Multifarious encryption tools	The encryption techniques used by ransomware are powerful enough to encrypt any kind of files including audio, video, image, etc.
3	Obfuscatory extensions	Once the victim's data is captured by ransomware, it scrambles the file names and extension to confuse the victim
4	Impenetrable control message	The ransomware seizes all the data of victim system and encrypts. Once all the data of victim is encrypted, it will issue a control message demanding ransom amount in the form of a text message or image. This control message is impossible to pass through or ignore until the ransom is paid. And the message is also be displayed in the victim's regional language
5	Untraceable payment mode	Usually ransom amount is demanded in bitcoins as bitcoin transactions cannot be traced
6	Time bound payment	The ransom payment has to be done within the time limit mentioned in the control message. If the victim fails to do so, either the amount will be increased or this data will be lost forever
7	Extendable threats	If the victim's system is connected in a network, all other systems connected in the network also have strong vulnerability of the same attack
8	Complex set of exfiltration techniques	Ransomwares apply a complex set of data exfiltration techniques to acquire password, usernames, mail ID and other sensitive data from the victim's system
9	Geographical targets	In some cases, the ransomware Attack is aimed at systems from particular geographical area

3.1 Sources of Research Data

The relevant research papers are collected from various research repositories such as IEEE Explore, Springer Link, ACM Digital Library, MDPI, Science Direct, Web of Science, Google Scholar, etc. However, only the abstract of the research works is available in some libraries, and the complete texts have to be obtained from the journal web pages.

Fig. 2 Search crititeria

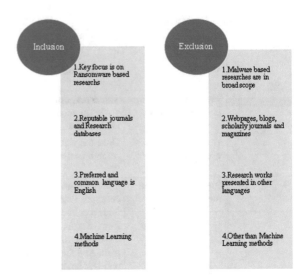

3.2 Search Key Phrase

Appropriate search strings have to be used to collect the research data. It will narrow down the search spectrum, and the works related to the objective will be collected. The key phrases include 'Ransomware', 'Ransomware + detection', 'Ransomware + Machine Learning' and 'Ransomware + Automated detection'. These phrases help to retrieve the necessary research data efficiently from the voluminous data sources.

3.3 Search Criteria

The research data collected using the key phrases will include all types of data, some of which may be less useful and out of the scope of the study. Therefore, search criteria have to be decided to avoid this. The search criteria describe Inclusion and Exclusion principles for segregating the collected data. These principles are shown in the diagram (see Fig. 2).

In some cases, the search criteria for inclusion and exclusion have to be refined further to eliminate unpublished works, duplicate copies, and irrelevant contents, i.e., the contents of the work may be beyond the scope of the study.

Table 2 Machine Learning based ransomware research

S/N	Authors	Methods	Findings
1	Homayoun et al. [56]	ML-J48, RF, Bagging, MLP	Detection Classification
2	Daku et al. [42]	J48 DT	Classification
3	Medhal et al. [79]	ML based framework	Detection
4	Alhawi et al. [9]	Network Traffic	Detection
5	Su et al. [100]	SVM,DT, RF and LR	Detection
6	Azmoodeh et al. [20]	KNN,NN, SVM and RF	Detection
7	Harikrishnan and Soman [53]	Supervised learning	Detection Classification
8	Cusack et al. [40]	ML based SDN	Early detection
9	Verma et al. [106]	SVM, LDA, QDA,Complex tree, KNN.	Real time detection
10	Lachtar et al. [70]	RF, SVM, SVM,KNN, ANN	Detection
11	Lee et al. [71]	KNN, DT Ensemble, NN	Detection
12	Adamu et al. [1]	ML Techniques	Detection Classification
13	Bae et al. [21]	RF, LR, NB,SGD, KNN,SVM	Detection Classification
14	Zhang et al. [113]	DT,RF, KNN, NB GBDT	Detection Classification
15	Poudyal et al. [85]	NB, LR, SVM, RF and DT	Detection Framework
16	Qian et al. [36]	ML	Early detection
17	Li et al. [72]	GAN	To build a resilient Machine Learning model for ransomware detection, they generated malicious samples and measured the performance of the Machine learning methods. The non-linear boundary of SVM-radial -CNN detects the adversarial samples effectively
18	Chen Li et al. [35]	Augmented SVM	Detection
19	Ahmed et al. [3]	DT, KNN, LR, RF and SVM	Detection
20	Zuhair et al. [115]	Multitier analytics	Detection Classification
21	Turner et al. [102]	Bitcoin Intelligence framework	Detection Classification
22	Lorenzo et al. [47]	OC-SVM	Detection Classification

(continued)

Table 2 (continued)

S/N	Authors	Methods	Findings
23	Hwang et al. [59]	ML	Early detection
24	Kok et al. [67]	ML	Early prediction
25	Manzano et al. [78]	ML	Empirical results show that random forest (RF) achieved a 96% accuracy in classifying ransomware, higher than decision tree (DT) and K-nearest neighbor (KNN)
26	Zuhair et al. [114]	ML	The empirical findings highlight the overlooked issues along with the perspectives that will boost future achievements in this domain
27	Dion et al. [44]	RF, GBDT, NN using MLP and SVM	Random Forest, GBDT and SVM have shown optimal results in detection of ransomware
28	Fernando et al. [46]	ML	The experiments identified possibledefects in the ML methods and suggested that the longevity shouldbe taken into consideration when ML algorithms are trained
29	UshaRani et al. [105]	RF, TB, SVM	Classification
30	Reddy et al. [87]	ML	Early detection
31	Borah et al. [30]	Ensemble	Early detection

3.4 Data Collection

Once the search criteria are set, the necessary research data are obtained from the research repositories [80] mentioned in Sect. 3.1. The primary considerations are the search key phrases, time frame, and awareness of unbiased works during the research data retrieval. The data has been retrieved for a particular time (2017–2021). In total, 184 research papers were collected exclusively for Ransomware. Ransomware is a notable kind of Malware, and Malware is an umbrella term covering different types of malicious logic. Hence, the term malware is excluded in the search criteria, and the critical term 'ransomware' is explicitly added.

3.5 Selection of Relevant Research Works

The data collected contains both relevant and redundant works. The redundant and obsolete works have to be eliminated from the collection by further refinement. During the data collection phase, 184 research papers are selected that fulfill this study's search criteria and objective. Irrelevant content, other language research works, scholarly research journals, and redundant research works are eliminated to refine this collection. The focus of this work is on Machine Learning-based research works for ransomware mitigation. Hence those research papers are collected (34 papers in total).

Further refinement is to select only those papers from reputed journals such as Springer, Elsevier, IEEE, ACM, Taylor and Francis, MDPI, etc., to improve the quality of the collection. Hence the count of such papers comes down to 32 (see Fig. 3). The papers are collected from the year 2017 to the mid of the year 2021. Hence the number of papers for the year 2021, will be less when compared with the count of the year 2020. These selected papers are taken for study and analysis about the Machine Learning-based ransomware detection model.

4 Results and Discussion

4.1 Ransomware Detection Models

Ransomware is a severe threat, and immediate mitigation is the need of the hour. Ransomware can be mitigated in three ways, namely Ransomware Prediction, Detection, and Prevention [13, 14, 55, 56, 68]. Each method involves a set of strategies and procedures. In some research works, Prediction and Detection are considered interchangeable. In the Prediction method, the various possibilities for ransomware attacks are studied and identified in a system.

The objective of this method is to keep the system under the scan always for monitoring suspicious activities or triggers in the system. Specific software models are built to capture the ransomware activities before the extortion starts in the detection method. These detection models are trained and tested with the data and signatures of previous attacks to make accurate detection successfully. Many such models are depicted in Fig. 3. Finally, in the Preventions method, the system is secured tightly to stay away from vulnerabilities by a periodical update of software, including Operating system, anti-virus kits, patch management, firewall, and periodical backup. Among the three methods, detection is more significant, and many research works are on it.

To detect Ransomware in the early stages, many researchers employ different forms of detection parameters [52, 98]. The detection parameters are the sensitive entities in systems that the Ransomware misuses to execute their attacks. The most commonly used detection parameters are API calls [29, 60, 86, 95, 98, 104] and log

RESEARCH SURVEY

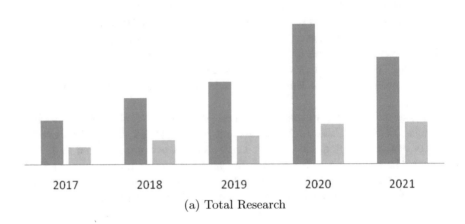

(a) Total Research

PAPERS FROM REPUTED JOURNALS

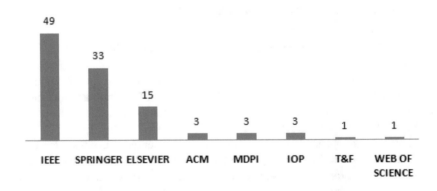

(b) ML based research

Fig. 3 Results of search criteria

files [71]. The Ransomware executes its sequence of instructions by taking the API under its control. Hence in many ransomware detection models, the API call is used as a detection parameter. The log files are the major contributor for tracing the ransomware trails in a system as it exposes the behavioral changes [8, 16, 36, 42, 88]. The other essential parameters are Registry Key, System File Process [16], Directory

Actions, Application folders, Control Panel Settings, System File Locations, Command and Control Server, Windows Volume Shadow (vssadmin.exe and WMIC.exe), File Fingerprint, Directory Listing Queries, File Extensions, Files Names, Portable Executable Header [9, 19, 52], Embedded Resources, Shannon's Entropy, Portable Executable Signature, Embedded Scripts and Fuzzy Hashing. These are the potential traits for the successful execution of a ransomware attack. Hence as a reverse engineering approach, these traits are primary detection parameters in a ransomware detection model. Other than the potential traits, some other detection parameters are also used. They are weblogs, profiling kernel activities, storage, opcode density, ransom footprint, remote desktop protocol, sensor data, payment transaction data, file paths, network activity, network traffic [9, 31, 78, 111], reused code, dropped file, etc.

The figure (see Fig. 4) shows the ransomware detection models developed in the time interval of 2017–2021. The models are developed using any one of the methodologies, which include particle swarm optimization, profile kernaling, OSS technologies, behavioral model, reused code analysis, SDN based model, file storage based model [22, 23], PR header-based model, entropy model [41, 71], parallel processing model, salp swarm technique [45, 99], OP code density analysis model [24, 62, 112, 113], PE file analysis, blockchain method, web search log, situational awareness model, API logs, sensor-based model, remote desktop protocol based model [109] and Machine Learning based models. Among all other models, Machine Learning based models are commonly developed for ransomware detection due to its versatility and robustness.

4.2   Machine Learning Based Detection Models

Machine Learning is a blooming field of Computer Science, and it employs a set of algorithms for the automation of a process in a system. It paves the way for a paradigm shift in the process of programming a problem. In traditional programming languages, the program will be coded manually by a programmer, and it is fed into the system with the set of inputs. However, in Machine Language, the system will be given a set of inputs and corresponding outputs. The program will be generated in the form of the working model by the system. The Machine Learning field achieves this with the help of the popular field of Artificial Intelligence. In this programming method, the explicit requirement is data. Therefore, data is a vital component in any Machine Learning model.

Machine Learning methods can be applied in any industry to automate complex processes, including the business world. Wherever the data flow is abundant, the Machine Learning methods can be utilized. There are three broad types of Machine Learning, namely Supervised Learning, Unsupervised Learning, and Reinforcement Learning. Under the category of Supervised Learning, there is one subcategory, viz., Semi-supervised Learning which is partially supervised and unsupervised learning.

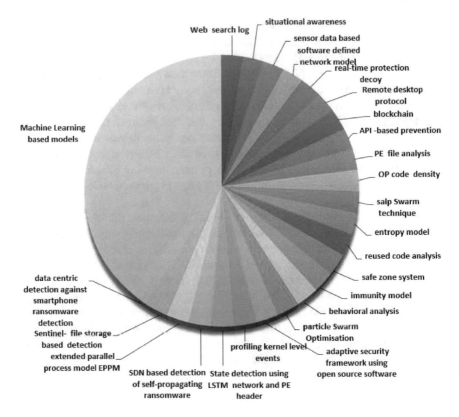

Fig. 4 Ransomware detection models [2017 to mid of 2021]

Supervised Learning is guided Learning in which the system processes the target data based on the labeled data. The labeled data is given to the system to learn about the data. This is called training the model. Based on the knowledge learned from this training, the model has to process the unlabelled data. In Unsupervised Learning, the data will not be labeled in a training session. The model has to learn it on its own. In Reinforcement learning, the unlabelled data will be given; if the model correctly processes the data, it will be encouraged by offering rewards. For correct responses, rewards will be offered and for each wrong result. Based on the rewards and penalties, the model has to learn the processes correctly.

Supervised Learning consists of two different methods, namely Regression and Classification. Linear Regression, Regression Trees, Non-Linear Regression, Bayesian Linear Regression, Polynomial Regression are examples of Regression methods. Random Forest, Decision Trees, Logistic Regression, Support Vector Machines are examples of classification.

K-means clustering, KNN (k-nearest neighbors), Hierarchal clustering, Anomaly detection, Neural Networks, Principle Component Analysis, Independent Component Analysis, Apriori algorithm, Singular value decomposition are examples of

																										ML FRAME-WORK
1	Homayoun et al. [56]	✓	✓							✓										✓						
2	May Medhal et al. [79]																									✓
3	Alhawi et al. [9]																									
4	Harikrishnan and Soman [53]																								✓
5	Daku et al. [42]		✓	✓																						
6	Verma et al. [106]					✓	✓		✓				✓												
7	Su et al. [100]		✓	✓		✓	✓		✓																
8	Azmoodeh et al. [20]		✓	✓		✓	✓			✓															
9	Cusack et al. [40]																								
10	Adamu et al. [1]																								
11	Lachtar et al. [70]		✓	✓		✓	✓		✓											✓					
12	Bae et al. [21]		✓	✓		✓	✓		✓				✓												
13	Lee et al. [71]			✓		✓			✓				✓			✓									
14	Poudyal et al. [85]		✓	✓		✓	✓		✓				✓												
15	Li et al. [72]																	✓							
16	Zhang et al. [113]		✓	✓		✓	✓																		
17	Qian et al. [36]																									✓

(continued)

Table 3 (continued)

S no	Author	ML-J48	RF	MLP	SVM	DT	LDA	KNN	LR	QDA	ANN	NB	GD	SMO	SGD	AB	GB	XGB	BAGGING	ENSEMBLE	ML FRAME-WORK
18	Chen Li and Chih Yuan et al. [35]				✓																
19	Zuhair et al. [115]			✓																	✓
20	Kok et al. [67]																				✓
21	Turner et al. [102]																				✓
22	Ahmed et al. [3]		✓		✓	✓		✓	✓												
23	Dion et al. [44]		✓	✓														✓			
24	Fernandez et al. [47]				✓																✓
25	Fernando et al. [46]																				✓
26	Hwang et al. [59]																	✓			
27	UshaRani et al. [105]		✓		✓																
28	Reddy et al. [87]																				✓
29	Borah et al. [30]																			✓	

Unsupervised Learning methods. Markov Decision Process and Q-learning are examples of Reinforcement learning. In the table (Table 3), various research works involving Machine Learning methods are listed. The most widely applied Machine Learning methods are SVM [24, 54], RF [63, 94] and DT.

4.3   Architecture of ML Based Ransomware Detection

The machine learning method is implemented in a sequence of steps [67], which are essential for building a model.

Gathering Data Machine Learning method is applied in a problem involving a large amount of data. The first step is to collect all sources of data relevant to the problem to be solved. This is a time-consuming step as all the data has to be collected.

Preparing that Data Once the data is collected, it must be prepared for processing by cleaning and pre-processing. Only then the results of the process will be accurate.

Choosing a Model Once the data is pre-processed, it is ready for processing. So the next step is to select an appropriate model.

Hyperparameter Tuning Hyperparameters are vital to control the overall behavior of a Machine Learning model. The combination of hyperparameters can obtain the best results of the model.

Evaluation After the model is constructed, its results have to be evaluated to check whether it produces accurate results.

Prediction Once the model is evaluated on test data, it must be verified with the actual data. The block diagram of a Machine Learning based ransomware detection model is depicted in Fig. 5.

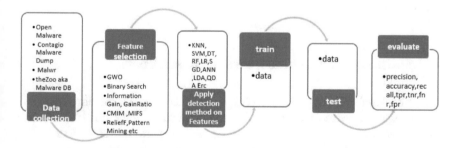

Fig. 5 ML based detection model

4.4 Performance Metrics for ML Based Ransomware Detection Models

The performance of the Machine Learning model can be evaluated by a set of parameters [39, 76] such as confusion matrix, Root-Mean-Squared Error (RMSE),

AUC-ROC, Accuracy, Precision, R-Call, TPR, TNR, FPR and FNR. A confusion matrix is a table of correct and incorrect values of the attribute under consideration. TPR (True Positive Rate) is the number of positive results which are correctly identified and TNR (True Negative Rate) is the number of negative results which are correctly identified.

FPR (False Positive Rate) is the number of negative results which are incorrectly identified as positive and FNR (False Negative Rate) is the number of positive results which are incorrectly identified as false. The performance metrics used in various Machine Learning based ransomware detection models are given in the Table 4.

4.5 Datasets for Ransomware Detection Models

There are many datasets available for implementing and testing ransomware detection models. The most commonly used datasets [98] are VirusTotal, VirusShare, Honeypot [82], Malwr.com, the Zoo, WEKA, openMalware, malware traffic-analysis.net, RISS of ICL ML data repository, samples generated by Sangfor Technologies Inc, Wallet Explorer data, http://en.softonic.com, Anzhi Market, Contagio Malware Dump, etc.

In addition, some researchers generate their own data sets for validating their ransomware detection models based on their detection criteria and parameters used. Many websites also offer Ransomware related details for research works. The following Table 5 depicts the usage statistics of various datasets. WEKA tool is also used for generating datasets for ransomware detection models. The data from the table shows that the VirusShare and VirusTotal datasets are widely used in research works.

5 Limitations of Machine Learning Based Tools

Many ML-based anti-ransomware tools have been developed using different detection strategies. These tools face some shortcomings while being applied for detecting various Ransomware of different families. Some tools perform well with the detection process in the learning phase. Though they are highly accurate in semi-realistic platforms [39], they fail to provide comprehensive detection in natural environments. The reasons for such crunch factors are listed in the Table 6.

Table 4 Performance metrics

S/N	Author	Accuracy	Precision	F-measure	FPR	TPR	Elapsed time	Miss rate	Auc	Recall
1	Homayoun et al. [56]	✓		✓	✓	✓	✓			
2	May Medhal et al. [79]	✓	✓	✓	✓					
3	Alhawi et al. [9]					✓				
4	Harikrishnan and Soman [53]	✓	✓							
5	Daku et al. [42]	✓								
6	Verma et al. [106]	✓								
7	Su et al. [100]	✓	✓	✓	✓					
8	Azmoodeh et al. [20]	✓	✓	✓						✓
9	Cusack et al. [40]	✓	✓	✓						✓
10	Adamu et al. [1]	✓								
11	Lachtar et al. [70]	✓								
12	Bae et al. [21]	✓	✓	✓			✓			
13	Lee et al. [71]								✓	
14	Poudyal et al. [85]	✓								
15	Li et al. [72]									
16	Zhang et al. [113]	✓		✓						
17	Qian et al. [36]	✓								

(continued)

Table 4 (continued)

S/N	Author	Accuracy	Precision	F-measure	FPR	TPR	Elapsed time	Miss rate	Auc	Recall
18	Chen Li and Chih Yuan et al. [35]	✓			✓					
19	Zuhair et al. [115]	✓					✓	✓		
20	Kok et al. [67]	✓								
21	Turner et al. [102]									
22	Ahmed et al. [3]	✓	✓	✓	✓		✓			
23	Dion et al. [44]									
24	Fernandez et al. [47]	✓	✓				✓			
25	Fernando et al. [46]				✓					
26	Hwang et al. [59]				✓		✓			
27	UshaRani et al. [105]	✓				✓				
28	Reddy et al. [87]	✓								
29	Borah et al. [30]	✓								

Table 5 Data sets usage statistics

S/N	Research works	Virus total	WEKA	Open malware	Contagio malware	Malwr	TheZoo	Virus share	Malware traffic	RISS of ICL	Sangfor	Wallet explorer	Softonic	Honey pot	Anzhi
1	Homayoun et al. [56]	✓													
2	May Medhal et al. [79]														
3	Alhawi et al. [9]	✓	✓												
4	Harikrishnan and Soman [53]	✓		✓	✓	✓	✓	✓							
5	Daku et al. [42]	✓													
6	Verma et al. [106]					✓		✓							
7	Su et al. [100]	✓													✓
8	Azmoodeh et al. [20]														
9	Cusack et al. [40]								✓						
10	Adamu et al. [1]									✓					
11	Lachtar et al. [70]														
12	Bae et al. [21]	✓													
13	Lee et al. [71]						✓								
14	Poudyal et al. [85]	✓													
15	Li et al. [72]														

(continued)

Table 5 (continued)

S/N	Research works	Virus total	WEKA	Open malware	Contagio malware	Malwr	TheZoo	Virus share	Malware traffic	RISS of ICL	Sangfor	Wallet explorer	Softonic	Honey pot	Anzhi
16	Zhang et al. [113]														
17	Qian et al. [36]														
18	Chen Li and Chih Yuan et al. [35]	✓													
19	Zuhair et al. [115]														
20	Kok et al. [67]						✓								
21	Turner et al. [102]											✓			
22	Ahmed et al. [3]	✓						✓							
23	Dion et al. [44]														
24	Fernandez et al. [47]														
25	Fernando et al. [46]	✓													
26	Hwang et al. [59]							✓					✓		
27	UshaRani et al. [105]													✓	
28	Reddy et al. [87]														
29	Borah et al. [30]														

Table 6 Limitations and Suggestions

S/N	Factor	Description
1	Static traits	Many ML based tools have not incorporated the complete set of static traits for mitigating ransomware threats. In some cases, the ransomwares apply complicated techniques to intrude the victim's system which is a big challenge to the ML based anti-ransomware tools for detection by static traits
2	Dynamic traits	Anti-ransomware tools using dynamic trait are more efficient in detection than those with static traits. The dynamic traits are normally classified based on behavioral patterns of the systems, resource usage patterns, events execution patterns. In general, the successful ransomware attack forms a chain of infection in the above mentioned dynamic traits. This is the place where the ML based anti-ransomware tools failed to produce better results in real environment
3	Multi-descent versions	Ransomwares are evolving in different families with different signatures. The newer versions of ransomware are emerging with common traits which are also inter-related both in static and dynamic traits. Due to this multi-descent nature, the ransomware detection is complex as well as a challenging task. They also lead to false positive results in many cases which compromise the results accuracy of the Machine Learning models
4	Source of data	To get accurate detection results, the entire process solely depends on the training data fed into the detection model. If the data is not comprehensive in nature, it will compromise the prediction accuracy of the models. Many datasets are collected from the attributes of existing ransomware families and ransomware attacks; they fail to produce valid and reliable data to the training component of the model. This untrustworthy nature leads to false predictions in the detection of malware as well as benign software
5	Environment	Despite all the hurdles mentioned above, a well trained model may be designed to produce accurate results every time it is tested in the simulated environment. Though it produces high level accuracy in the semi real environments, there is a fair level of possibility of producing FNR and FPR when it is deployed in the real environments due to the high variants of ransomware. FNR is more serious threat to detection models as it may pave way to the ransomware attack in the corresponding environment

6 Conclusion and Future Scope

In this paper, systematic literature review [66] is conducted to study about the application of Machine Learning methods to detect the ransomware attacks in detail. The mitigation techniques for ransomware detection using various methods from the past research works are discussed in detail. The scope of Machine Learning in devising automated detection tool for ransomware is also described elaborately. The set of evaluation parameters used for a Machine Learning model is also discussed with the results from the past research works. Finally the challenges faced when the Machine Learning methods are applied in developing a successful ransomware detection tool, are also described. This work will give an idea for devising a new novel and efficient detection method for ransomware using Machine Learning mehods and it is suggested as a future work.

References

1. Adamu U, Awan I (2019) Ransomware prediction using supervised learning algorithms. In: 2019 7th international conference on future internet of things and cloud (FiCloud). IEEE, Istanbul, Turkey
2. Agrawal R, Stokes JW, Selvaraj K, Marinescu M (2019) Attention in recurrent neural networks for ransomware detection. ICASSP 2019–2019 IEEE international conference on acoustics, speech and signal processing (ICASSP). IEEE, Brighton, United Kingdom, pp 3222–3226
3. Ahmed YA, Koçer B, Huda S, Al-Rimy BAS, Hassan MM (2020) A system call refinement-based enhanced minimum redundancy maximum relevance method for ransomware early detection. J Netw Comput Appl 167:102753. https://doi.org/10.1016/j.jnca.2020.102753
4. Akcora CG, Li Y, Gel YR, Kantarcioglu M (2020) Bitcoinheist: topological data analysis for ransomware prediction on the bitcoin blockchain. In: Proceedings of the twenty-ninth international joint conference on artificial intelligence. Yokohama, Japan, international Joint Conferences on Artificial Intelligence Organization, pp 4439–4445
5. Al-Hawawreh M, Sitnikova E (2019) Leveraging deep learning models for ransomware detection in the industrial internet of things environment. 2019 military communications and information systems conference (MilCIS). IEEE, Canberra, Australia, pp 1–6
6. Al-rimy B, Maarof M, Mohd Shaid SZ (2019) Crypto-ransomware early detection model using novel incremental bagging with enhanced semi-random subspace selection. Future Gener Comput Syst. https://doi.org/10.1016/j.future.2019.06.005
7. Al-rimy B, Maarof M, Shaid S (2018) Ransomware threat success factors, taxonomy, and countermeasures: a survey and research directions. Comput Secur 74. https://doi.org/10.1016/j.cose.2018.01.001
8. Al-rimy BAS, Maarof MA, Shaid SZM (2018) A 0-day aware crypto-ransomware early behavioral detection framework. In: Saeed F, Gazem N, Patnaik S, Balaid A, Mohammed F (eds) Recent trends in information and communication technology, vol 5. Springer International Publishing, Cham, pp 758–766
9. Alhawi OM, Baldwin J, Dehghantanha A (2019) Leveraging machine learning techniques for windows ransomware network traffic detection. In: Dehghantanha A, Conti M, Dargahi T (eds) Cyber threat intelligence, vol 70. Springer International Publishing, Cham, pp 93–106
10. Almashhadani AO, Kaiiali M, Sezer S, O'Kane P (2019) A multi-classifier network-based crypto ransomware detection system: a case study of locky ransomware. IEEE Access 7:47053–47067. https://doi.org/10.1109/ACCESS.2019.2907485

11. Alotaibi FM, Vassilakis VG (2021) SDN-based detection of self-propagating ransomware: the case of badrabbit. IEEE Access 9:28039–28058. https://doi.org/10.1109/ACCESS.2021.3058897
12. Alrawashdeh K, Purdy C (2018) Ransomware detection using limited precision deep learning structure in FPGA. NAECON 2018–IEEE national aerospace and electronics conference. IEEE, Dayton, OH, pp 152–157
13. AlSabeh A, Safa H, Bou-Harb E, Crichigno J (2020) Exploiting ransomware paranoia for execution prevention. ICC 2020–2020 IEEE international conference on communications (ICC). IEEE, Dublin, Ireland, pp 1–6
14. Alshaikh H, Ramadan N, Hefny H (2020) Ransomware prevention and mitigation techniques. Int J Comput Appl 117:31–39. https://doi.org/10.5120/ijca2020919899
15. Alzahrani N, Alghazzawi D (2019) A review on android ransomware detection using deep learning techniques. In: Proceedings of the 11th international conference on management of digital ecosystems. ACM, Limassol Cyprus, pp 330–335
16. Arabo A, Dijoux R, Poulain T, Chevalier G (2020) Detecting ransomware using process behavior analysis. Procedia Comput Sci 168:289–296. https://doi.org/10.1016/j.procs.2020.02.249
17. Atapour-Abarghouei A, Bonner S, McGough AS (2019) A king's ransom for encryption: ransomware classification using augmented one-shot learning and bayesian approximation. 2019 IEEE international conference on big data. IEEE, Los Angeles, CA, USA, pp 1601–1606
18. Ayub MA, Continella A, Siraj A (2020) An i/o request packet (IRP) driven effective ransomware detection scheme using artificial neural network. IEEE, Las Vegas, NV, USA, pp 319–324
19. Azeez NA, Odufuwa OE, Misra S, Oluranti J, Damaševičus R (2021) Windows pe malware detection using ensemble learning. Informatics 8(1). https://www.mdpi.com/2227-9709/8/1/10
20. Azmoodeh A, Dehghantanha A, Conti M, Choo KKR (2018) Detecting crypto-ransomware in IoT networks based on energy consumption footprint. J Ambient Intell Human Comput 9. https://doi.org/10.1007/s12652-017-0558-5
21. Bae S, Lee G, Im EG (2019) Ransomware detection using machine learning algorithms. Concurr Comput: Pract Exp 32:e5422. https://doi.org/10.1002/cpe.5422
22. Baek S, Jung Y, Mohaisen A, Lee S, Nyang D (2018) SSD-insider: internal defense of solid-state drive against ransomware with perfect data recovery. 2018 IEEE 38th international conference on distributed computing systems (ICDCS). IEEE, Vienna, pp 875–884
23. Baek S, Jung Y, Mohaisen D, Lee S, Nyang D (2021) SSD-assisted ransomware detection and data recovery techniques. IEEE Trans Comput 70(10):1762–1776. https://doi.org/10.1109/TC.2020.3011214
24. Baldwin J, Dehghantanha A (2018) Leveraging support vector machine for opcode density based detection of crypto-ransomware. In: Dehghantanha A, Conti M, Dargahi T (eds) Cyber threat intelligence, vol 70. Springer International Publishing, Cham, pp 107–136
25. Bansal C, Deligiannis P, Maddila C, Rao N (2020) Studying ransomware attacks using web search logs. In: Proceedings of the 43rd international ACM SIGIR conference on research and development in information retrieval. ACM, Virtual Event China, pp 1517–1520
26. Berrueta E, Morato D, Magaña E, Izal M (2020) Open repository for the evaluation of ransomware detection tools. IEEE Access 8:65658–65669. https://doi.org/10.1109/ACCESS.2020.2984187
27. Bhateja V, Peng SL (2021) Suresh chandra satapathy. In: Zhang YD (ed) Evolution in computational intelligence: frontiers in intelligent computing: theory and applications (FICTA), vol 1, 1176. Springer, Singapore
28. Bibi I, Akhunzada A, Malik J, Ahmed G, Raza M (2019) An effective android ransomware detection through multi-factor feature filtration and recurrent neural network, pp 1–4. https://doi.org/10.1109/UCET.2019.8881884
29. Black P, Sohail A, Gondal I, Kamruzzaman J, Vamplew P, Watters P (2020) Api based discrimination of ransomware and benign cryptographic programs. In: Yang H, Pasupa K, Leung AS,

Kwok J, Chan J, King I (eds) Neural information processing, vol 12533. Springer International Publishing, Cham, pp 177–188

30. Borah P, Bhattacharyya DK, Kalita JK (2020) Cost effective method for ransomware detection-an ensemble approach. In: Distributed computing and internet technology, pp 203–219. Springer International Publishing. https://doi.org/10.1007/978-3-030-65621-8_13

31. Cabaj K, Gregorczyk M, Mazurczyk W (2016) Software-defined networking-based crypto ransomware detection using http traffic characteristics. Comput Electr Eng 66. https://doi.org/10.1016/j.compeleceng.2017.10.012

32. Castillo PA, Laredo JLJ, Fernández F (2020) Applications of evolutionary computation. In: Vega (ed) 23rd European conference, EvoApplications, held as part of EvoStar 2020. Proceedings, vol 12104. Springer International Publishing, Seville, Spain

33. Chadha S, Kumar U (2017) Ransomware: let's fight back! 2017 international conference on computing, communication and automation (ICCCA). IEEE, Greater Noida, pp 925–930

34. Chen J, Wang C, Zhao Z, Chen K, Du R, Ahn GJ (2018) Uncovering the face of android ransomware: characterization and real-time detection. IEEE Trans Inf Forensics Secur 13(5):1286–1300. https://doi.org/10.1109/TIFS.2017.2787905

35. Chen L, Yang CY, Paul A, Sahita R (2018) Towards resilient machine learning for ransomware detection. arXiv preprint arXiv:1812.09400

36. Chen Q, Islam SR, Haswell H, Bridges RA (2019) Automated ransomware behavior analysis—pattern extraction and early detection. In: Science of cyber security, pp 199–214. Springer International Publishing. https://doi.org/10.1007/978-3-030-34637-9_15

37. Cheng L, Leung ACS (2018) In: Ozawa S (ed) Neural information processing: 25th international conference, ICONIP 2018, Siem Reap. Proceedings, Part VI, vol 11306. Springer International Publishing, Cambodia

38. Cimitile A, Mercaldo F, Nardone V, Santone A, Visaggio CA (2018) Talos: no more ransomware victims with formal methods. Int J Inf Secur 17. https://doi.org/10.1007/s10207-017-0398-5

39. Connolly Y, Lena SD (2019) Wall."the rise of crypto-ransomware in a changing cybercrime landscape: taxonomising countermeasures." Comput Secur 87(101568). https://doi.org/10.1016/j.cose.2019.101568

40. Cusack G, Michel O, Keller E (2018) Machine learning-based detection of ransomware using sdn. In: Proceedings of the 2018 ACM international workshop on security in software defined networks & network function virtualization. ACM, Tempe, AZ, USA, pp 1–6

41. Cuzzocrea A, Martinelli F, Mercaldo F (2018) A novel structural-entropy-based classification technique for supporting android ransomware detection and analysis. 2018 IEEE international conference on fuzzy systems (FUZZ-IEEE). IEEE, Rio de Janeiro, pp 1–7

42. Daku H, Zavarsky P, Malik Y (2018) Behavioral-based classification and identification of ransomware variants using machine learning. In: 2018 17th IEEE international conference on trust, security and privacy in computing and communications/12th IEEE international conference on big data science and engineering (TrustCom/BigDataSE), pp 1560–1564. https://doi.org/10.1109/TrustCom/BigDataSE.2018.00224

43. Dargahi T, Dehghantanha A, Nikkhah P, Conti M, Bianchi G, Benedetto L (2019) A cyber-kill-chain based taxonomy of crypto-ransomware features. J Comput Virol Hacking Tech 15. https://doi.org/10.1007/s11416-019-00338-7

44. Dion Y, Brohi S (2020) An experimental study to evaluate the performance of machine learning algorithms in ransomware detection. J Eng Sci Technol 15:967–981

45. Faris H, Habib M, Almomani I, Eshtay M, Aljarah I (2020) Optimizing extreme learning machines using chains of salps for efficient android ransomware detection. Appl Sci 10(11). https://www.mdpi.com/2076-3417/10/11/3706

46. Fernando DW, Komninos N, Chen T (2020) A study on the evolution of ransomware detection using machine learning and deep learning techniques. IoT 1(2):551–604. https://www.mdpi.com/2624-831X/1/2/30

47. Fernández Maimó L, Huertas Celdrán A, Perales Gómez NL, García Clemente FJ, Weimer J, Lee I (2019) Intelligent and dynamic ransomware spread detection and mitigation in integrated clinical environments. Sensors 19(5). https://www.mdpi.com/1424-8220/19/5/1114

48. Ferrante A, Malek M, Martinelli F, Mercaldo F, Milosevic J (2017) Extinguishing ransomware-A hybrid approach to android ransomware detection. In: Imine A, Fernandez J, Marion JY, Logrippo L, Garcia-Alfaro J (eds) Lecture notes in computer science, vol 10723. Springer International Publishing, Cham, pp 242–258
49. Gharib A, Ghorbani A (2017) DNA-droid: a real-time android ransomware detection framework. In: Yan Z, Molva R, Mazurczyk W, Kantola R (eds) Lecture notes in computer science, vol 10394. Springer International Publishing, Cham, pp 184–198
50. Gowtham R, Menen A (2020) Automated dynamic approach for detecting ransomware using finite-state machine. Decis Support Syst 138:113400. https://doi.org/10.1016/j.dss.2020.113400
51. Gupta BB, Perez GM, Agrawal DP, Gupta D (eds) Handbook of computer networks and cyber security: principles and paradigms. Springer International Publishing, Cham
52. Hampton N, Baig Z, Zeadally S (2018) Ransomware behavioural analysis on windows platforms. J Inf Secur Appl 40:44–51. https://doi.org/10.1016/j.jisa.2018.02.008
53. Harikrishnan N, Soman K (2018) Detecting ransomware using gurls. 2018 second international conference on advances in electronics, computers and communications (ICAECC). IEEE, Bangalore, pp 1–6
54. Hasan MM, Rahman MM (2017) Ranshunt: a support vector machines based ransomware analysis framework with integrated feature set. 2017 20th international conference of computer and information technology (ICCIT). IEEE, Dhaka, pp 1–7
55. Herrera Silva JA, Barona L, Valdivieso L, Alvarez M (2019) A survey on situational awareness of ransomware attacks-detection and prevention parameters. Remote Sens 11:1168. https://doi.org/10.3390/rs11101168
56. Homayoun S, Dehghantanha A, Ahmadzadeh M, Hashemi S, Khayami R (2020) Know abnormal, find evil: frequent pattern mining for ransomware threat hunting and intelligence. IEEE Trans Emerg Top Comput 8(2):341–351. https://doi.org/10.1109/TETC.2017.2756908
57. Hu JW, Zhang Y, Cui YP (2020) Research on android ransomware protection technology. J Phys: Conf Ser 1584(012004). https://doi.org/10.1088/1742-6596/1584/1/012004
58. Humayun M, Jhanjhi N, Alsayat A, Ponnusamy V (2021) Internet of things and ransomware: evolution, mitigation and prevention. Egypt Inform J 22(1):105–117
59. Hwang J, Kim J, Lee S, Kim K (2020) Two-stage ransomware detection using dynamic analysis and machine learning techniques. Wirel Pers Commun 112:1–13. https://doi.org/10.1007/s11277-020-07166-9
60. J, Z, M, H, Y, K, A, I (2020) In: Evaluation to classify Ransomware variants based on correlations between APIs. In Proceedings of the 6th International conference on information systems Security and Privacy, vol 1, pp 465–472. https://doi.org/10.5220/0008959904650472
61. Kara I, Aydos M (2020) Cyber fraud: Detection and analysis of the crypto-ransomware. 2020 11th IEEE Annual ubiquitous computing, electronics & mobile communication conference (UEMCON). IEEE, New York, NY, USA, pp 0764–0769
62. Karimi A, Moattar MH (2017) Android ransomware detection using reduced opcode sequence and image similarity. 2017 7th international conference on computer and knowledge engineering (ICCKE). IEEE, Mashhad, pp 229–234
63. Khammas BM (2020) Ransomware detection using random forest technique. ICT Express 6(4):325–331
64. Khan F, Ncube C, Ramasamy LK, Kadry S, Nam Y (2020) A digital DNA sequencing engine for ransomware detection using machine learning. IEEE Access 8:119710–119719. https://doi.org/10.1109/ACCESS.2020.3003785
65. Kharraz A, Robertson W, Kirda E (2018) Protecting against ransomware: a new line of research or restating classic ideas? IEEE Secur Priv 16(3):103–107. https://doi.org/10.1109/MSP.2018.2701165
66. Kitchenham B, Pearl Brereton O, Budgen D, Turner M, Bailey J, Linkman S (2009) Systematic literature reviews in software engineering—A systematic literature review. Inf Softw Technol 51(1):7–15 (2009). https://doi.org/10.1016/j.infsof.2008.09.009. https://www.sciencedirect.com/science/article/pii/S0950584908001390 (special Section—Most Cited Articles in 2002 and Regular Research Papers)

67. Kok S, Abdullah A, Jhanjhi N (2020) Early detection of crypto-ransomware using pre-encryption detection algorithm. J King Saud Univ Comput Inf Sci
68. Kok S, Abdullah A, Zaman N, Supramaniam M (2019) Prevention of crypto-ransomware using a pre-encryption detection algorithm. Computers 8:79. https://doi.org/10.3390/computers8040079
69. Koli, J.D.: Randroid: Android malware detection using random machine learning classifiers. In: 2018 technologies for smart-city energy security and power (ICSESP). pp 1–6 (2018). https://doi.org/10.1109/ICSESP.2018.8376705
70. Lachtar N, Ibdah D, Bacha A (2019) The case for native instructions in the detection of mobile ransomware. IEEE Lett Comput Soc 2(2):16–19. https://doi.org/10.1109/LOCS.2019.2918091
71. Lee K, Lee SY, Yim K (2019) Machine learning based file entropy analysis for ransomware detection in backup systems. IEEE Access 7:110205–110215. https://doi.org/10.1109/ACCESS.2019.2931136
72. Li Z, Rios ALG, Trajkovic L (2020) Detecting internet worms, ransomware, and blackouts using recurrent neural networks. 2020 IEEE international conference on systems, man, and cybernetics (SMC). IEEE, Toronto, ON, Canada, pp 2165–2172
73. Lokuketagoda B, Weerakoon MP, Kuruppu UM, Senarathne AN, Abeywardena KY (2018) R-killer: an email based ransomware protection tool. In: 2018 13th international conference on computer science & education (ICCSE). Colombo. IEEE
74. Lu T, Zhang L, Wang S, Gong Q (2017) Ransomware detection based on v-detector negative selection algorithm. 2017 international conference on security, pattern analysis, and cybernetics (SPAC). IEEE, Shenzhen, pp 531–536
75. Luhach AK, Kosa JA, Poonia RC (2020) Xiao-zhi Gao. In: Singh D (ed) First international conference on sustainable technologies for computational intelligence: proceedings of ICTSCI 2019, vol 1045. Springer, Singapore, Singapore
76. Maigida AM, Abdulhamid SM, Olalere M, Alhassan JK (2019) Haruna chiroma, and emmanuel gbenga dada."systematic literature review and metadata analysis of ransomware attacks and detection mechanisms." J Reliab Intell Environ 5(2):67–89. https://doi.org/10.1007/s40860-019-00080-3
77. Manavi F, Hamzeh A (2020) A new method for ransomware detection based on PE header using convolutional neural networks. 2020 17th international ISC conference on information security and cryptology (ISCISC). IEEE, Tehran, Iran, pp 82–87
78. Manzano C, Meneses C, Leger P (2020) An empirical comparison of supervised algorithms for ransomware identification on network traffic. 2020 39th international conference of the chilean computer science society (SCCC). IEEE, Coquimbo, Chile, pp 1–7
79. Medhat M, Gaber S, Abdelbaki N (2018) A new static-based framework for ransomware detection. In: 2018 IEEE 16th International conference on dependable, autonomic and secure computing, 16th international conference on pervasive intelligence and computing, 4th international conference on big data intelligence and computing and cyber science and technology congress(DASC/PiCom/DataCom/CyberSciTech), pp 710–715. https://doi.org/10.1109/DASC/PiCom/DataCom/CyberSciTec.2018.00124
80. Misra S, A step by step guide for choosing project topics and writing research papers in ICT related disciplines, vol 1350. Springer, Cham
81. Mohammad A (2020) Ransomware evolution, growth and recommendation for detection. Modern Appl Sci 14:68. https://doi.org/10.5539/mas.v14n3p68
82. Ng C, Rajasegarar S, Pan L, Jiang F, Zhang L (2020) Voterchoice: a ransomware detection honeypot with multiple voting framework. Concurr Comput: Pract Exp 32. https://doi.org/10.1002/cpe.5726
83. Pastor A, Mozo A, Vakaruk S, Canavese D, López DR, Regano L, Gómez-Canaval S, Lioy A (2020) Detection of encrypted cryptomining malware connections with machine and deep learning. IEEE Access 8:158036–158055. https://doi.org/10.1109/ACCESS.2020.3019658
84. Pont J, Oun OA, Brierley C, Arief B, Hernandez-Castro J (2019) A roadmap for improving the impact of anti-ransomware research. In: Askarov A, Hansen R, Rafnsson W (eds) Secure IT systems, vol 11875. Springer International Publishing, Cham, pp 137–154

85. Poudyal S, Dasgupta D, Akhtar Z, Gupta KD (2019) A multi-level ransomware detection framework using natural language processing and machine learning
86. Qin B, Wang Y, Ma C (2020) API call based ransomware dynamic detection approach using textCNN. 2020 international conference on big data, artificial intelligence and internet of things engineering (ICBAIE). IEEE, Fuzhou, China, pp 162–166
87. Reddy BV, Krishna GJ, Ravi V, Dasgupta D (2020) Machine learning and feature selection based ransomware detection using hexacodes. In: evolution in computational intelligence, pp 583–597. Springer Singapore. https://doi.org/10.1007/978-981-15-5788-056
88. Rosli MS, Syahirah R, Yassin W, Faizal MA, Nur W (2020) Ransomware behavior attack construction via graph theory approach. Int J Adv Comput Sci Appl 11
89. Rouka E, Birkinshaw C, Vassilakis VG (2020) SDN-based malware detection and mitigation: the case of expetr ransomware. 2020 IEEE international conference on informatics, IoT, and enabling technologies (ICIoT). IEEE, Doha, Qatar, pp 150–155
90. Roy K, Chen Q (2021) Deepran: attention-based bilstm and crf for ransomware early detection and classification. Inf Syst Front 23. https://doi.org/10.1007/s10796-020-10017-4
91. Vinayakumar R, Jolfaei MA, Jolfaei A, Soman KP, Poornachandran P (2019) Ransomware triage using deep learning: twitter as a case study. 2019 cybersecurity and cyberforensics conference (CCC). IEEE, Melbourne, Australia, pp 67–73
92. Saeed S, Jhanjhi N, Naqvi M, Humayun M, Ahmed S (2020) Ransomware: a framework for security challenges in internet of things. 2020 2nd international conference on computer and information sciences (ICCIS). IEEE, Sakaka, Saudi Arabia, pp 1–6
93. Sahay SK, Goel N (2020) Vishwas patil. In: Jadliwala M (ed) Secure Knowledge Management. In: Artificial Intelligence Era: 8th international conference, SKM 2019. Proceedings, vol 1186. Springer, Singapore, Goa, India, pp 1–6
94. Saleh MA, Rass A, Evaluation of supervised machine learning classifiers for detecting ransomware based on naïve bayes, svm, knn, c 4.5, and random forest algorithms. Int J Innov Sci Res Technol 5(1):10
95. Scalas M, Maiorca D, Mercaldo F, Visaggio CA, Martinelli F, Giacinto G (2019) On the effectiveness of system API-related information for android ransomware detection. Comput Secur 86:168–182. https://doi.org/10.1016/j.cose.2019.06.004
96. Sechel: Sergiu."a comparative assessment of obfuscated ransomware detection methods. Inform Econ 23(2):45–62. https://doi.org/10.12948/issn14531305/23.2.2019.05
97. Shaukat K, Luo S, Chen S, Liu D (2020) Cyber threat detection using machine learning techniques: a performance evaluation perspective. 2020 international conference on cyber warfare and security (ICCWS). IEEE, Islamabad, Pakistan, pp 1–6
98. Sheen S, Yadav A (2018) Ransomware detection by mining API call usage. 2018 international conference on advances in computing, communications and informatics (ICACCI). IEEE, Bangalore, pp 983–987
99. Song J, Meng Q, Luo C, Naik N, Xu J (2020) An immunization scheme for ransomware. Comput Mater Continua 64(2):1051–1061. https://doi.org/10.32604/cmc.2020.010592
100. Su D, Liu J, Wang X, Wang W (2019) Detecting android locker-ransomware on chinese social networks. IEEE Access 7:20381–20393. https://doi.org/10.1109/ACCESS.2018.2888568
101. Sultan NA, Thanoon KH, Ibrahim OA (2020) Ethical hacking implementation for lime worm ransomware detection. J Phys: Conf Ser 1530(012078). https://doi.org/10.1088/1742-6596/1530/1/012078
102. Turner AB, McCombie S, Uhlmann AJ (2020) Discerning payment patterns in bitcoin from ransomware attacks. J Money Laund Control 23(3):545–589. https://doi.org/10.1108/JMLC-02-2020-0012
103. Uandykova M, Lisin A, Stepanova D, Baitenova L, Mutaliyeva L (2020) Serhat yuksel, and hasan dincer. "the social and legislative principles of counteracting ransomware crime." Entrep Sustain Issues 8(2):777–798. https://doi.org/10.9770/jesi.2020.8.2(47)
104. Ullah F, Javaid Q, Salam A, Ahmad M, Sarwar N (2020) Dilawar shah, and muhammad abrar. "modified decision tree technique for ransomware detection at runtime through API calls." Sci Program 2020:1–10. https://doi.org/10.1155/2020/8845833

105. Usharani S, Bala P, Mary MJ (2021) Dynamic analysis on crypto-ransomware by using machine learning: gandcrab ransomware. J Phys: Conf Ser 1717(012024). https://doi.org/10.1088/1742-6596/1717/1/012024
106. Verma M, Kumarguru P, Deb SB, Gupta A (2018) Analysing indicator of compromises for ransomware: leveraging IOCS with machine learning techniques. 2018 IEEE international conference on intelligence and security informatics (ISI). IEEE, Miami, FL, pp 154–159
107. Vinayakumar R, Soman K, Velan K, Ganorkar S (2017) Evaluating shallow and deep networks for ransomware detection and classification. 2017 international conference on advances in computing, communications and informatics (ICACCI). IEEE, Udupi, pp 259–265
108. Wan YL, Chang JC, Chen RJ, Wang SJ (2018) Feature-selection-based ransomware detection with machine learning of data analysis. 2018 3rd international conference on computer and communication systems (ICCCS). IEEE, Nagoya, Japan, pp 85–88
109. Wang Z, Liu C, Qiu J, Tian Z, Cui X, Su S (2018) Automatically traceback RDP-based targeted ransomware attacks. Wirel Commun Mob Comput 2018:1–13. https://doi.org/10.1155/2018/7943586
110. Wani A, Revathi S (2020) Ransomware protection in IoT using software defined networking. Int J Electr Comput Eng (IJECE) 10(3). https://doi.org/10.11591/ijece.v10i3.pp3166-3175
111. Xia T, Sun Y, Zhu S, Rasheed Z, Shafique K (2018) Toward a network-assisted approach for effective ransomware detection. In: ICST Trans Secur Safety 168506. https://doi.org/10.4108/eai.28-1-2021.168506
112. Zhang B, Xiao W, Xiao X, Sangaiah AK, Zhang W, Zhang J (2020) Ransomware classification using patch-based CNN and self-attention network on embedded N-grams of opcodes. Fut Gener Comput Syst 110:708–720. https://doi.org/10.1016/j.future.2019.09.025
113. Zhang H, Xiao X, Mercaldo F, Ni S, FabioMartinelli AKS (2019) Classification of ransomware families with machine learning based N-gram of opcodes. Fut. Gener. Comput. Syst. 90:211–221. ISSN 0167:739X. https://doi.org/10.1016/j.future.2018.07.052
114. Zuhair H, Selamat A, An empirical analysis of machine learning efficacy in anti-ransomware tools. AUE Int Res Conf/Dubai 8
115. Zuhair H, Selamat A, Krejcar O (2020) A multi-tier streaming analytics model of 0-day ransomware detection using machine learning. Appl Sci 10(9). https://doi.org/10.3390/app10093210
116. Zhou J, Hirose M, Kakizaki Y, Inomata A (2020) Evaluation to classify ransomware variants based on correlations between APIs. In: Proceedings of the 6th international conference on information systems security and privacy. vol 1, pp 465–472. https://doi.org/10.5220/0008959904650472

Printed in the United States
by Baker & Taylor Publisher Services